Lecture Notes in Physics

Monographs

Springer

Berlin
Heidelberg
New York
Barcelona
Hong Kong
London
Milan
Paris
Singapore
Tokyo

Physics and Astronomy

ONLINE LIBRARY

http://www.springer.de/phys/

The Editorial Policy for Monographs

The series Lecture Notes in Physics reports new developments in physical research and teaching - quickly, informally, and at a high level. The type of material considered for publication in the monograph Series includes monographs presenting original research or new angles in a classical field. The timeliness of a manuscript is more important than its form, which may be preliminary or tentative. Manuscripts should be reasonably self-contained. They will often present not only results of the author(s) but also related work by other people and will provide sufficient motivation, examples, and applications.

The manuscripts or a detailed description thereof should be submitted either to one of the series editors or to the managing editor. The proposal is then carefully refereed. A final decision concerning publication can often only be made on the basis of the complete manuscript, but otherwise the editors will try to make a preliminary decision as definite as they can on the basis of the available information.

Manuscripts should be no less than 100 and preferably no more than 400 pages in length. Final manuscripts should be in English. They should include a table of contents and an informative introduction accessible also to readers not particularly familiar with the topic treated. Authors are free to use the material in other publications. However, if extensive use is made elsewhere, the publisher should be informed. Authors receive jointly 30 complimentary copies of their book. They are entitled to purchase further copies of their book at a reduced rate. No reprints of individual contributions can be supplied. No royalty is paid on Lecture Notes in Physics volumes. Commitment to publish is made by letter of interest rather than by signing a formal contract. Springer-Verlag secures the copyright for each volume.

The Production Process

The books are hardbound, and quality paper appropriate to the needs of the author(s) is used. Publication time is about ten weeks. More than twenty years of experience guarantee authors the best possible service. To reach the goal of rapid publication at a low price the technique of photographic reproduction from a camera-ready manuscript was chosen. This process shifts the main responsibility for the technical quality considerably from the publisher to the author. We therefore urge all authors to observe very carefully our guidelines for the preparation of camera-ready manuscripts, which we will supply on request. This applies especially to the quality of figures and halftones submitted for publication. Figures should be submitted as originals or glossy prints, as very often Xerox copies are not suitable for reproduction. For the same reason, any writing within figures should not be smaller than 2.5 mm. It might be useful to look at some of the volumes already published or, especially if some atypical text is planned, to write to the Physics Editorial Department of Springer-Verlag direct. This avoids mistakes and time-consuming correspondence during the production period.

As a special service, we offer free of charge LaTeX and TeX macro packages to format the text according to Springer-Verlag's quality requirements. We strongly recommend authors to make use of this offer, as the result will be a book of considerably improved technical quality.

For further information please contact Springer-Verlag, Physics Editorial Department II, Tiergartenstrasse 17, D-69121 Heidelberg, Germany.

Series homepage – http://www.springer.de/phys/books/lnpm

Jorge Berger Jacob Rubinstein (Eds.)

Connectivity
and Superconductivity

 Springer

Editors

Jorge Berger
Ort Braude College
21982 Karmiel, Israel

Jacob Rubinstein
Department of Mathematics
Technion
32000 Haifa, Israel

Library of Congress Cataloging-in-Publication Data

Connectivity and superconductivity / Jorge Berger, Jacob Rubinstein, eds.
    p. cm. -- (Lecture notes in physics. Monographs, ISSN 0940-7677 ; vol. m62)
  Includes bibliographical references.
  ISBN 3540679324 (alk. paper)
    1. Superconductivity--Mathematics. 2. Connections (Mathematics) 3. Mathematical
physics. I. Berger, Jorge, 1948- II. Rubinstein, Jacob, 1955- III. Lecture notes in
physics. New series m, Monographs ; v. m62.

QC611.92 .C65 2000
537.6'23'0151--dc21

00-063784

ISSN 0940-7677 (Lecture Notes in Physics. Monographs)
ISBN 3-540-67932-4 Springer-Verlag Berlin Heidelberg New York

Springer-Verlag Berlin Heidelberg New York
a member of BertelsmannSpringer Science+Business Media GmbH

© Springer-Verlag Berlin Heidelberg 2000
Printed in Germany

Typesetting: Camera-ready by the authors/editors
Cover design: *design & production*, Heidelberg

Printed on acid-free paper
SPIN: 10780149   55/3141/du - 5 4 3 2 1 0

Preface

The motto of *connectivity and superconductivity* is that the solutions of the Ginzburg–Landau equations are qualitatively influenced by the topology of the boundaries. Special attention is given to the "zero set", the set of the positions (usually known as "quantum vortices") where the order parameter vanishes. The paradigm of connectivity and superconductivity is the Little–Parks effect, discussed in most textbooks on superconductivity.

This volume is intended to serve as a reference book for graduate students and researchers in physics or mathematics interested in superconductivity, or in the Schrödinger equation as a limiting case of the Ginzburg–Landau equations.

The effects considered here usually become important in the regime where the coherence length is of the order of the dimensions of the sample. While in the Little–Parks days a lot of ingenuity was required to achieve this regime, present microelectronic techniques have transformed it into a routine. Moreover, measurement and visualization techniques are developing at a pace which makes it reasonable to expect verification of distributions, and not only of global properties.

Activity in the field has grown and diversified substantially in recent years. We have therefore invited experts ranging from experimental and theoretical physicists to pure and applied mathematicians to contribute articles for this book. While the skeleton of the book deals with superconductivity, micronetworks and generalizations of the Little–Parks situation, there are also articles which deal with applications of the Ginzburg–Landau formalism to several fundamental topics, such as quantum coherence, cosmology, and questions in materials science.

The sequence of the chapters in the book follows similarity of subjects rather than authors' disciplines, so that articles by physicists and by mathematicians are intermixed. We have made an effort to have all authors express themselves in a common language, but the reader will still identify differences in their styles.

Haifa,
July 2000

Jorge Berger
Jacob Rubinstein

Contents

5 Zero Set of the Order Parameter, Especially in Rings

6 Persistent Currents in Ginzburg–Landau Models

List of Contributors

Luís Almeida
Université de Nice
06108 Nice Cédex 02, France
luis@math.unice.fr

Jorge Berger
Ort Braude College
21982 Karmiel, Israel
phr76jb@tx.technion.ac.il

Fabrice Bethuel
Université Pierre et Marie Curie
75252 Paris Cédex 5, France
bethuel@ann.jussieu.fr

Carlos Bolech
Rutgers University
Piscataway, NJ 08854, USA
bolech@pion.rutgers.edu

Vital Bruyndoncx
Katholieke Universiteit Leuven
3001 Leuven, Belgium
Vital.Bruyndoncx@
fys.kuleuven.ac.be

Gustavo Buscaglia
Centro Atómico Bariloche
8400 Bariloche, Argentine
gustavo@cab.cnea.gov.ar

José Castro
Universidad Nacional de San Juan
5400 San Juan, Argentine
jicastro@
server.ffha.unsj.edu.ar

Pierre-Gilles De Gennes
ESPCI
75005 Paris, France
Pierre-Gilles.DeGennes@
espci.fr

Guy Deutscher
Tel Aviv University
69879 Tel Aviv, Israel
chava@post.tau.ac.il

Sanatan Digal
Universitat Bielefeld
33615 Bielefeld, Germany
digal@Physik.Uni-Bielefeld.DE

Bernard Helffer
Université Paris-Sud
91405 Orsay, France
Bernard.Helffer@math.u-psud.fr

Maria Hoffmann-Ostenhof
Universität Wien
1090 Vienna, Austria
mho@nelly.mat.univie.ac.at

Thomas Hoffmann-Ostenhof
Universität Wien
1090 Vienna, Austria
Thomas.Hoffmann.Ostenhof@
esi.ac.at

Charles Kuper
Technion – Israel Institute of
Technology
3200 Haifa, Israel
charles@physics.technion.ac.il

Antony Leggett
University of Illinois at Urbana-
Champaign
Urbana, IL 61801-3080, USA
tony@cromwell.physics.uiuc.edu

Arturo López
Centro Atómico Bariloche
8400 Bariloche, Argentine
alopez@cab.cnea.gov.ar

Victor Moshchalkov
Katholieke Universiteit Leuven
3001 Leuven, Belgium
victor.moshchalkov@
fys.kuleuven.ac.be

Mark Owen
Technische Universität Wien
1090 Vienna, Austria
mowen@wiener.fam.tuwien.ac.at

Rajarshi Ray
Institute of Physics

751005 Bhubaneswar, India
rajarshi@iopb.res.in

Jacob Rubinstein
Technion – Israel Institute of
Technology
32000 Haifa, Israel
koby@math.technion.ac.il

Supratim Sengupta
Institute of Physics
751005 Bhubaneswar, India
supratim@iopb.res.in

Ajit Srivastava
Institute of Physics
751005 Bhubaneswar, India
ajit@iopb.res.in

Peter Sternberg
Indiana University
Bloomington, IN 47405, USA
sternber@indiana.edu

Lieve Van Look
Katholieke Universiteit Leuven
3001 Leuven, Belgium
lieve.vanlook@
fys.kuleuven.ac.be

In the Memory of Shlomo Alexander

Pierre-Gilles de Gennes

ESPCI, 10 rue Vauquelin, 75005 Paris, France

My first recollection of superconducting clusters is based on work by Guy Deutscher and coworkers in Tel Aviv, where they probed certain random arrays of microscopic metallic particles. Coming to Israel soon after, I mentioned the amusing properties of "wire networks", where some superconducting threads (simple loops, or connected periodic networks) are exposed to a magnetic field.

Shlomo was immediateley interested by these questions. With Amnon Aharony and others he had fully understood the question of classical diffusion on percolation networks — what I had called the "ant problem". He was immediately able to transpose these concepts to the quantum mechanical problem of a Landau–Ginzburg equation with a vector potential, and to predict the magnetic behavior.

In another direction, he kept an active eye on self similar fractals, such as the Serpinskii gasket, where he constructed a very interesting set of wave functions. As usual, I was deeply impressed by all he had brought within a short time...

Recently, many years later, I had another contact with random clusters and their spectral dimension: the field here is very different. We talk about flexible branched polymer chains, and how they can be sucked into a nanopore. Shlomo discussed this with us at great length. One amusing feature was that a very simple calculation, based on Flory's ideas, was able to predict the spectral dimension of the clusters. He enjoyed this.

Now, if we find a difficult problem, or a strange concept, we cannot come to him and ask for help. We have lost a great man.

P.G. de Gennes, April 1999

1 Topological Considerations in Superconductivity

Jacob Rubinstein

Department of Mathematics, Technion, 32000 Haifa, Israel

1.1 Introduction

Some of the fundamental phenomena of superconductivity are observed in samples with a nontrivial topology. The purpose of this book is to assemble evidence ranging from experimental physics to pure mathematics that supports this assertion. The task is not easy. The different communities represented in the book have very different perspectives. They often even talk in different languages. But we feel that the gap between them has narrowed in recent years, and we hope that the book will serve to narrow it further.

In this introductory chapter, I review results by my colleagues and myself over the last decade. They include several examples in which topology plays a crucial role in the mathematical analysis of the Ginzburg Landau model, and where physics is deeply affected too. I shall not try to survey the entire field. This is the purpose of the book as a whole. Rather, I shall try to highlight some of the main features of topological considerations, particularly from the mathematical view point.

Following a brief introduction to the Ginzburg Landau model, I proceed to investigate one of the earliest fascinating patterns discovered in superconductivity: permanent currents. I show that these currents persist even without an external magnetic field. A nontrivial topology (a toroidal domain, for example) is essential for maintaining the currents. Permanent currents are of course well known in physics. It is interesting to note, though, that the mathematics needed to study them is quite novel.

I continue with the Little Parks effect, which is closely associated with the Aharonov Bohm effect. Here the topology manifests itself in the phase transition diagram. The effect becomes even more involved when we consider arbitrary graphs. Now the topology (related to the cycles of the graph) has to be coupled with the geometry (the connectivity of the graph).

Vortices are arguably the most important pattern in superconductivity. The very notion of a vortex is topological in nature, since a vortex is associated with a degree - the circulation of the order parameter phase around the vortex core. There is an important difference, though, between the topological argument related to vortices and those appearing in the examples in the previous paragraphs: in the study of vortices the topology is built into the solution through the initial data and the external conditions, such as the applied magnetic field. It is not required for the sample to have unusual geometry to have vortices. In addition to the built-in topological feature of a vortex through degree considerations, the vortex line may have a nontrivial spatial shape. It can form loops, knots and links. The phenomena of vortex line entanglement plays a prominent role in turbulence in superfluids. Therefore, a natural question is whether such patterns can arise naturally also in superconductivity. While I do not have a complete answer, I present below a partial result pertaining to a simplified model, in which magnetic fields are neglected. In this toy model one can show that complex patterns formed by superconducting vortices tend to unknot themselves. Hence vortex entanglement is not expected to be typical in superconductivity, unless it is forced through unusual boundary and initial conditions.

Vortex formation and the Little Parks effect are shown to be related to each other in asymmetric domains with nontrivial topology. We examine this point by studying the zero set of the order parameter in thin networks. The problem of the linear GL model on graphs is further elaborated by Castro and López in Chap. 2. The case of a two dimensional setup, i.e. networks with finite thickness, is considered in Chap. 3. The analysis of the zero set in networks is used to introduce another interesting topological tool - the double covering of a manifold. Moreover, a nontrivial topology can imply a zero set of codimension 1, which is quite unusual for complex valued functions. Finally we point out a connection between a ring topology and the Josephson effect.

1.2   Introduction to the Ginzburg–Landau Model

The Ginzburg Landau (GL) model for superconductivity is based on the Landau theory for second order phase transitions. The model consists of an energy functional depending on a complex order parameter Ψ and the magnetic vector potential A. In dimensional units, the energy density is given by

$$a(T)|\Psi|^2 + b(T)|\Psi|^4 + \frac{1}{4m}|\hbar\nabla\Psi - 2ieA\Psi|^2 + \frac{\mu_M}{2}H\cdot(H - 2H_e) \quad (1.1)$$

where T is the temperature, H_e is the applied magnetic field, $b(T)$ is a positive temperature dependent material function, the material function $a(T)$ is positive for $T > T_c$, and negative for $T < T_c$, μ_M is the magnetic permeability, m is a free parameter, taken in general to be the electron mass, and T_c is the critical transition temperature. A material is associated with two fundamental length scales: the coherence length (characterizing the order parameter) $\xi = \hbar/2\sqrt{m|a|}$, and the penetration length (characterizing the magnetic potential) $\lambda = mb/2|a|\mu_M e^2$.

Several kinds of nondimensional units are often used. One of them is based on scaling all lengths with respect to λ:

$$\Psi \to \frac{\sqrt{|a|}}{b}\psi, \; H \to |a|\sqrt{\frac{2}{b\mu_M}}H, \; x \to \lambda x. \quad (1.2)$$

In these units the GL functional takes the form

$$F_{GL} = \int_{\Omega} |(\frac{1}{\kappa}\nabla - iA)\psi|^2 + (|\psi|^2 - 1)^2 + \int_{R^3} (\nabla \times A - H_e)^2. \quad (1.3)$$

Here we introduced the nondimensional GL parameter $\kappa = \lambda/\xi$, and completed the magnetic contribution to a full square by adding a constant to the energy.

Another kind of scaling is based on using the coherence length as the sale length [26]. Scaling by either the coherence length or by the penetration length has the deficiency that the length unit is temperature dependent. This situation is undesired when studying phase transitions. Thus we sometimes restore to a different scaling in which the sample scale is chosen to scale lengths in the problem [7]. Throughout this chapter, and, in fact, throughout the book, different scalings will be used depending on the problem under consideration. The GL functional and its related Euler Lagrange equations have been extensively studied from the pure mathematical view point. We refer to [14] and [26] for further references. In addition we refer to the three classical texts by de Gennes [13], by Abrikosov [1] and by Tinkham [33] for a comprehensive treatment of superconductivity in general.

1.3 Permanent Currents

One of the classical quantized objects is provided by a permanent current circulating in domains with non trivial topology, such as tori. While in the setting we shall consider there are no vortices, the permanent current is closely related to the issue of quantized fluxoid that we define below. Obviously, permanent currents are well known and understood both theoretically and experimentally from the physical view point. Nevertheless, it takes somewhat unusual mathematics to establish their existence in arbitrary multiply connected domains. We note that we are investigating permanent superconducting current *without* any driving mechanism such as an external magnetic field. Clearly these patterns can be only local minima of the GL functional.

Denote by Ω the domain in R^3 occupied by the superconducting material. Assume furthermore that Ω is multiply connected. More precisely, we assume that the fundamental group of Ω is nontrivial. Consider now smooth (Lipschitz) mappings from Ω into the unit circle S^1, and define for each of them a homotopy type, determined by its circulation along the 1-skeleton of Ω. We can thus divide the smooth mappings from Ω into S^1 into equivalent classes defined by their homotopy types. The solid torus is a concrete example that we shall use in the sequel. In this case the set of homotopy class is exactly Z - the set of integers.

The material is not subjected to an external magnetic field. Therefore we write its energy in the form

$$F_{\kappa}(\psi, A) = \int_{\Omega} |(\nabla - iA)\psi|^2 + \kappa^2(|\psi|^2 - 1)^2 + \int_{R^3} |\nabla \times A|^2, \quad (1.4)$$

where the coherence length scaling is used. The first term is the energy associated with the superconducting electrons, which are confined to Ω. The second term is the energy of the magnetic field, which is defined over the entire space. We have to choose an appropriate function space for the magnetic field A. The natural space seems to be the Sobolev space $H^1(\mathbb{R}^3; \mathbb{R}^3)$, which can be taken as the closure of C_0^∞ vector fields in the norm $\int_{\mathbb{R}^3} |\nabla v|^2$. Recalling the gauge invariance of the GL energy, we impose the additional constraint $\nabla \cdot A = 0$; thus we consider for A the space $H^1_{div}(\mathbb{R}^3; \mathbb{R}^3)$, which is the closure of divergence free C_0^∞ vector fields under the $\int_{\mathbb{R}^3} |\nabla v|^2$ norm.

We shall show that for each homotopy type m and for κ sufficiently large, there exists a local minimizer $(\psi_\kappa^m, A_\kappa^m)$ of (1.4). We first introduce, for each m, an auxiliary pair (ψ^m, A^m). For this purpose we introduce the Sobolev space $H^1(\Omega, S^1)$ of mappings between Ω and the unit circle. Considering the functional F_κ for functions in this space, we observe that the middle term on the right hand side of (1.4) vanishes identically; thus we define the functional

$$F_0(\psi, A) = \int_\Omega |(\nabla - iA)\psi|^2 + \int_{\mathbb{R}^3} |\nabla \times A|^2. \tag{1.5}$$

A crucial question is how to associate homotopy types with H^1 functions (that are not necessarily smooth). The solution to this problem is provided in two parts. The first is a theorem by Bethuel, stating that the smooth functions are dense in $H^1(\Omega, S^1)$ [10]. The second ingredient is a theorem by White [34], indicating that two smooth functions that are sufficiently close in the H^1 norm must have the same homotopy type. It follows that non-smooth functions can be associated with a unique homotopy type through approximations by smooth functions. Moreover, White has shown that homotopy type is preserved under weak convergence in H^1. Thus we can partition the space $H^1(\Omega, S^1)$ into subspaces $H^1_m(\Omega, S^1)$ of functions with the same homotopy class.

Using these results, and the direct method of calculus of variations, it can be shown [31] that, for each m, there exists a pair (ψ, A) in $H^1_m(\Omega, S^1) \times H_{div}(\mathbb{R}^3)$, that is a local minimizer of (1.5) in the space $H^1(\Omega, S^1) \times H_{div}(\mathbb{R}^3)$. We denote this minimizer by (ψ^m, A^m).

We can now state a theorem about the homotopy classification of the minimizers for the full functional (1.4). A result along this line was first proved in [19] for the special case of symmetric domains. An abstract formulation that applies for arbitrary domains is provided in [31]:

Theorem 1. *For each homotopy type m there exists κ_0, such that for all $\kappa > \kappa_0$, the functional F_κ posseses a local minimizer $(\psi_\kappa^m, A_\kappa^m)$. Moreover, the sequence $(\psi_\kappa^m, A_\kappa^m)$ converges to (ψ^m, A^m) as $\kappa \to \infty$ in $H^1(\Omega, \mathbf{C}) \times H_{div}(\mathbb{R}^3)$.*

Proof

Fix a homotopy type m, and minimize F_κ over the set

$$\{(\psi, A) \in H^1(\Omega, \mathbf{C}) \times H_{div}(\mathbb{R}^3) \; : \; ||\psi - \psi^m|| \leq \gamma_m\},$$

where γ_m is chosen so that ψ^m minimizes F_0 in a ball of that radius in $H^1(\Omega, S^1)$. Clearly, the following chain of inequalities hold

$$F_0(\psi_\kappa^m, A_\kappa^m) \leq F_\kappa(\psi_\kappa^m, A_\kappa^m) \leq F_\kappa(\psi^m, A^m) = F_0(\psi^m, A^m). \qquad (1.6)$$

Hence $\int_\Omega V(\psi_\kappa^m) \leq \kappa^{-2} F_0(\psi^m, A^m)$, which implies $|\psi_\kappa^m| \to 1$ pointwise a.e. as $\kappa \to \infty$. Also, H^1 compactness implies the convergence (possibly of subsequences) of ψ_κ^m to some function U, with $|U| = 1$, and of A_κ^m to some vector field A, weakly in $H^1(\Omega, \mathbf{C}) \times H_{div}(\mathbb{R}^3)$, respectively. The lower semicontinuity of the H^1 norm implies

$$F_0(U, A) \leq \liminf F_0(\psi_\kappa^m, A_\kappa^m). \qquad (1.7)$$

Combining (1.6) and (1.7) we find that U satisfies

$$F_0(U, A) \leq F_0(\psi^m, A^m), \quad \text{and} \quad ||U - \psi^m|| \leq \gamma_m. \qquad (1.8)$$

Therefore $U = e^{i\alpha} \psi^m$, and $A = A^m$ for some constant α that we can take without loss of generality to be zero. This implies the convergence of the L^2 norms of $\nabla \psi_\kappa^m$ and ∇A_κ^m to those of $\nabla \psi^m$ and ∇A^m, respectively. Therefore the convergence is in the strong sense.

An interesting question is whether the topology for the convergence can be upgraded to C^1, or at least to C^0. This is important from the physical view point, since it will imply that there are no vortices in Ω when κ is sufficiently large. Such an extension was indeed obtained recently in [11] within a more general framework. The general abstract mathematical question of understanding the critical points of the GL functional in multiply connected domains, and their dependence on the domain topology is addressed by Almeida and Bethuel [2].

The homotopy classification relates to the phenomena of *quantized fluxoids*. Assume again that Ω is a solid torus. Writing $\psi = y e^{i\phi}$, the superconducting current can be written in the form $J_s = y^2(\nabla \phi - A)$. Let σ be a closed curve in Ω homotopic to its 1-skeleton, and denote by Σ any surface bounded by σ. Then the fluxoid is defined as

$$FL = \int_\Sigma H d\Sigma + \int_\sigma \frac{J_s}{y^2} d\sigma.$$

The relation $\nabla \times A = H$ and Stokes theorem imply

$$FL = \int_\sigma \nabla \phi.$$

Therefore the fluxoid must be an integer times 2π. This integer is precisely the homotopy type that was used to classify the local minimizers.

The stable configurations we found are obviously associated with the existence of *permanent currents*. These currents are experimentally observed to circulate around superconducting rings *even in the absence of an external magnetic field*. Some driving force, though, was needed to create these currents, i.e. to drive the system from a quiescent equilibrium (which is the global minimum) to the vicinity of one of the local minimizers. For example, a strong magnetic field can be used to bring the system into the domain of attraction of one the local minimizers with large homotopy type. The approach presented here enables one to predict the structure of permanent currents in domains with various topological properties.

1.4   The Little–Parks Paradigm

The second main topological aspect of superconductivity is the Little Parks effect. Little and Parks observed in 1961 [21] that the phase transition temperature in long and thin cylindrical shells is essentially a periodic function of the axial magnetic flux through the cylinder. This is obviously a quantum mechanical effect, actually a manifestation of the Aharonov-Bohm effect. More precisely, the $T_c(\Phi)$ curve, where T_c is the critical temperature and Φ is the flux, is of the form of a periodic function superimposed on a parabola. The pure effect (i.e. without the parabolic background) has a simple explanation in terms of one dimensional considerations. The parabolic background was shown by Groff and Parks [15] to be a consequence of the finite thickness of the shell. The theoretical work was based on the assumption that the order parameter has essentially constant amplitude.

With the introduction of advanced fabrication methods, modern experiments are performed on essentially two dimensional domains (i.e. the shell is actually a flat ring). For a long time people found excellent agreement when they compared experimental results with theoretical calculations (e.g. [15]). A closer look, however reveals problems. As an example we mention the experiment of Zhang and Price [35]. They measured $dI/d\Phi$ as a function of Φ, where I is the current flowing in the ring, and Φ is the flux through a disc defined by the average radius of the ring. Near the critical temperature the graphs have strong positive peaks for flux values that are approximately integer plus half (in normalized units). These peaks are unaccounted for by the usual theory (e.g. [33]).

To understand these peaks, a more refined theory is needed. In fact, Berger and Rubinstein [5] observed already before this experiment that even a slight deviation from uniform thickness implies an unusual behavior at flux values near $Z + 1/2$. They showed that in this situation, and in a one dimensional setup, the order parameter has a zero whenever the flux is exactly in the set $Z + 1/2$. Moreover, the assumption of uniform amplitude for the order parameter breaks down in a temperature interval near T_c, and for flux values near $Z + 1/2$. They have further shown [6] that the new theory can explain, at least qualitatively, the anomalies in the Zhang-Price experiment. We refer to Chap. 5 for an extensive review of this subject.

Recent theoretical progress [7], [16] revealed that the key term is *symmetry breaking*. The nonuniformity of the thickness of a ring indeed gives rise to a specific symmetry breaking. But it is only a special case of a more general situation. In the next two sections we shall examine the general GL theory on graphs, and introduce precisely the issue of asymmetry in a one dimensional setup. Extensions to two dimensions are discussed in detail in [7], [16] and in Chap. 3.

We comment that the classical literature on Little Parks oscillations assumes that the superconducting sample is multiply connected, typically with a ring-like geometry. Nevertheless, an oscillatory $T_c(\Phi)$ curve can also be obtained for simply connected domains. This has been shown experimentally by Buisson et al. [12] who considered a small disc. Later Moshchalkov, Bruyndoncx and their coworkers (see Chap. 4 for a comprehensive review) measured the transition temperature $T_c(\Phi)$ for a mesoscopic square. Both groups reported oscillatory phase boundary superimposed on a linear background. The heuristic reasoning for the oscillations is that as the flux increases, the wave function concentrates near the boundary, and the sample appears effectively as a thin ring. This gives rise to a topological quantization constraint on the phase of the wave function and hence the Little Parks oscillations. The analogy with the experiments in thin shells is not perfect, though. In the case of a simply connected sample under strong fields, the width of the 'effective' ring is not fixed. Rather, the width shrinks as the applied field increases. Further theoretical study of this issue is provided in [9], [18] and in Chap. 7.

1.5   Ginzburg–Landau Model on Graphs

The theoretical framework for the magnetic Schrödinger operator on graphs was laid down by de Gennes and Alexander (dGA) about 20 years ago. The theory was further developed by Castro, López and their coworkers (see Chap. 2). A key idea was to transform the underlying ordinary differential equations for the order parameter ψ into equations for the values of ψ at the nodes of the graph. While the idea is limited to linear problems, it enabled the calculations of many useful quantities, including the phase transition diagram for rather complicated geometries.

The problem is somewhat more involved in the nonlinear version of the dGA theory, where the quartic term in the GL energy functional is taken into account. One possible approach to the problem is to integrate the phase equations. This can be done over each of the independent cycles of the graph. The consequence is that, instead of solving an equation for the *complex* order parameter, one needs to solve an equation for the *real* amplitude of ψ, and, in addition, to find a number of constants: the total phase change across each independent cycle in the graph. We shall show in some detail how this integration is performed.

We start by introducing fundamental graph theoretical notions. We shall use them to express compactly the integrated equations. An oriented finite graph M is described by a set $\mathcal{V}$ of vertices and a family $\mathcal{E}$ of edges. The edge-node incidence matrix of a graph M is the $|\mathcal{V}| \times |\mathcal{E}|$ matrix $\mathcal{A}$ defined by $\mathcal{A}_{vj} = +1$ if v is the origin of edge j and not its end, $\mathcal{A}_{vj} = -1$ if v is the end of edge j and not its origin, and 0 otherwise. Each edge of M is numbered by $j \in \{1, \ldots, |\mathcal{E}|\}$, and each vertex is numbered by $v \in \mathcal{V}$; the number of vertices is denoted by $|\mathcal{V}|$ and the number of edges is denoted by $|\mathcal{E}|$.

A path in M is a list of vertices $\{v_1, \ldots, v_n\}$ such that for each j, (v_j, v_{j+1}) or (v_{j+1}, v_j) belongs to the set of edges $\mathcal{E}$. A path is elementary if all its vertices are distinct, and it is simple if all its edges are distinct. A cycle is a path whose end vertex v_n coincides with the origin vertex v_1. It is elementary if all its vertices are distinct, except for the coincidence between the first and last vertex.

A cycle can also be seen as an algebraic sum of edges, affected with the $+$ sign if their orientation coincides with that of the cycle, with the $-$ sign if it is opposite and with the 0 sign if the edge does not belong to the cycle; therefore, the sum of cycles can be defined, as well as the 0 cycle, and the multiplication of a cycle by any relative integer. The graph possesses a basis of independent simple and elementary cycles. Notice that there are several such bases. Denoting by $\mathcal{C}$ an arbitrary family of simple cycles, we may define the edge-cycle incidence matrix: it is the $|\mathcal{C}| \times |\mathcal{E}$ matrix $\mathcal{B}(\mathcal{C}, M)$ such that $\mathcal{B}_{lj}(\mathcal{C}, M) = +1$ if edge j belongs to cycle l and has the same orientation, $\mathcal{B}_{lj}(\mathcal{C}, M) = -1$ if edge j belongs to cycle l and has the opposite orientation, and $\mathcal{B}_{lj}(\mathcal{C}, M) = 0$ otherwise.

Assume that we can embed the graph M in $\mathbb{R}^2$. This means that the vertices are points of $\mathbb{R}^2$ and that the edges are curves of $\mathbb{R}^2$ parameterized by a smooth map ρ_j from an interval (a_j, b_j) to $\mathbb{R}^2$; we assume that ρ'_j is bounded away from 0 over $[a_j, b_j]$. Thus, without loss of generality, the parameter is the arc length. It is convenient to denote by $M_j = \rho_j((a_j, b_j))$ the arc of M indexed by j.

We need to describe the arcs leaving or entering any vertex of M so that we can write down Kirchhoff-like transmission conditions. This is done by introducing for each $v \in \mathcal{V}$ the set $J(v)$ defined as follows:

$$J(v) = \{(j, a_j, +1) : \rho_j(a_j) = v\} \cup \{(j, b_j, -1) : \rho_j(b_j) = v\}. \qquad (1.9)$$

If ζ belongs to $J(v)$, its components are denoted $(\zeta[1], \zeta[2], \zeta[3])$. There are $|J(v)|$ curves which start or end at any vertex $v \in \mathcal{V}$; an arc might start and end at v if it is a loop.

We can now write down the one dimensional GL functional on M:

$$\mathcal{H}(\psi) = \sum_{j=1}^{|\mathcal{E}|} \int_{a_j}^{b_j} \left(|i\psi_j' + A_j\psi_j|^2 + \mu[(|\psi_j|^2 - 1)^2/2] \right) ds. \tag{1.10}$$

Here ψ is the complex-valued order parameter, A (which is a real valued function on M) is the tangential component of the magnetic vector potential $\mathbf{A}$ corresponding to the (given) applied magnetic field $\mathbf{H}_e$, and $\mu = \frac{R^2}{\xi(0)^2}\frac{T_c-T}{T_c}$, where R is a length scale characterizing the perimeter of the graph, $\xi(0)$ is a material length scale (an effective zero temperature coherence length) and T_c is the phase transition temperature in the absence of external magnetic fields. A word of caution – we work in units in which the fundamental flux unit is 2π. Therefore the special flux values (referred to in the previous section), in which ψ might have zeros are now given by $k\pi$ where k is an odd integer. The one dimensional problem of minimizing $\mathcal{H}(\psi)$ is the nonlinear version of the dGA model. It was shown [27] to be the rigorous limit of the two dimensional GL functional in thin structures centered around M.

The set $\mathbb{R}^2 \setminus M$ has a finite number of bounded connected components which we denote by F_l. The ordered boundaries of the F_l's are simple elementary cycles of M: they constitute a basis of cycles of M which will be denoted by $\mathcal{C}$. The flux of the applied magnetic field $\mathbf{H}_e$ through F_l is denoted by Φ_l: it follows from Stokes formula that

$$\Phi_l = \sum_{j \in \mathcal{E}} \mathcal{B}_{lj}(\mathcal{C}, M) \int_{a_j}^{b_j} A_j \, ds. \tag{1.11}$$

The Euler Lagrange equations associated with $\mathcal{H}$ are (for every $j \in \{1, \ldots, |\mathcal{E}|\}$):

$$\left(i\frac{d}{ds} + A_j \right)^2 \psi_j + \mu(|\psi_j|^2 - 1)\psi_j = 0, \quad s \in (a_j, b_j). \tag{1.12}$$

In addition ψ is continuous at the vertices, and the following Kirchhoff transmission conditions hold there:

$$\sum_{\kappa \in J(v)} \kappa[3] \left(i\psi_{\kappa[1]}' + A_{\kappa[1]}\psi_{\kappa[1]} \right)(\kappa[2]) = 0. \tag{1.13}$$

It is convenient at this point to define the absolute value and the phase of the order parameter: $\psi = ye^{i\phi}$. At each point where $\psi_j \neq 0$, y_j and ϕ_j satisfy the following pair of differential equations, obtained by taking the real part and the imaginary part of (1.12):

$$-y_j'' + (A_j - \phi_j')^2 y_j + \mu y_j(y_j^2 - 1) = 0, \tag{1.14}$$

$$2(A_j - \phi_j')y_j' + (A_j' - \phi_j'')^2 y_j = 0. \tag{1.15}$$

Multiplying equation (1.15) by y_j and integrating, we obtain

$$(\phi'_j - A_j)y_j^2 = Constant. \tag{1.16}$$

The constant on the right hand side of (1.16) is the branch current I_j. Thus, we write more explicitly

$$(\phi'_j - A_j)y_j^2 = I_j. \tag{1.17}$$

At each vertex where ψ does not vanish, condition (1.13) is equivalent to the two conditions

$$\sum_{\kappa \in J(v)} \kappa[3]y'_{\kappa[1]}(\kappa[2]) = 0, \tag{1.18}$$

and

$$\sum_{\kappa \in J(v)} \kappa[3]I_{\kappa[1]} = 0. \tag{1.19}$$

If I_j does not vanish, we have the relation:

$$\int_{a_j}^{b_j} (\phi'_j - A_j)\,\mathrm{d}s = I_j \int_{a_j}^{b_j} \frac{\mathrm{d}s}{y_j^2}. \tag{1.20}$$

Standard results in the theory of ordinary differential equations imply that if ψ_j vanishes on (a_j, b_j), then either it vanishes identically, or it vanishes at isolated points. Furthermore, if ψ vanishes at an isolated point s_0 of $[a_j, b_j]$, then, by uniqueness of solutions of ordinary differential equations, $\psi'_j(s_0)$ cannot vanish. Therefore, we have

$$\psi_j(s) = \psi'_j(s)(s - s_0) + \ldots \tag{1.21}$$

which implies

$$y_j(s_0 + 0)e^{i\phi_j(s_0+0)} = -y_j(s_0 - 0)e^{i\phi_j(s_0-0)}. \tag{1.22}$$

Therefore, the phase jumps by an odd number times π at the zeros of ψ_j; thus if p_j is the number of zeros of ψ_j over the open subset (a_j, b_j), we combine this information with (1.17) to obtain:

$$\phi_j(b_j) - \phi_j(a_j) = p_j\pi + \int_{a_j}^{b_j} A_j(s)\,\mathrm{d}s \mod 2\pi. \tag{1.23}$$

Any cycle C_m of M is a linear combination of the ∂F_l's, which are the oriented boundaries of the F_l's:

$$C_m = \sum_{l=1}^{|C|} \epsilon_{ml}\partial F_l \tag{1.24}$$

with coefficients $\epsilon_{ml} \in Z$; thus we can define for every cycle a flux $\varPhi_m$ through the oriented region bounded by this cycle through

$$\varPhi_m = \sum_{l=1}^{|\mathcal{C}|} \epsilon_{lm} \varPhi_l. \tag{1.25}$$

Define $\widehat{M}$ to be the subgraph of M whose vertex set is $\mathcal{V}$, and whose edges are those edges of M on which the branch current does not vanish. The edge-node incidence matrix for $\widehat{M}$ is the matrix $\widehat{\mathcal{A}}$ obtained by erasing from $\mathcal{A}$ the columns indexed by $j \notin \widehat{\mathcal{E}}$, where $\widehat{\mathcal{E}}$ is the set of edges of $\widehat{M}$. Let $\widehat{\mathcal{C}}$ be a basis of simple and elementary cycles of $\widehat{M}$. Since ψ does not vanish along any cycle $C_m \in \widehat{\mathcal{C}}$, the expression

$$\sum_{j \in \widehat{\mathcal{E}}} \mathcal{B}_{mj}(\widehat{\mathcal{C}}, \widehat{M}) \int_{a_j}^{b_j} \phi_j' \, ds \tag{1.26}$$

is well defined, and it is the phase difference along a cycle $C_m \in \widehat{\mathcal{C}}$. As ψ is a continuous function, the expression (1.26) must be a multiple of 2π, which we denote by $2\pi N_m$:

$$\sum_{j \in \widehat{\mathcal{E}}} \mathcal{B}_{mj}(\widehat{\mathcal{C}}, \widehat{M}) \int_{a_j}^{b_j} \phi_j' \, ds = 2\pi N_m. \tag{1.27}$$

Define now for each real-valued function z on (a_j, b_j)

$$\Lambda_j(z) = \int_{a_j}^{b_j} z^{-2} \, ds, \tag{1.28}$$

with the convention that $\Lambda_j(z) = +\infty$ if z^{-2} is not integrable. For all functions y on M, let $\Lambda(y)$ be the $|\mathcal{E}| \times |\mathcal{E}|$ diagonal matrix

$$\Lambda_{ij}(y) = \delta_{ij}\Lambda_j(y_j). \tag{1.29}$$

For all $C_m \in \widehat{\mathcal{C}}$, we multiply (1.20) by $\mathcal{B}_{mj}(\widehat{\mathcal{C}}, \widehat{M})$ and sum with respect to $j \in \mathcal{E}$; with the help of relations (1.11), (1.27) and (1.29), we see that

$$2\pi N_m = \varPhi_m + \left(\mathcal{B}(\widehat{\mathcal{C}}, \widehat{M})\Lambda I\right)_m \tag{1.30}$$

This expression can be given an alternative formulation: Observe that the above defined basis $\widehat{\mathcal{C}}$ of the cycle space of $\widehat{M}$ consists of cycles of M. Let $\mathcal{B}(\widehat{\mathcal{C}}, M)$ be the edge-cycle incidence matrix for $\widehat{\mathcal{C}}$ in M. Then the edge-cycle incidence matrix $\mathcal{B}(\widehat{\mathcal{C}}, \widehat{M})$ is simply obtained by erasing from $\mathcal{B}(\widehat{\mathcal{C}}, M)$ all the columns indexed by $j \notin \widehat{\mathcal{E}}$.

Define further the $|\widehat{\mathcal{E}}| \times |\widehat{\mathcal{E}}|$ matrix $\widehat{\Lambda}(y)$ by

$$\big(\widehat{\Lambda}(y)\big)_{jk} = \delta_{jk}\Lambda_j(y), \quad \forall j, k \in \widehat{\mathcal{E}}. \tag{1.31}$$

We write

$$\widehat{I} = (I_j)_{j\in\widehat{\mathcal{E}}}, \quad \widehat{\xi} = 2\pi\widehat{N} - \widehat{\Phi} = (2\pi N_m - \Phi_m)_{m\in\widehat{\mathcal{C}}}. \tag{1.32}$$

With notations (1.32), relation (1.30) can be rewritten as

$$\widehat{\xi} = \widehat{\mathcal{B}}\widehat{\Lambda}(y)\widehat{I}. \tag{1.33}$$

We observe that if the edge j is a loop starting and ending at v, the terms containing I_j cancel out in (1.19); there are no contributions of $\mathcal{E}\backslash\widehat{\mathcal{E}}$ to (1.19). Thus relation (1.19) can be rewritten

$$\forall v \in \mathcal{V}, \quad \sum_{j\in\widehat{\mathcal{E}}} \widehat{\mathcal{A}}_{vj}\widehat{I}_j = 0, \tag{1.34}$$

or in a more concise form

$$\widehat{\mathcal{A}}\widehat{I} = 0. \tag{1.35}$$

We now show how to take into account the edges where the branch current vanishes. Indeed, if C_m is a cycle, we denote by $Z_m(\psi)$ the number of zeros along it, counted algebraically. Then, since, according to (1.23) and (1.20)

$$\int_{a_j}^{b_j} \phi_j' \, \mathrm{d}s = \Lambda_j I_j + \int_{a_j}^{b_j} A_j(s) \, \mathrm{d}s, \tag{1.36}$$

whenever $I_j \neq 0$, while

$$\int_{a_j}^{b_j} \phi_j' \, \mathrm{d}s = p_j \pi + \int_{a_j}^{b_j} A_j(s) \, \mathrm{d}s \tag{1.37}$$

if $I_j = 0$, we see that (1.30) generalizes to

$$2\pi N_m = \big(\mathcal{B}\Lambda I\big)_m + \Phi_m + Z_m(\psi)\pi, \tag{1.38}$$

where $Z_m(\psi)$ is the number of zeros of ψ in the cycle C_m.

The solution of the algebraic system (1.33), (1.35) is characterized by the following result:

Lemma 1. *For any choice of $\widehat{\mathcal{E}} \subset \mathcal{E}$ and for any diagonal $|\widehat{\mathcal{E}}| \times |\widehat{\mathcal{E}}|$ matrix $\widehat{\Lambda}$ with strictly positive diagonal terms, there exists a unique solution of the system (1.35), (1.33); moreover, this solution is given by*

$$\widehat{I} = \widehat{\mathcal{B}}^T \big(\widehat{\mathcal{B}}\,\widehat{\Lambda}(y)\,\widehat{\mathcal{B}}^T\big)^{-1}\widehat{\xi}. \tag{1.39}$$

This lemma enables us to give an expression of $\mathcal{H}(\psi)$ in terms of y: we have

$$\left|i\psi_j' + A_j\psi_j\right|^2 = \left|y_j'\right|^2 + y_j^2\left(\phi_j' - A_j\right)^2, \tag{1.40}$$

and thanks to (1.17),

$$\left|i\psi_j' + A_j\psi_j\right|^2 = \left|y_j'\right|^2 + I_j^2 y_j^{-2}. \tag{1.41}$$

Therefore, we obtain the identity:

$$\sum_{j\in\mathcal{E}}\int_{a_j}^{b_j}\left|i\psi_j' + A_j\psi_j\right|^2 \,\mathrm{d}s = \sum_{j\in\mathcal{E}}\int_{a_j}^{b_j}\left|y_j'\right|^2\,\mathrm{d}s + \sum_{j\in\widehat{\mathcal{E}}}\Lambda_j(y)I_j^2. \tag{1.42}$$

We now infer from Lemma 1

$$\sum_{j\in\widehat{\mathcal{E}}}I_j^2\Lambda_j(y) = \widehat{\xi}^T\left(\widehat{\mathcal{B}}\,\widehat{\Lambda}(y)\,\widehat{\mathcal{B}}^T\right)^{-1}\widehat{\xi}. \tag{1.43}$$

This implies:

Lemma 2. *Let ψ be a minimizer of $\mathcal{H}$ over $H^1(M)$ which does not vanish at the vertices of M, and let $y = |\psi|$; let $\widehat{\mathcal{E}}$ be the set of edge indices for which the branch current I_j does not vanish and let $\widehat{\mathcal{C}}$ be a maximal set of independent cycles of M on which the branch current does not vanish. Then there exists an integer vector $\widehat{N} \in Z^{\widehat{\mathcal{C}}}$, such that*

$$\mathcal{H}(\psi)=\sum_{j\in\mathcal{E}}\int_{a_j}^{b_j}\left[\left|y_j'\right|^2+\mu(y_j^2-1)^2/2\right]\,\mathrm{d}s+\left(2\pi\widehat{N}-\widehat{\Phi}\right)^T\left(\widehat{\mathcal{B}}\,\widehat{\Lambda}(y)\,\widehat{\mathcal{B}}^T\right)^{-1}\left(2\pi\widehat{N}-\widehat{\Phi}\right)^T \tag{1.44}$$

An elementary analysis indicates that when ψ has a zero on a branch j, then the associated Λ_j is infinite. This enables us to consider the formulation described in Lemma 2, while removing the "hats", and taking the (continuous) limit $\Lambda_j \to \infty$. Recalling the phase jump across zeros (1.23), we can summarize now our analysis in the following theorem:

Theorem 2. *Consider the functional*

$$\mathcal{K}(y,\xi,\Lambda) = \sum_{j\in\mathcal{E}}\int_{a_j}^{b_j}\left(\left|y_j'\right|^2 + \mu(y_j^2-1)^2/2\right)\,\mathrm{d}s + \left(\xi\right)^T\left(\mathcal{B}\Lambda\mathcal{B}^T\right)^{-1}\left(\xi\right), \tag{1.45}$$

where Λ is a diagonal matrix with diagonal entries belonging to $(0, +\infty]$. Let $\Lambda(y)$ be defined by (1.28) and (1.29). Then

$$\min\{\mathcal{H}(\psi),\quad \psi \in H^1(M;C)\} = \tag{1.46}$$

$$\min\{\mathcal{K}(y,\xi+\pi Z(y),\Lambda(y)),\quad y \in H^1(M),\quad \xi \in 2\pi Z^{|\mathcal{C}|} - \Phi\}.$$

Thus the phase equations were integrated up to a set of integers $\{N_l, \ l = 1, \ldots, |\mathcal{C}|\}$. From the definition of ξ_l and (1.45) we see at once that the minimizers are periodic with respect to each Φ_l with a (non-dimensional) period of 2π. This is of course well known experimentally. If the matrix $(\mathcal{B}^T \Lambda \mathcal{B})^{-1}$ is diagonal, then when there are no zeros, for any value of Λ the energy is minimized when every ξ_l^2 is minimized. Therefore each N_l is the closest integer to $\Phi_l/2\pi$. This is the case, for example, when every edge of M is a closed loop. For arbitrary M, however, the determination of the $\{N_l\}$ cannot be decoupled from the amplitude equations. In fact, it is possible to construct examples in which some of the $\{\xi_l^2\}$ are not minimized. Such examples, and further theoretical considerations regarding the functional $\mathcal{K}$ are provided in [28].

The dGA functional $\mathcal{H}$ is valid in the limit of very thin networks. Actually, the derivation is performed under the assumption that, as the network thickness d shrinks to zero, all other parameters in the problem are fixed. It is useful to consider one dimensional models under other preferred scaling. For example, [24] consider the limit in which the applied magnetic field is of the form $d^{-1}H_e$. The resulting one dimensional limit has an additional factor, quadratic in H_e. This model can be used, for example, to study Little Parks oscillations with the parabolic background.

Another interesting limit is when the GL parameter scales like $O(d^p)$, for some positive parameter p. Computing this limit enables the analysis of phase transitions in mesoscopic samples with small GL parameter. It is well known that in bulk samples, the transition between the normal state and the superconducting state is of type I (discontinuous) if the GL parameter is lower than $1/\sqrt{2}$. Nevertheless, type II (continuous) phase transitions are observed experimentally in Al mesoscopic samples, even though the GL parameter for this material is quite small. This effect was considered in [8] and [25] where one dimensional models were derived for canonical scaling of the GL parameter. It is interesting to note that the exponent p depends on the domain connectivity. Consider, for example, a narrow two dimensional strip. If the strip is open, i.e. it is not closed to form a ring, then the phase transition is continuous if $\kappa > C_1 d$ for some $O(1)$ constant. On the other hand, if the strip forms an annular region, the necessary condition for smooth phase transition is $\kappa > C_2 d^{1/2}$. The heuristic reasoning behind the quantitatively different behaviors in closed or open strips, is that, in fact, the nature of the phase transition is determined by the Meissner effect. Strong Meissner current favors discontinuous phase transition. In thin open strips the superconducting currents are weak (since they have nowhere to go), while in a ring the current may circulate around the hole bounded by it.

1.6   The Zero Set

One of the fascinating subjects in the study of the GL model in multiply connected domains is the zero set of the order parameter. Several chapters in this volume address this question from different perspectives. We shall consider it now for the case of one dimensional models, with a geometry similar to the preceding section. We have shown in this section how to handle zeros while integrating the phase equations. But do zeros indeed occur? Consider, for example, a single uniform loop. It is natural to expect a solution with a uniform order parameter. However, it was shown in [5] that even slight deviations from uniformity imply (for certain flux values) large variations in the amplitude of the order parameter. The question is closely related to the nature of the transition between different types of circulation along cycles in the graph. As the flux through a cycle j is varied, we reach a value where the total circulation, characterizing the j-th entry of the vector ξ changes to an adjacent integer. The question is whether this change is continuous (and thus accompanied by a zero of ψ somewhere along the cycle), or discontinuous.

A partial answer was given in [6] for single narrow rings whose cross section deviates slightly from a constant. In the limit case of the one dimensional model it was found that in general a zero will form at some point along the loop whenever the flux is an odd multiple of π. A general theoretical framework to study this zero formation, together with extensions to two dimensional situations was developed later in [7] (see also Chap. 5). One consequence of this theory is that under generic asymmetry assumption, a zero will form for the critical flux values even for rings with arbitrary cross section. The crucial point is to define an appropriate notion of asymmetry in this context, since the definition used in [7] is neither natural nor easy to check. The problem was resolved by Helffer et al. in [16] (see also Chap. 3 for a survey of recent progress by this group). They proposed to consider the GL equations over the double covering of the ring. In this space one can find a convenient gauge, and the equations simplify considerably.

I shall apply the method of Helffer et al. to the problem of zeros in graphs. I assume that the flux through each F_l is an odd multiple of π. Furthermore, to simplify the exposition, I shall limit myself to linear problems. Physically, this amounts to considering the phase transition from the normal state, where $\psi \equiv 0$, to a superconducting state. In graphs this phase transition is always continuous, taking the form of a bifurcating branch. The bifurcation occurs at the lowest μ where the zero solution is no longer a local minimum. We denote this critical value by μ_c. Calculating the second variation of the GL functional [4] one finds that μ_c is exactly the ground energy of the magnetic Schrödinger operator on M. The superconducting order parameter is proportional to the ground state. Therefore the problem we shall consider now is whether the ground state has zeros.

The double covering of M is denoted by $\widetilde{M}$. In fact there are many double coverings, and we have to choose an appropriate one. For this purpose we lift both ψ and A into $\widetilde{M}$. We denote a function f on M lifted to $\widetilde{M}$ by $\tilde{f}$. We pick a covering such that the integral of $\tilde{A}$ over every cycle in $\widetilde{M}$ is $k\pi$ for some even $k \in Z$. Denoting the primitive of $\tilde{A}$ by ζ, our choice of $\widetilde{M}$ guarantees that $e^{i\zeta}$ is a single valued function on $\widetilde{M}$. This is the crucial observation that facilitates the analysis. It is here that we need the condition that the flux through every cycle of M is an odd integer times π. We can therefore define the new gauge

$$\tilde{v} = e^{-i\zeta}\tilde{\psi}. \tag{1.47}$$

In this gauge the GL functional over $\widetilde{M}$ reads

$$\widetilde{F}_{GL} = \int_{\widetilde{M}} \tilde{w}(|\tilde{v}'|^2 + \frac{1}{2}\mu(|\tilde{v}|^2 - 1)^2)ds, \tag{1.48}$$

where we introduced a weight $\tilde{w}$ that models nonuniformities in the network thickness. The nonuniformity will enable us to investigate symmetry breaking even in single loops.

Consider further the mapping $G : \widetilde{M} \to \widetilde{M}$ which sends every point to its corresponding point on the other copy of M. Since $\tilde{\psi}$ is G symmetric by construction, then $\tilde{v}$ is G antisymmetric. The weight $\tilde{w}$, on the other hand, is lifted into $\widetilde{M}$ from the weight w in M, and thus it is G symmetric. Therefore the variational problem we consider is to minimize $\widetilde{F}_{GL}$ over complex G antisymmetric functions in $\widetilde{M}$. The advantage of the new formulation is that a complicated problem on M was reduced to a simpler problem on $\widetilde{M}$, except that the minimization is now taken over a special class of functions. The critical temperature is determined by the eigenvalue problem of minimizing $\tilde{\mu}_c = \int_{\widetilde{M}} \tilde{w}|\tilde{v}'|^2$ over all complex G antisymmetric functions $\tilde{v}$ such that $\int_{\widetilde{M}} \tilde{w}|\tilde{v}|^2 = 1$. The minimal value is the eigenvalue $\mu_c = \tilde{\mu}_c$, associated with a ground state $\tilde{v}$.

Instead of minimizing over complex valued functions, we consider the eigenvalue problem of minimizing $\tilde{\mu}_c^s = \int_{\widetilde{M}} \tilde{w}\tilde{u}'^2$ over all *real* G antisymmetric functions $\tilde{u}$ such that $\int_{\widetilde{M}} \tilde{w}\tilde{u}^2 = 1$. Assume $\tilde{\mu}_c^s$ is a simple eigenvalue, associated with the eigenfunction $\tilde{u}_c^s$. We argue that in this case $\tilde{\mu}_c = \tilde{\mu}_c^s$, and $\tilde{v}_c = \tilde{u}_c^s$. For suppose, in contradiction, that $\tilde{\mu}_c < \tilde{\mu}_c^s$. Since the spectral problem for $\tilde{\mu}_c$ is invariant under complex conjugation, both the real and imaginary parts of $\tilde{v}$ are real G antisymmetric eigenfunctions. Hence our assumption on $\tilde{\mu}_c^s$ implies that they are proportional to each other and to $\tilde{u}$.

Consider now the special example where M is a single loop, and assume that $\tilde{\mu}_c^s$ is simple (this is why we need the nonuniformity; when $w = 1$, the eigenspace of $\tilde{\mu}_c^s$ has dimension two). Clearly a G antisymmetric function must have at least one zero on each of the copies of M that comprise $\widetilde{M}$. Using G antisymmetry again, it is easy to verify that there cannot be two zeros on each copy, and simple surgery argument implies that an eigenfunction associated with a smallest simple eigenvalue cannot have three or more zeros.

An interesting question is whether the zero(s) occur only in the linear bifurcating solution. It was shown in [7] (and in more detail in Chap. 3) that, in fact, if $\widetilde{\mu}_c^s$ is simple, there is an interval $(\mu_c, \mu_c + \delta)$ (physically, a temperature interval) where the solution to the full nonlinear GL problem has a zero for $\Phi = k\pi$ for odd k. Another important question regards the nature of the zero in a truly two dimensional domain Ω with holes. The important aspect in our construction was our ability to "gauge out" the magnetic potential. This can be done whenever the potential is a gradient of some function. In one dimension this is always true. But in two dimensions we need the compatibility condition $\nabla \times A = 0$ to hold. This condition is equivalent to assuming that the magnetic field vanishes in Ω (although it should not vanish in the holes bounded by Ω, in order to guarantee the desired flux value). Indeed, it can be shown (Chap. 3) that the conclusion regarding the identification of $\widetilde{v}$ with $\widetilde{u}$, in the case where $\widetilde{\mu}_c^s$ is simple, holds in two dimensional multiply connected domains, under the special assumption that $H_e \equiv 0$ in Ω. Again the zero set of ψ is the nodal set of $\widetilde{u}$ – the leading G antisymmetric eigenfunction of the Laplacian on $\widetilde{\Omega}$. Thus the zero set is of codimension one. When H_e does not vanish in Ω, there may still be a smooth transition between circulations along closed loops in Ω. But now the transition is mediated by vortices (see Chap. 5 for a detailed investigation of this effect).

We have shown that when M is a loop and $\widetilde{\mu}_c^s$ is simple, there is exactly one zero in the order parameter. We now set $w \equiv 1$, i.e. assume uniform thickness, and consider a canonical version of the model in an arbitrary graph. Given a graph M we face two questions . First, is $\widetilde{\mu}_c^s$ simple? Then, if it is simple, what is the size and structure of the zero set? Clearly there will be at least one zero, but in sufficiently complex graphs we could expect more zeros. To make the discussion more concrete, let us analyze the special case of symmetric ladders. An n symmetric ladder consists of n identical squares in a row. Parks conjectured that the order parameter will have zeros in an n ladder if and only if n is even. The conjecture was verified experimentally for $n = 1, 2, 3$. For the case of symmetric ladders, it can be checked that when n is odd, the symmetries of the ladder imply that $\widetilde{\mu}_c^s$ is not simple. On the other hand, it can be verified that $\widetilde{\mu}_c^s$ is simple for n even. Thus we have a simple justification of Park's conjecture in this case. We emphasize, though, that the original conjecture did not take into account the strict symmetry requirement. By this I mean that ψ can have zeros even in a 3 ladder, say, if we replace the squares by unequal loops, or if we introduce nonuniformities in M. Returning to the question of estimating the size of the zero set, it is easy to check that the number of zeros is at least $n/2$ (for even n). I suspect that for symmetric ladders this is also the actual number of zeros, but I have not verified that. I refer to [29] for an extensive discussion on double covering of graphs and estimates on the number of zeros of G antisymmetric functions there.

1.7   Other Problems with Topological Flavor

We finally mention briefly two more examples in which topology plays an interesting role in superconductivity.

1.7.1   Josephson Junctions

Our first example concerns loops with impurities. The impurities can be modeled in several ways. For example, Chap. 5 discusses impurities in the form of very narrow contrictions in the loop. Another way to model impurities is through a modified GL model. The idea is to recall that the Landau phase transition theory is based on an energy functional with a quadratic term, $a(T)|\Psi|^2$, in (1.1), such that a becomes negative as T decreases past a critical value T_c. An impurity can be modeled by a quadratic term with a coefficient that remains positive even at low T. The model was introduced long time ago (e.g. [3], [20]). It has been recently reconsidered in [17] who incorporated the screening effect of the impurity into the topological constraint on the phase in a closed loop.

To describe the model in some detail, consider a ring parameterized by $0 \leq s \leq 2\pi$. The ring consists of two sets: The normal part $\{0 < s < d\}$ and the superconducting part $\{d < s < 2\pi\}$. The Modified Ginzburg Landau (MGL) model then takes the form

$$F_{MGL}(\psi) = \int_0^d (|(\frac{d}{ds} - iA)\psi|^2 + \frac{\alpha^2}{d^p}|\psi|^2)ds$$
$$+ \int_d^{2\pi} (|(\frac{d}{ds} - iA)\psi|^2 + \frac{\mu^2}{2}(|\psi|^2 - 1)^2)ds, \qquad (1.49)$$

where p is a positive parameter that controls the 'strength' of the impurity and μ is again proportional to $\frac{T_c-T}{T_c}$. The model is somewhat similar to the 'toy model' of Section 3.6, except that the impurity contribution is not exactly in the form of a potential added to the regular GL functional.

The model (1.49) can be investigated along the lines of Chap. 5. One of the interesting quantities to compute is the dependence of the critical temperature T_c (via μ) on the flux Φ enclosed by the loop. Recall that the critical temperature is defined as the smallest value of μ for which the zero solution is no longer a minimizer. The function $\mu_c(\Phi)$ can be computed explicitly in the limit of small d for one or more impurities. Its shape depends crucially on the impurity strength, defined as the L_1 norm of the impurity potential. For weak impurities ($p < 1$), the loop behaves to leading order as a 'clean ring'. When $p = 1$, one obtains a complex dependence of μ_c on Φ. For strong impurities ($p > 1$) the order parameter vanishes to leading order at $s = 0$, and the supercurrent satisfies the Josephson formula $I \sim \sin(\Phi)$.

1.7.2   Vortex Lines

So far we have discussed topological effects that are associated with the geometry confining the superconducting sample. Considering superconductors in $\mathbb{R}^3$, we may encounter additional interesting topological entities, since the zero set of the order parameter might form nontrivial loops, links and knots. This problem received a lot of attention in the study of superfluids, where the entanglement of vortex lines is associated with turbulence. The standard time dependent GL model in superconductivity is dissipative, so a different behavior is expected here. Nevertheless, one wonders if the action of dissipation may not act just to simplify the form of the vortex line, without altering its basic topological identity. For example, suppose that the initial form of the vortex line is a knot of some kind; will the knot simplify in due course to a canonical knot, or will the knot change its form completely, or even vanish while evolving under the flow?

The problem is still open as far as the full GL model is concerned. But an answer is available in the special mathematical model where the magnetic vector potential is neglected [30]. Let $u(x,t)$ be the solution to

$$u_t = \Delta u + u(1 - |u|^2|) \tag{1.50}$$

in $\mathbb{R}^3 \times \mathbb{R}_+$, subjected to the initial data $u(x,0) = u_0(x)$. Then, under mild smoothness assumptions on u_0, $u(x,t)$ converges as $|x| + t \to \infty$ to some constant unit vector e. Suppose now that the zero set (line vortices) of the initial data has a complex pattern of loops and knots. The result quoted above implies that these complex shapes disappear in finite time. Therefore the flow associated with the model (1.50) cannot stabilize or even support nontrivial topological patterns in its zero set.

Acknowledgments

This research was supported by the Israel Science Foundation. I wish to thank my colleagues J. Berger, M. Schatzman and P. Sternberg for their help and advice.

References

1. A.A. Abrikosov: *Fundamentals of the Theory of Metals*, North Holland, 1988.
2. l. Almeida, F. Bethuel: J. Math. Pure Appl. **77**, 1 (1998)
3. A. Baratoff, J.A. Blackburn, B.B. Schwarz: Phys. Rev. Lett. **16**, 1096 (1970)
4. P. Bauman, D. Phillips, Q. Tang: Arch. Rat. Mech. Anal. **142**, 1 (1998)
5. J. Berger, J. Rubinstein: Phys. Rev. Lett. **75**, 320 (1995)
6. J. Berger, J. Rubinstein: Phys. Rev. B. **56**, 5124 (1997)
7. J. Berger, J. Rubinstein: Comm. Math. Phys. **202**, 621 (1999)
8. J. Berger, J. Rubinstein: "Continuous phase transitions in mesoscopic systems", ZAMP, in press

9. A. Bernoff, P. Sternberg: J. Math. Phys. **39**, 1272 (1998)
10. F. Bethuel: Acta Math. **167**, 153 (1991)
11. H. Brezis, Y. Li, P. Mironescu, L. Nirenberg: "Degree and Sobolev Spaces", preprint.
12. O. Buisson, P. Gandit, R. Rammal, Y.Y. Wang B. Pannetier: Phys. Lett. **150**, 36 (1990).
13. P.G. de Gennes: *Superconductivity in metals and Alloys*, Addison Wesley (1989).
14. Q. Du, M.D. Gunzburger, J.S. Peterson: SIAM Review **34**, 54, (1992)
15. R.P. Groff, R.D. Parks: Phys. Rev. **176**, 567 (1968)
16. B. Helffer, M. Hoffmann-Ostenhof, T. Hoffmann-Ostenhof, M. Owen: Comm. Math. Phys. **202**, 629, (1999)
17. E. Hill, P. Sternberg, J. Rubinstein: "The modified Ginzburg Landau model for Josephson junctions in a ring", preprint
18. H.T. Jadallah, J. Rubinstein, P. Sternberg: Phys. Rev. Lett. **82** 2935, (1999)
19. S. Jimbo, Y. Morita: SIAM J. Math. Anal. **27**, 1360 (1996)
20. K. Likharev: Rev. Mod. Phys. **51**, 101 (1979)
21. W.A. Little, R.D. Parks: Phys. Rev. Lett **9**, 9 (1962)
22. V.V. Moshchalkov, L. Gielen, C. Strunk, R. Jonckheere, X. Qiu, C. Van Haesendonck, Y. Bruynseraede: Nature (London) **373**, 319 (1995)
23. B. Pannetier: "Superconducting wire networks", In *Quantum Coherence in Mesoscopic Systems* B. Kramer, Ed. (Plenum Press 1991), 457-484
24. G. Richardson, J. Rubinstein: Proc. Royal Soc. London **455**, 2549 (1999)
25. G. Richardson, J. Rubinstein: "Effective equations for thin superconductors with small Ginzburg-Landau parameter", SIAM J. Appl. Math., to appear.
26. J. Rubinstein: "Six lectures on superconductivity", in *Boundaries, Interfaces and Transitions*, M. Delfour Ed. (American Mathematical Society 1998), 163-184
27. J. Rubinstein, M. Schatzman: J. Math. Pure Appl. **77**, 801 (1998)
28. J. Rubinstein, M. Schatzman: "Variational problems on multiply connceted thin strips III: Integration of the Ginzburg Landau equations over graphs", preprint
29. J. Rubinstein, M. Schatzman: "Variational problems on multiply connceted thin strips IV: Zero set for the linearized Ginzburg Landau equations", in preparation
30. J. Rubinstein, M. Schatzman, P. Sternberg: C. R. Acad. Sci. Paris. **322**, 31 (1996)
31. J. Rubinstein, P. Sternberg: Comm. Math. Phys. **179**, 257 (1996)
32. C. Strunk, V. Bruyndoncx, V.V. Moshchalkov, C. Van Haesendock, Y. Bruynseraede, R. Jonckheere: Phys. Rev. **B 54**, R12701 (1996)
33. M. Tinkham: *Introduction to Superconductivity*, McGraw Hill, 1996.
34. B. White: Acta Math. **160**, 1 (1988)
35. X. Zhang, J.C. Price: Phys. Rev. B **55** 3128 (1997)

2 The de Gennes–Alexander Theory of Superconducting Micronetworks

José I. Castro[1] and Arturo López[2]

[1] Universidad Nacional de San Juan, Argentina
[2] Centro Atómico Bariloche and Instituto Balseiro, Argentina

2.1 Introduction

The technology of the second half of the XX century was largely based on the applications of the quantum physics of the solid state, which explains the electronic properties of matter. Technology of the XXI century will surely be based on the *macroscopic* quantum properties of matter, that is to say, on the coherent macroscopic states of physical systems, some of their best known examples are the *laser, superfluidity* and *superconductivity*. The present review is dedicated to a presentation of the bibliography and the more relevant results referred to *superconducting micronetworks*, that is to say, circuits made of superconducting material, fabricated by photolitography or carved using particle beams, electrons or ions, on insulating substrates. The circuits to which we refer have connections with characteristic lengths of the order of $1\ \mu m$ and transverse sections of about $0.01\ \mu m^2$. This allows to identify them as *mesoscopic*, since, from the microscopic point of view, they are large relative to the superconducting properties characterized by the coherence length of the Cooper pairs, but small relative to the characteristic dimensions of macroscopic superconductivity. It was de Gennes [9,10] who first introduced these structures in 1981 as a model for *inhomogeneous* superconductors. His interest was centered in the analysis of the magnetic susceptibility of a kind of "spaghetti and sauce" superconductor, where the spaghetti were made up of superconducting material, and the sauce was some insulating material. The thermodynamic and electrodynamic properties of these systems are more related to their topology (i.e. their connectivity) than to their geometry. To study this kind of system in the vicinity of the normal-superconductor transition, de Gennes devised the theory which we are going to present below. The importance of percolation phenomena in this systems lead Shlomo Alexander [1,3] in 1983 to the study of certain systems which are paradigmatic in percolation theory, like the so called Sierpinsky gasket. The possibility of building *regular* superconducting networks lead other authors to the study of these latter structures, as models of artificial systems in which quantum coherence plays an important role [12,23,25,26] due to fluxoid quantization along intertwined loops. Later on several papers applied the theoretical formulation developed by de Gennes and Alexander to micronetworks of varied geometric and topological characteristics (infinite networks, planar and spatial networks, fractal and disordered networks, etc.) obtaining results which are important not only because of the physics underlying them but also because of possible future applications.

The de Gennes-Alexander theory was very successful in explaining the description of the second order phase transition between the normal and superconducting phases and is the subject of this review. The properties of these same systems in the superconducting state far from the phase transition boundary cannot be described by this theory, which is a linear approximation to the Ginzburg-Landau theory. In Section 2.4 we review nonlinear aspects of the Ginzburg-Landau theory, allowing for the exploration of the phase diagram in the whole range of validity of the GL theory.

2.1.1   The Ginzburg–Landau Theory

The Ginzburg-Landau [20,15] theory of the superconductor-normal phase - transition in a bulk material, in the presence of an applied magnetic field, describes the superconductor thermodynamics assuming that:

1) There is a complex order parameter $\psi(\mathbf{r})$ characterizing the electronic condensation in the superconducting phase at position $\mathbf{r}$ inside the material; ψ goes to zero continuously and vanishes when the material is in the normal state; $|\psi|^2$ is proportional to the density of superconducting electrons.

2) The free energy density F_S can be expanded in powers of $|\psi|^2$ and $|\nabla\psi|^2$, with coefficients which are regular functions of the temperature T.

Taking into account the gauge invariance condition for F_S, the total free energy of the system is an integral of the free energy density over all space, in the form

$$F_S = F_N + \int \left[\alpha\,(T)\,|\psi|^2 + \frac{\beta(T)}{2}\,|\psi|^4 + \frac{1}{2m_e}\left|(i\hbar\nabla - 2\frac{e}{c}\mathbf{A})\psi\right|^2 + \frac{H_i^2}{8\pi} \right] d^3\mathbf{r}.$$
$$(2.1)$$

Here F_N is the normal state free energy (when $\psi \equiv 0$), $\mathbf{H}_i$ is the magnetic field induced by the supercurrents and $\mathbf{A}$ is the magnetic vector potential, i.e. $\nabla \times \mathbf{A} = \mathbf{H}_e + \mathbf{H}_i$, where $\mathbf{H}_e$ is the applied field. The magnetic potential is usually taken in the Coulomb gauge, $\nabla \cdot \mathbf{A} = 0$. The normal phase is assumed non magnetic.

The coefficients $\alpha(T)$ and $\beta(T)$ are assumed to be of the form:

$$\alpha(T) = C(T - T_c) \ , \ \ \beta(T) = \beta(T_c) = C' \qquad (2.2)$$

where T_c is the zero field critical temperature, C and C' being constants.

Furthermore, it is possible to relate $\alpha(T)$ and $\beta(T)$ to the zero field order parameter $\psi_\infty(T)$ and to the thermodynamic critical field $H_c(T)$, both defined for the bulk material under Meissner conditions (complete field expulsion):

$$\frac{\alpha}{\beta} = -|\psi_\infty|^2 \quad , \quad 4\pi\frac{\alpha^2}{\beta} = H_c^2. \tag{2.3}$$

Defining the "GL coherence length" $\xi = \hbar/\sqrt{2m_e|\alpha|}$, the free energy difference between the superconducting and normal states for a volume V of material can be written

$$
\begin{aligned}
F_{GL} &= F_S - F_N \\
&= \frac{H_c^2}{4\pi}\int\left[-|\psi|^2 + \frac{|\psi|^4}{2} + \xi^2\left|\left(i\nabla - \frac{2\pi}{\Phi_0}\mathbf{A}\right)\psi\right|^2\right]d^3\mathbf{r} + \frac{1}{8\pi}\int H_i^2 d^3\mathbf{r}\ .
\end{aligned}
\tag{2.4}
$$

In this equation and from now on, we use a normalized order parameter $\psi(\text{new}) \equiv \psi(\text{old})/\psi_\infty$, $\Phi_0 = ch/2e$ is the magnetic flux quantum (or "fluxon") and

$$\xi = \frac{\xi(0)}{\sqrt{1 - T/T_c}} \tag{2.5}$$

The condition for F_{GL} to have an extremum under variations of the order parameter and the magnetic vector potential implies that the fields $\psi = \psi(\mathbf{r})$ and $\mathbf{A} = \mathbf{A}(\mathbf{r})$ must satisfy the well known GL differential equations:

$$\xi^2\left(i\nabla - \frac{2\pi}{\Phi_0}\mathbf{A}\right)^2\psi + (|\psi|^2 - 1)\psi = 0 \tag{2.6}$$

$$\frac{c}{4\pi}\nabla \times (\nabla \times \mathbf{A}) = \mathbf{j} \tag{2.7}$$

where the superconducting current is

$$\mathbf{j} = \frac{c\Phi_0}{(4\pi)^2\lambda^2}\left[\psi^*\left(i\nabla - \frac{2\pi}{\Phi_0}\mathbf{A}\right)\psi + (c.c.)\right] \tag{2.8}$$

and $\lambda = \sqrt{m_e c^2/16\pi e^2\,|\psi_\infty|^2}$ is the "GL penetration depth". The equation (2.6) has the form of Schroedinger equation for a particle in a magnetic field and with a scalar potential energy proportional to $|\psi|^2$; equation (2.7) is Maxwell equation for the magnetic field in terms of the superconducting current density.

The coherence length ξ and the penetration depth λ are the characteristic distances for changes in the fields ψ and $\mathbf{A}$ respectively. The approximate form

$$\lambda = \frac{\lambda(0)}{\sqrt{1 - T/T_c}} \tag{2.9}$$

is generally assumed, $\lambda(0)$ being a constant expressed in terms of the normal state electronic parameters. It is useful to observe that

$$2\pi\sqrt{2}\lambda\xi H_c = \Phi_0 \tag{2.10}$$

and

$$\kappa = \frac{\lambda}{\xi} = \frac{\lambda(0)}{\xi(0)} = \frac{m_e\Phi_0}{2\pi\hbar^2}\sqrt{\frac{\beta}{2\pi}} \tag{2.11}$$

are quantities independent of T in this theory; the dimensionless ratio κ is usually called the "GL constant".

Equations (2.6) and (2.7) must be solved with the boundary conditions following from the variational procedure applied to the free energy functional F_{GL}.

Assuming $\psi = |\psi|\,e^{i\varphi}$ in (2.7), the current density can be expressed as:

$$\mathbf{j} = -\frac{2c\Phi_0}{(4\pi\lambda)^2}\,|\psi|^2\left(\nabla\varphi + \frac{2\pi}{\Phi_0}\mathbf{A}\right) = -\frac{2c\Phi_0}{(4\pi\lambda)^2\xi}\,|\psi|^2\,\mathbf{Q}, \tag{2.12}$$

where we have defined the supercurrent "velocity" $\mathbf{Q} = \xi(\nabla\varphi + \frac{2\pi}{\Phi_0}\mathbf{A})$. The negative electron charge makes $\mathbf{j}$ and $\mathbf{Q}$ opposite vectors. Integrating this expression along a closed path C inside the material, and recalling that the phase φ can vary only by integer multiples of 2π when returning to the starting point, the important relation

$$\frac{4\pi\lambda^2}{c}\oint_C \frac{\mathbf{j}}{|\psi|^2}\cdot d\mathbf{r} + \Phi = m\,\Phi_0 \ (m\ \text{integer}) \tag{2.13}$$

is obtained, where $\Phi = \oint_C \mathbf{A}\cdot d\mathbf{r}$ is the total magnetic flux across a surface with contour C. The left hand side in this formula is called the "fluxoid" and relation (2.13) is called the "fluxoid quantization condition". The integer m is also called the phase winding number.

Finally, using the GL equations (2.6) and (2.7) and the boundary conditions in the free energy F_{GL} the thermodynamic equilibrium free energy is found to be:

$$F_{eq} = -\frac{H_c^2}{8\pi}\int_V |\psi|^4\,d^3\mathbf{r} + \frac{1}{8\pi}\int_{(\infty)} H_i^2 d^3\mathbf{r}. \tag{2.14}$$

In this expression the interplay between the negative superconductive condensation energy, depending of $|\psi|^2$), and the positive magnetic energy, depending of the induced supercurrents $\mathbf{j}$ in the material, can be clearly seen.

2.1.2   de Gennes–Alexander Linear Approach for Micronetworks

At the beginning of the 80's, Pierre G. de Gennes [9,10] applied the GL theory to filamentary structures of superconducting material at the submicron

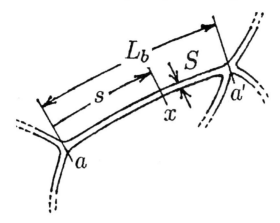

Fig. 2.1. Branch $b \equiv (a, a')$ with cross section area S and length L_b. The point x has curvilinear coordinate s

level; the theory was devised to deal with randomly distributed filaments as a model for random heterogenous systems; these multiply connected structures are called "superconducting micronetworks". Using the GL equations in the limit when $\psi \to 0$, de Gennes studied the normal-superconductor second order phase transition boundary on the (H_e, T) plane; in this limit, H_e (the externally applied magnetic field) can be identified with $|\nabla \times \mathbf{A}|$ (the total local magnetic field), because the induced superconducting current approaches zero: $\mathbf{j} \to 0$. The first system to be studied was a loop connected to an open branch, that de Gennes called "the lasso". He then made an statistical extrapolation to a "soup" of lassos as a model of a large randomly connected filamentary system. Later Shlomo Alexander [1,3] extended the theory to regular complex systems (square network, triangular Sierpinsky gasket), emphasizing the algebraic equations for the order parameter at the nodes; Fink, López and Maynard [12] worked out the case when external currents are fed into the micronetwork and Rammal, Lubensky and Toulouse [25] applied the theory to ladder structures (See also Ref. [26], [27]). A brief presentation of de Gennes-Alexander (dGA) theory follows.

A superconducting micronetwork is a collection of thin wires with uniform cross section S, connected at nodes n; a wire connecting two nodes (a, a') is called a branch b with length L_b (Fig. 2.1). A loop l is a closed path on the network including a certain number of branches only once; an arbitrary network with B branches and N nodes has $L = B - N + 1$ independent loops that uniquely describe the superconducting current distribution. The whole network is assumed to lie on an insulating substrate and to be in the presence of an uniform externally applied magnetic field.

We assume thin wires meaning that the following conditions are satisfied

$$a) \; \xi^2(0) < S < \xi^2$$

$$b) \; L_b \geq \xi$$

$$c) \; S \leq \lambda^2$$

The lower limit for S given by a) ensures the validity of GL theory. Conditions a) and b) imply that the order parameter can be considered uniform on the wire cross section, varying only along the curvilinear coordinate s defined along each branch: $\psi = \psi(s)$. Condition c) in the limit when $\psi \to 0$, implies that there is no magnetic shielding inside the wire: the network is "transparent" to the magnetic field. The vanishingly small current density $\mathbf{j}$ is uniform on S. Because of the geometry, only the component of $\mathbf{A}$ along the branches is relevant. At a given point of coordinate s on a branch we take $\mathbf{A} = A(s)\mathbf{s}$, $\mathbf{s}$ being the unit vector directed along the branch at that point.

Near the phase transition boundary, $\psi \to 0$, $\mathbf{j} \to 0$ and $\mathbf{A} \to \mathbf{A}_e$, where $\mathbf{A}_e$ is the externally applied magnetic potential at the transition. Taking into account that $\psi = \psi(s)$ and $\mathbf{A} = A(s)\mathbf{s}$ and neglecting terms of order higher than $|\psi|^2$, the free energy (2.4) for the network becomes

$$F_{GL} = \frac{H_c^2 S}{4\pi} \sum_b \int_0^{L_b} \left[-|\psi(s)|^2 + \xi^2 \left| \left(i\frac{d}{ds} - \frac{2\pi}{\Phi_0} A_e(s) \right)\psi(s) \right|^2 \right] ds \quad (2.15)$$

where $\sum_b$ is the sum over all branches. In this expression, $\psi(s)$ is the only quantity to be determined variationally. The condition for the extremum of F_{GL} leads to the GL linearized equation on each branch

$$\xi^2 (i\frac{d}{ds} - \frac{2\pi}{\Phi_0} A_e(s))^2 \psi(s) - \psi(s) = 0 \quad (2.16)$$

and to the conditions at each node of the network

$$\sum_{(a')} \left[(i\frac{d}{ds} - \frac{2\pi}{\Phi_0} A_e(s))\psi(s) \right]_a = 0 \; , \; (a = 1, ..., n) \quad (2.17)$$

Here $\sum_{(a')}$ is the sum over the branches connecting the node a with its first neighbor nodes a'. This conditions imply the validity of Kirchhoff laws at each node. Equation (2.16) is formally identical to Schroedinger equation for a free electron in a domain with the network symmetry, in presence of a magnetic field; ξ^{-2} plays the role of the eigenvalue for the Hamiltonian operator $(i\frac{d}{ds} - \frac{2\pi}{\Phi_0} A_e(s))^2$. The phase transition boundary is determined by the lowest eigenvalue ξ^{-2}, which we call μ_0:

$$\mu_0 = \xi^{-2} = (\xi(0))^{-2}\left(1 - \frac{T_0}{T_c}\right), \tag{2.18}$$

T_0 being the temperature at which the transition occurs when the considered external field is applied.

The solution of (2.16) on branch $b \equiv (a, a')$ is

$$\psi_0(s) = \frac{\exp(-i\gamma_0(s))}{\sin(\sqrt{\mu_0}L_b)}[\psi_0(a)\sin(\sqrt{\mu_0}(L_b - s))$$
$$+ \psi_0(a')\exp(i\gamma_0(L_b))\sin(\sqrt{\mu_0}s)] \tag{2.19}$$

with

$$\gamma_0(s) = \frac{2\pi}{\Phi_0}\int_0^s A_e(s')\,ds' \tag{2.20}$$

Substituting (2.19) in (2.17) we obtain a linear system of n equations for the n order parameters ($\psi_0(a)$, $a = 1, ..., n$) at the nodes of the network, that are our unknowns; these are the "Alexander nodal equations":

$$\psi_0(a)\sum_{(a')}\cot(\sqrt{\mu_0}L_b) - \sum_{(a')}\frac{\psi_0(a')\exp(i\gamma_0(L_b))}{\sin(\sqrt{\mu_0}L_b)} = 0 \ (a = 1, ..., n) \tag{2.21}$$

The non trivial solution of this homogeneous system determines the relative magnitude of the order parameter at each node, whereas the condition of zero characteristic determinant gives a relation between μ_0 and γ_0 corresponding to the $H_e = H_e(T_0)$ line for the second order phase transition boundary in the (H, T) plane.

The superconducting current density in each branch b, j_{0b}, can be obtained from (2.8) and (2.19):

$$j_{0b} = \frac{2c\Phi_0}{(4\pi\lambda)^2}\,\text{Re}\left[\psi_0^*(s)(i\frac{d}{ds} - \frac{2\pi}{\Phi_0}A_e(s))\psi_0(s)\right]_{s=0}$$
$$= J_b\frac{(\sqrt{\mu_0}L_b)}{\sin(\sqrt{\mu_0}L_b)}|\psi_0(a)|\,|\psi_0(a')|\sin[(\varphi_a - \varphi_{a'}) - \gamma_0(L_b)] \tag{2.22}$$

where $J_b = (2c\Phi_0)/(4\pi\lambda)^2L_b$. Notice that charge conservation implies that j_{0b} is independent of the coordinate s along the branch.

It is necessary to note here that, being (2.16) a linear equation, the amplitude of $\psi(s)$ is undetermined, and the equations (2.21) allows us to find only the relative magnitude of the order parameter; being j_{0b} proportional to $|\psi_0|^2$, also j_{0b} is undetermined in amplitude. The amplitude can be obtained considering the non linear terms of the GL theory that are neglected in the de Gennes-Alexander approximation.

2.1.3   Symmetry Analysis in the dGA Theory

Although the physical properties of networks are basically determined by the topology, symmetry properties are very helpful in analyzing the solutions to the linearized equations [4]. Artificially fabricated superconducting micronetworks can have geometrical symmetries by design, useful to predict the type of "modes" that the nodal order parameter will show when searching for the solutions of Alexander equations (2.21); these modes correspond to the different superconductive condensation and currents distributions in the wires, and their relative stabilities depend on the field and temperature ranges considered. Being a linear theory, it is possible to analyze the solutions applying the theory of symmetry groups; this allows for the classification of modes according to the transformation properties of the order parameter under the group operations.

The external field $\mathbf{H}_e$ reduces the number of symmetry elements of the geometrical space group. In fact, being the magnetic field a pseudovector, the system remains invariant only under those point operations that keep $\mathbf{H}_e$ invariant (as an example, the mirror planes parallel to the field *are not symmetry elements of the system*). Using the symmetric gauge $\mathbf{A}_e = \mathbf{H}_e \times \mathbf{r}/2$, the only acceptable point groups for regular superconducting micronetworks, referred to the axis defined by $\mathbf{H}_e$, are C_n (rotation around an axis by $2\pi/n$), C_{nh} (same as C_n plus a symmetry plane normal to the axis; C_{1h} is also called C_s), and S_n (similar to C_{nh}, but the symmetry includes *both rotation and reflection*) in Schönflies notation.

Taking the nodal order parameters $\psi_0(a)$ as a basis, we can construct the representation $\mathbf{\Gamma}_N$ of the symmetry group. Defining the N-uple vector $\psi_0 \equiv [\psi_0(a)]_{a=1}^{N}$, a symmetry operation S produces an equivalent configuration characterized by the transformed vector $\psi_0' = \overset{\wedge}{\mathbf{S}}\,\psi_0$ where $\overset{\wedge}{\mathbf{S}}$ is the matrix of S in the representation $\mathbf{\Gamma}_N$. If $\mathbf{\Gamma}_i$ are the irreducible representations of the group, we have

$$\mathbf{\Gamma}_N = \sum_i a_i \mathbf{\Gamma}_i \tag{2.23}$$

a_i being the multiplicity of $\mathbf{\Gamma}_i$ in $\mathbf{\Gamma}_N$. The coefficients a_i can be calculated from a corollary of the "great orthogonality theorem" of group theory:

$$a_i = \frac{1}{g} \sum_c g_c \, \chi_c(i) \, \chi_c(N) \tag{2.24}$$

where $\chi_c(i)$ and $\chi_c(N)$ are the characters of class c in the representations $\mathbf{\Gamma}_i$ and $\mathbf{\Gamma}_N$, g is the number of operations of the group and g_c is the number of operations of class c; for the groups we are dealing here, $g_c \equiv 1$. The characters $\chi_c(N)$ can be found using the non shifted nodes technique. The condensation modes are classified according to the irreducible representation with the same transformation properties under the symmetry operations of

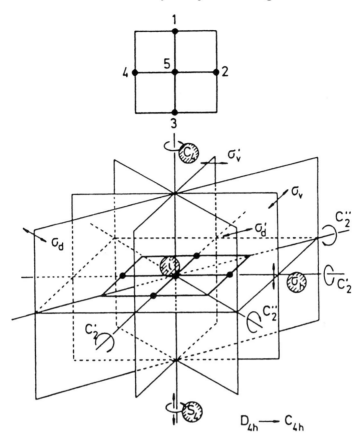

Fig. 2.2. Shown are the symmetry elements of a simple planar network. The presence of a magnetic field in the C_4 axis direction leaves only the symmetry elements shown as hatched. (Taken from Ref. [4])

the group. They are grouped in sets with the numerical proportion provided by the coefficients a_i.

An example can clearly show this. Consider the four identical square loops network studied by Rammal et al. [25], drawn in Fig. 2.2; when there is no external field, the point group of space symmetry is D_{4h}. When an external uniform magnetic field is applied, parallel to the C_4 axis, the planes σ_v, σ_d and the C_2 axes are eliminated and the point space group is the subgroup C_{4h}, whose symmetry operations are E, C_4, C_2, C_4^3, I, S_4^3, σ_h and S_4, with

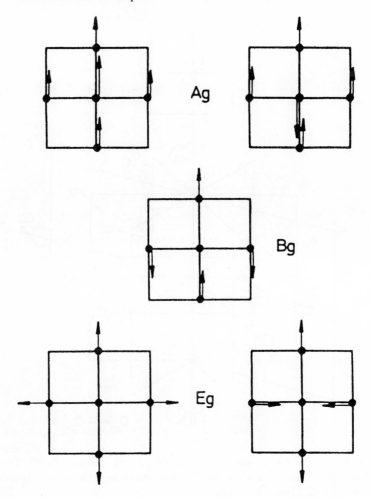

Fig. 2.3. Shown are the five modes of the order parameter for the lattice of Fig. 2.2. The arrows indicate the phase of the order parameter

the character table

C_{4h}	E	C_4	C_2	C_4^3	I	S_4^3	σ_4	S_4
A_g	1	1	1	1	1	1	1	1
B_g	1	−1	1	−1	1	−1	1	−1
E_g	1	i	−1	$-i$	1	i	−1	$-i$
E_g	1	$-i$	−1	i	1	$-i$	−1	i
A_u	1	1	1	1	−1	−1	−1	−1
B_u	1	−1	1	−1	−1	1	−1	1
E_u	1	i	−1	$-i$	−1	$-i$	1	i
E_u	1	$-i$	−1	i	−1	i	1	$-i$

If $\psi_0 \equiv [\psi_0(1), \psi_0(2), \psi_0(3), \psi_0(4), \psi_0(5)]$, the C_4 operation applied to it gives

$$\psi_0' = \hat{C}_4 \, \psi_0 \equiv [\psi_0(4), \psi_0(1), \psi_0(2), \psi_0(3), \psi_0(5)]$$

and this implies that

$$\hat{C}_4 = \begin{bmatrix} 0 & 0 & 0 & 1 & 0 \\ 1 & 0 & 0 & 0 & 0 \\ 0 & 1 & 0 & 0 & 0 \\ 0 & 0 & 1 & 0 & 0 \\ 0 & 0 & 0 & 0 & 1 \end{bmatrix}$$

The remaining matrices are obtained in similar manner:

$$\hat{E} = \overset{\wedge}{\sigma}_h = \begin{bmatrix} 1 & 0 & 0 & 0 & 0 \\ 0 & 1 & 0 & 0 & 0 \\ 0 & 0 & 1 & 0 & 0 \\ 0 & 0 & 0 & 1 & 0 \\ 0 & 0 & 0 & 0 & 1 \end{bmatrix} , \quad \hat{C}_4 = \hat{S}_4 = \begin{bmatrix} 0 & 0 & 0 & 1 & 0 \\ 1 & 0 & 0 & 0 & 0 \\ 0 & 1 & 0 & 0 & 0 \\ 0 & 0 & 1 & 0 & 0 \\ 0 & 0 & 0 & 0 & 1 \end{bmatrix}$$

$$\hat{C}_2 = \hat{I} = \begin{bmatrix} 0 & 0 & 1 & 0 & 0 \\ 0 & 0 & 0 & 1 & 0 \\ 1 & 0 & 0 & 0 & 0 \\ 0 & 1 & 0 & 0 & 0 \\ 0 & 0 & 0 & 0 & 1 \end{bmatrix} , \quad \hat{C}_4^3 = \hat{S}_4^3 = \begin{bmatrix} 0 & 1 & 0 & 0 & 0 \\ 0 & 0 & 1 & 0 & 0 \\ 0 & 0 & 0 & 1 & 0 \\ 1 & 0 & 0 & 0 & 0 \\ 0 & 0 & 0 & 0 & 1 \end{bmatrix}$$

In this case $g = 8$ and

$$\chi_E(n) = \chi_{\sigma_h}(n) = 5,$$

$$\chi_{C_4}(n) = \chi_{S_4}(n) = \chi_{C_2}(n) = \chi_I(n) = \chi_{C_4^3}(n) = \chi_{S_4^3}(n) = 1;$$

using 2.23 and 2.24, and the character table for C_{4h} it follows that

$$\Gamma_N = 2\, \Gamma_{Ag} + \Gamma_{Bg} + 2\, \Gamma_{Eg}$$

In Fig. 2.3 we plot the five modes of this network, with its corresponding classification; the directions of the arrows represent the phase the order parameter and the length, the relative modules; the normal-superconducting boundary is determined by only four of the modes because one of the A_g modes is more energetic, as a consequence of the strong gradient of the order parameter between the central node and the external ones. The order parameter has a zero in each of the internal branches.

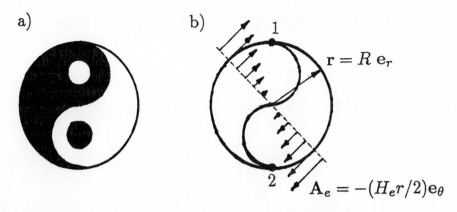

Fig. 2.4. a) The yin-yang (YY). b) Geometry of the associated superconducting micronetwork

2.2 Some Simple Applications of the dGA Theory to Symmetric Micronetworks

2.2.1 The Yin-Yang

The yin-yang (YY) (see Fig. 2.4) is a superconducting micronetwork with three branches of equal length and two loops of equal area [5]. Its symmetry is C_{2h}. Due to these relations between the length of different branches, the equations are quite simple for this example. Some extensions of this example are discussed in Section 5.2.

Taking the symmetric gauge $\mathbf{A}_e = -(H_e r/2)\mathbf{e}_\theta$, where $\mathbf{e}_\theta$ is the cylindrical unit vector, the flux in the basic loop σ is

$$\Phi_e = \int\int_\sigma \mathbf{H}_e \cdot d\sigma = \oint \mathbf{A}_e \cdot d\mathbf{r} = H_e(\pi R^2/2) \; ;$$

by the symmetry of the network, it is obvious that the circulation of $\mathbf{A}_e$ along the external branches amounts

$$\gamma_e = \frac{2\pi}{\Phi_0} \int_1^2 \mathbf{A}_e \cdot d\mathbf{r} = \frac{2\pi\Phi_e}{\Phi_0} = 2\pi\phi_e$$

where we have defined the reduced flux $\phi_e = \Phi_e/\Phi_0$. The circulation along the central branch is $\gamma_e' = 0$. Calling $L = \pi R$ the common length of the branches, the nodal equations are

$$3\,\cos(\sqrt{\mu_0}L)\,\psi_0(1) - (1 + 2\,\cos\gamma_e)\,\psi_0(2) = 0$$
$$3\,\cos(\sqrt{\mu_0}L)\,\psi_0(2) - (1 + 2\,\cos\gamma_e)\,\psi_0(1) = 0.$$

There are two types of solutions: $\psi_0(2) = \pm\psi_0(1)$ corresponding to the irreducible representations A_g (symmetric) and B_u (antisymmetric) of the C_{2h}

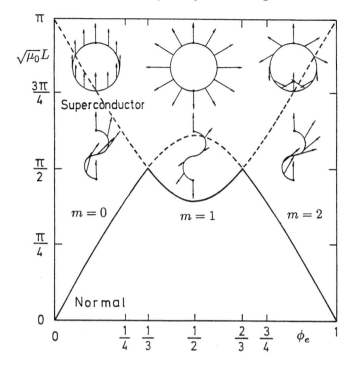

Fig. 2.5. Phase diagram for the yin-yang: the solid line is the N-S second order phase transition boundary . Shown is also the phase of the order parameter along the external ring and also along the central branch for the modes $m \doteq 0, 1, 2$

group. The phase boundary (see Fig. 2.5) is given in each case by the characteristic equations

$$\cos(\sqrt{\mu_0}L) = \pm \frac{[1 + 2 \cos(2\pi\phi_e)]}{3}$$

In these expressions the $+$ $(-)$ sign corresponds to an even (odd) number of flux quanta for the fluxoid on the external ring of the YY; this is equivalent to characterize the modes of the order parameter by means of its "phase winding number" $m = -\frac{1}{2\pi} \oint_C \nabla\varphi \cdot d\mathbf{r}$ on the external ring: if m is even (odd), the mode is A_g (B_u). The phase boundary is determined by the mode $m = 0$ if $0 \le \phi_e < 1/3$, by the mode $m = 1$ if $1/3 < \phi_e < 2/3$, and by the mode $m = 2$ if $2/3 < \phi_e \le 1$.

Using 2.19 it is simple to find the order parameter on the branches starting from the node denoted by 1, for which we impose $\psi_0(1) = 1$. For the left and right branches we have

$$\psi_0(s) = \frac{\exp(-i\gamma_e s/L)}{\sin(\sqrt{\mu_0}L)} \left[\sin\left[\sqrt{\mu_0}\,(L-s)\right] \pm e^{i\gamma_e} \sin\left(\sqrt{\mu_0}s\right)\right] \text{ (right branch)}$$

$$\psi_0(s) = \frac{\exp(i\gamma_e s/L)}{\sin(\sqrt{\mu_0}L)} \left[\sin\left[\sqrt{\mu_0}\,(L-s)\right] \pm e^{-i\gamma_e} \sin\left(\sqrt{\mu_0}s\right)\right] \text{ (left branch)}$$

For the central branch:

$$\psi_0(s) = \exp(-i\gamma_e'(s)) \frac{\cos\left[\sqrt{\mu_0}(L/2-s)\right]}{\cos\left[\sqrt{\mu_0}L/2\right]} \text{ (central branch, } m \text{ even)}$$

$$\psi_0(s) = \exp(-i\gamma_e'(s)) \frac{\sin\left[\sqrt{\mu_0}(L/2-s)\right]}{\sin\left[\sqrt{\mu_0}L/2\right]} \text{ (central branch, } m \text{ odd)}$$

where

$$\gamma_e'(s) = \begin{cases} \phi_e\left[\sin(2\pi s/L)/2 + \pi s/L\right] & (0 \le s \le \frac{L}{2}) \\ \phi_e\left[\pi(1 - s/L) - \sin(2\pi s/L)/2\right] & (\frac{L}{2} \le s \le L) \end{cases}$$

It is interesting to note that $[\psi_0(s)]_{\text{left}} = (-1)^m\,[\psi_0(L-s)]_{\text{right}}$, as it must be by the C_{2h} symmetry in the different modes, and that the order parameter is maximum at the center of the network for the even modes, and it vanishes at that point for the odd modes.

The current density in the external branches is

$$j_0 = J\,\frac{\sqrt{\mu_0}L}{\sin(\sqrt{\mu_0}L)}\,\sin\left[2\pi(\frac{m}{2} - \phi_e)\right]$$

and $j_0 = 0$ in the central branch. When ϕ_e is a half-integer number, the currents vanish for whatever value of m. In the case $m = 0$, the order parameter vanishes for $\phi_e = 1/2$ at the center of the *external branches* allowing for the phase to rearrange in the following even mode $m = 2$. The transition between the modes $m = 0$ and $m = 2$ proceeds via two "zeros" of the order parameter in the external branches, where the superconducting circuit opens and interrupts the current in the external loop.

2.2.2   The Double Yin-Yang

The symmetry properties of the YY extend to the "double yin-yang (2YY)" (see Fig. 2.6), a network topologically identical to that of Fig. 2.2, but simpler to study. The symmetry group of the 2YY is C_{4h} and the presence of the magnetic field along the C_4 axis does not modify the group. It is again convenient to use the magnetic vector potential in the symmetric gauge $\mathbf{A}_e = -(H_e r/2)\mathbf{e}_\theta$. The flux across the basic loop is

$$\Phi_e = \int\int_\sigma \mathbf{H}_e \cdot d\sigma = \oint \mathbf{A}_e \cdot d\mathbf{r} = H_e(\pi R^2/4)\,.$$

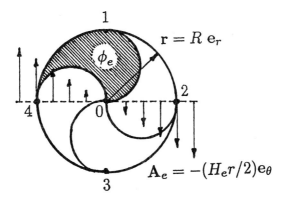

Fig. 2.6. Geometry of the "double yin-yang" (2YY). This network has the same topology and, under the external magnetic field, the same symmetry C_{4h} as the network of Fig. 2.2

From the symmetry of the network it is easy to see that the circulation of $\mathbf{A}_e$ along the branches takes the values

$$\gamma_e = \frac{2\pi}{\Phi_0}\int_1^2 \mathbf{A}_e \cdot d\mathbf{r} = \frac{2\pi\Phi_e}{\Phi_0} = 2\pi\phi_e, \gamma_e' = \frac{2\pi}{\Phi_0}\int_1^0 \mathbf{A}_e \cdot d\mathbf{r} = \frac{\pi\Phi_e}{\Phi_0} = \pi\phi_e = \frac{\gamma_e}{2}$$

With $L = \pi R/2$, the nodal equations are

$$\begin{cases} 4 \, \cos(\sqrt{\mu_0}L) \, \psi_0(0) - e^{-i\gamma_e/2} \displaystyle\sum_{a=1}^{4} \psi_0(a) = 0 \text{ (central node 0)} \\[2mm] 3 \, \cos(\sqrt{\mu_0}L) \, \psi_0(a) - e^{i\gamma_e}\psi_0(a+1) - e^{-i\gamma_e}\psi_0(a-1) - e^{i\gamma_e/2}\psi_0(0) = 0 \\ \hspace{6cm} (a = 1, 2, 3, 4) \end{cases}$$

for which there are five solutions:

a) Two of type A_g with

$$\exp(i\pi\phi_e) \, \cos(\sqrt{\mu_0}L) \, \psi_0(0) = \psi_0(2) = \psi_0(3) = \psi_0(4) = \psi_0(1).$$

We have fixed the complex phase taking $\psi_0(1)$ to be real and normalized such that $\psi_0(1) = 1$; for these two modes the characteristic determinant gives for the phase boundary

$$\cos(\sqrt{\mu_0}L) = \frac{\cos(2\pi\phi_e)}{3} \pm \sqrt{\frac{\cos^2(2\pi\phi_e)}{9} + \frac{1}{3}} \, .$$

The less energetic solution corresponds to the plus sign and determines the N-S transition when the applied external flux is close to an integer multiple of Φ_0; the solution corresponding to the minus sign has higher energy because the gradient of the order parameter on the branches is also higher than for the other mode.

b) One of type B_g with

$$\psi_0(0) = 0, \psi_0(2) = \psi_0(4) = -\psi_0(1), \psi_0(3) = \psi_0(1).$$

For this mode the phase boundary is determined by

$$\cos(\sqrt{\mu_0}L) = -\frac{2}{3}\cos(2\pi\phi_e);$$

this solution determines the N-S transition when the applied external flux is in the neighborhood of half-integer multiples of Φ_0.

c) Two of type E_g with

$$\psi_0(0) = 0, \psi_0(2) = \mp i\psi_0(1), \psi_0(3) = -\psi_0(1), \psi_0(4) = \pm i\psi_0(1).$$

The second order boundary is in this case given by

$$\cos(\sqrt{\mu_0}L) = \pm\frac{2}{3}\sin(2\pi\phi_e).$$

Fig. 2.7 shows the thermodynamical phase diagram resulting from the solutions a), b) and c); the solid line breaks at certain "critical" magnetic flux values where the mode determining the N-S phase boundary changes ($\phi_{ec} \doteq .216, .375, .625, .784$, and so on).

The current can be calculated using equation (2.22). The central branches carry no currents; the peripheral branches support the same current

$$j_0 = -J\frac{\sqrt{\mu_0}L}{\sin(\sqrt{\mu_0}L)} \ |\psi_0(1)|^2 \sin\left[2\pi(\phi_e - \frac{m}{4})\right].$$

The number m takes the value 0 for both A_g modes; $m = 2$ for the mode B_g and m is 1or 3 for the modes E_g. The current j_0 as a function of ϕ_e for the different modes shows inversions and jumps that allows for the fluxoid quantization condition to be satisfied. This behavior is typical for systems with complex phase coherence in multiply connected domains.

Whenever the currents vanish, the critical temperature for the N-S transition shows a relative maximum. This effect appears because the absence of currents lowers the free energy of the system; in the absence of currents the superconducting state remains stable up to higher temperatures.

2.2.3   Wheatstone Bridge

Figure 2.8a represents a network with four nodes, equal length branches and equal area minimal loops; it has no central node. This system can be called

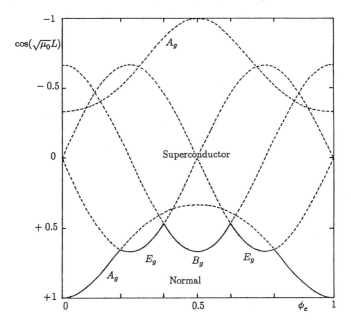

Fig. 2.7. Phase diagram for the "double yin-yang"

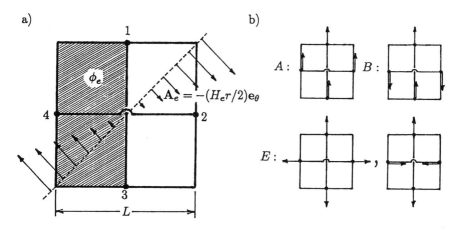

Fig. 2.8. a) Wheatstone bridge network: its symmetry is S_4. b) The four possible modes for the order parameter

a "superconducting Wheatstone bridge". It is really a three dimensional network. The corresponding symmetry space group is S_4. The nodal equations are

$$3\cos(\sqrt{\mu_0}L)\,\psi_0(a)-e^{-i\gamma_e}\psi_0(a-1)-e^{i\gamma_e}\psi_0(a+1)-\psi_0(a+2) = 0(a = 1, 2, 3, 4)$$

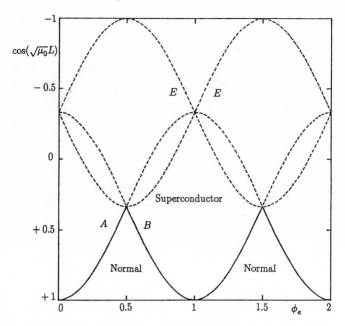

Fig. 2.9. Phase diagram for the Wheatstone bridge network

For these equations we can expect four solutions: one of type A, one of type B, and two of type E. The solutions are (see Fig. 2.8b)

$$\psi_0(a) = \psi_0(1) \exp\left[-i\pi(m-1)(a-1)/2\right] (m, \ a \doteq 1, 2, 3, 4)$$

where the subscript m takes on the values $m = 1$ for the mode A, $m = 3$ for the mode B, and $m \doteq 2, 4$ for the modes E. The characteristic determinant for each mode gives

$$\cos(\sqrt{\mu_0}L) = \frac{2 \, \cos\left[\pi\left(\phi_e - (m-1)/2\right)\right] + (-1)^{m-1}}{3}.$$

This allows us to plot the phase diagram shown in Fig. 2.9. The N-S boundary is determined by modes A and B. The modes E have higher energy. For half-integer values of ϕ_e the normal-superconductor phase boundary is given by a threefold degenerated eigenvalue.

The analysis of the fluxoid quantization and the phase twisting number on each loop of the network can be seen in Fig. 2.10, where we have displayed mode B.

2.2.4   The Basic Triangular Gasket

The number of papers in the literature on simple micronetworks is quite large, so we have selected as a final example of this kind of systems one with

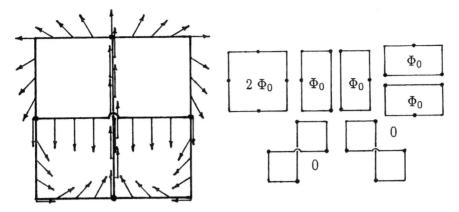

Fig. 2.10. Fluxoid quantization on the seven loops shown at right, for the B mode. The scheme at left shows the order parameter on the branches

three nodes, useful to analyze the fractal "triangular Sierpinsky gasket" [16]. The system we refer to is depicted in Fig. 2.11b. It is topologically equivalent to the basic building block of the gasket (Fig. 2.11a) but with equal length branches and equal area minimal loops, a condition that makes it simple to deal with. As shown in the figure it has three symmetrical nodes, placed a distance l from one another, with circumference arcs of length $L = 1.043\ l$; the minimal loop area is $2/5$ of the area of the triangles in Fig. 2.11a. Taking the same gauge as in the previous examples, the circulation of $\mathbf{A}_e$ on the branches is given by

$$\gamma_e = \frac{8}{3}\pi\phi_e \text{ and } \gamma_e' = \frac{2}{3}\pi\phi_e$$

for the external and internal branches respectively.

This system has, in the presence of an externally applied magnetic field, two symmetry elements: the plane σ_h containing the network and the axis normal to the plane, C_3; the space point group is C_{3h} and we must expect three solutions for the nodal equation, one corresponding to the A' irreducible representation and two corresponding to the E' one. The nodal equations are

$$4\cos(\sqrt{\mu_0}L)\,\psi_0(a) - (e^{i\gamma_e} + e^{i\gamma_e'})\psi_0(a+1) - (e^{-i\gamma_e} + e^{-i\gamma_e'})\psi_0(a-1) = 0$$
$$(a = 1,2,3)$$

with solutions

$$\text{for mode } A' \ \left\{ \psi_0(1) = \psi_0(2) = \psi_0(3) \right.$$

$$\text{for mode } E' \ \left\{ \begin{array}{l} \psi_0(2) = \psi_0(1)\,e^{\mp i2\pi/3} \\ \psi_0(3) = \psi_0(1)\,e^{\pm i2\pi/3} \end{array} \right. .$$

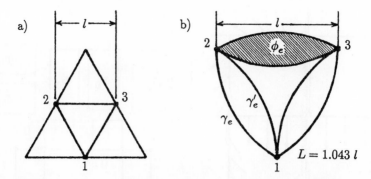

Fig. 2.11. a) Basic triangular Sierpinsky gasket network. b) Network topologically equivalent to the previous network and with the same symmetry (C_{3h})

The characteristic determinant gives in each case

$$\cos(\sqrt{\mu_0}L) = \cos(5\pi\phi_e/3) \; \cos(\pi\phi_e) \qquad \text{mode } A'$$

$$\cos(\sqrt{\mu_0}L) = \cos\left[\tfrac{5}{3}\pi(\phi_e \mp \tfrac{2}{5})\right] \cos(\pi\phi_e) \quad \text{mode } E'$$

The resulting phase diagram is shown in Fig. 2.12. Notice the similarity with the phase diagram experimentally obtained by Gordon et al. [16] for the Sierpinsky gasket (Fig. 2.13). The experiment shows a repetition of the basic pattern up to order four; the ideal gasket, being a fractal structure, would show a repetition of infinite order. Fig. 2.14 shows the stable mode E' when $\phi_e = 1$.

The currents on the external and internal branches, are

$$j_0 = -J \, \frac{\sqrt{\mu_0}L}{\sin(\sqrt{\mu_0}L)} \; |\psi_0(1)|^2 \sin\left[\frac{8}{3}\pi(\phi_e - \frac{m}{4})\right]$$

and

$$j_0' = -J \, \frac{\sqrt{\mu_0}L}{\sin(\sqrt{\mu_0}L)} \; |\psi_0(1)|^2 \sin\left[\frac{2}{3}\pi(\phi_e - m)\right]$$

respectively, where $m \doteq 0, 1, 2$ for the modes A' and E'; the "distribution" of currents among external and internal branches is determined by the ratio

$$\frac{j_0}{j_0'} = \frac{\sin\left[\frac{8}{3}\pi(\phi_e - \frac{m}{4})\right]}{\sin\left[\frac{2}{3}\pi(\phi_e - m)\right]}$$

When $\phi_e \to$ integer, it is easy to see that $j_0/j_0' \to 4$, a consequence of fluxoid quantization. In fact, in that limit we have $\sqrt{\mu_0}L \to 0$ and, according to equation (2.19), $\psi_0(s) \to \psi_0(1) \exp(-i\gamma_e(s))$, $|\psi_0(s)|^2 \to |\psi_0(1)|^2$ and the order parameter becomes uniform over the network; the fluxoid quantization

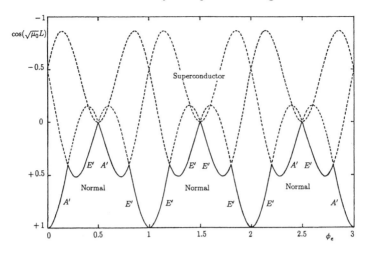

Fig. 2.12. Phase diagram for the network of Fig. 2.11b). The period of the modes is $3\Phi_0$; the period of the observable quantities $|\psi|$ and $\mathbf{Q}$ is Φ_0

equation (2.13) for a loop with uniform current is

$$\frac{4\pi\lambda^2}{c\Phi_0}\frac{j_0}{|\psi_0(1)|^2}\cdot\oint_C ds + \phi_e = m$$

and can be applied to the large loop formed by the external branches and also to the small loop formed by the internal branches, providing $j_0 = 4j_0'$ as desired. In Fig. 2.15 we plot the currents for each mode; for certain values of flux, the current in the loop formed by the external branches flows in opposite direction to that flowing in the loop formed by the internal branches to satisfy fluxoid quantization on each loop.

2.3 The dGA Theory for Extended Micronetworks

2.3.1 Geometrical Properties of Ladder Systems

The regular square ladder (see Fig. 2.16a) has been intensively studied in references [13,26] in the framework of the dGA theory; it is a network with equal length branches and equal area minimal loops that facilitate its analytical solution. A network with the same properties is the regular rhombic ladder (Fig. 2.16b), that reduces to the square ladder for $\delta = \pi/2$. If we consider a rhombic ladder with $\delta = \pi/3$ and attach new rectilinear branches joining nodes a^+ and $(a+1)^-$, we obtain a regular triangular ladder (Fig. 2.16c), which also has equal length branches and equal area minimal loops. In the presence of an uniform external applied magnetic field, the space symmetry group is $T_1(L)\otimes C_{2h}$ for the three ladders (square, rhombic and triangular).

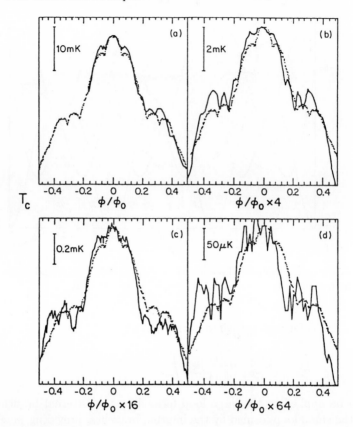

Fig. 2.13. a) The superconducting transition temperature as a function of the normalized flux through an elementary triangle of the gasket. The solid line is experimental data and the points represent the theory for a second-order gasket scaled to fit the data at $\phi/\phi_0 = \frac{1}{2}$. In b)-d) the field axes have been expanded 4, 16, and 64 times, respectively, in order to highlight the small-field structure in $T_c(H)$. The self-similar nature of the curve is made apparent by our superimposing the second-order theory on each plot. (Taken from Ref. [16]; note that $\phi/\phi_0 \equiv \Phi_e/\Phi_0$)

Here $T_1(L)$ is the translation group and C_{2h} is the point group; the symbol $\otimes$ indicates direct product. The essential difference between the triangular and rhombic or square ladders lies on the topology of the network; the increased connectivity at each node changes the "frustration" [24] conditions of the order parameter.

We will concentrate on the triangular ladder, but simultaneously we will show the corresponding results for the rhombic ladder in order to analyze similarities and differences. Choosing the magnetic vector potential in Landau gauge $\mathbf{A}_e = H_e y \mathbf{e}_x$, the circulations of $\mathbf{A}_e$ along longitudinal and transverse

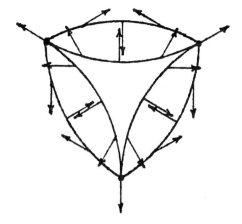

Fig. 2.14. The stable E' mode for the network of Fig. 2.11b), with $\phi_e = \Phi_e/\Phi_0 = 1$

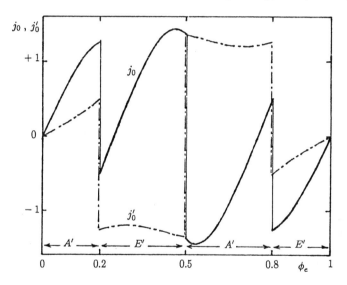

Fig. 2.15. The solid (dotted) line shows the current densities on the external (internal) branches for the modes that determine the N-S boundary in Fig. 2.12. Note the changes in the relative direction of the currents when comparing external and internal branches

branches turn out to be

$$\gamma_e(s) = \pi\phi_e \tfrac{s}{L} \quad \gamma_e'(s) = \pi\phi_e(\tfrac{s}{L} - 1)\tfrac{s}{L}\cos\delta \rightarrow \text{rhombic}$$

$$\gamma_e(s) = 2\pi\phi_e \tfrac{s}{L} \quad \gamma_e'(s) = \pi\phi_e(\tfrac{s}{L} - 1)\tfrac{s}{L} \qquad \rightarrow \text{triangular}$$

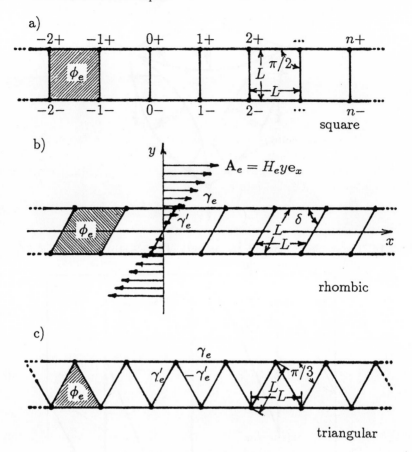

Fig. 2.16. a) The square ladder. b) The rhombic ladder. c) The triangular ladder. The same Landau gauge is used for the three cases

Note that, to compare the results, the fluxes across the minimal loops must be "scaled" by a factor 2.

2.3.2 Nodal Equations and Phase Transition Boundaries

Alexander equations for nodes a^+ and a^- are

$$3\cos(\sqrt{\mu_0}L)\psi_{a+} - e^{i\gamma_e}\psi_{(a+1)+} - e^{-i\gamma_e}\psi_{(a-1)+} - \psi_{a-} = 0$$

$$3\cos(\sqrt{\mu_0}L)\psi_{a-} - e^{-i\gamma_e}\psi_{(a+1)-} - e^{i\gamma_e}\psi_{(a-1)-} - \psi_{a+} = 0$$

rhombic   (2.25)

and

$$4\cos(\sqrt{\mu_0}L)\psi_{a+} - e^{i\gamma_e}\psi_{(a+1)+} - e^{-i\gamma_e}\psi_{(a-1)+} - \psi_{a-} - \psi_{(a+1)-} = 0$$

triangular

$$4\cos(\sqrt{\mu_0}L)\psi_{a-} - e^{-i\gamma_e}\psi_{(a+1)-} - e^{i\gamma_e}\psi_{(a-1)-} - \psi_{a+} - \psi_{(a-1)+} = 0 \quad (2.26)$$

with $\gamma_e = \gamma_e(L)$. These equations are equivalent to the Schroedinger equation for an electron in the same geometry, in the tight binding approximation.

Since the space group is $T \otimes C_{2h}$, the translational invariance under the symmetry operations of T gives solutions periodic in a (Floquet theorem). These solutions can be cast into the form

$$\psi_{a\sigma} = f_\sigma(q) \exp(iqa) \quad (\sigma \doteq +, -) \quad (2.27)$$

where q is a "mode index" and $f_\sigma(q)$ is an amplitude to be determined. Replacing (2.27) in the nodal equations (2.25) and (2.26) we obtain for $f_\sigma(q)$ the equations

$$\begin{cases} \beta_+ f_+(q) - f_-(q) = 0 \\ \beta_- f_-(q) - f_+(q) = 0 \end{cases} \quad (2.28)$$

Here

$$\beta_\pm = 3\cos(\sqrt{\mu_0}L) - 2\cos(\gamma_e \pm q) \qquad \text{rhombic}$$

$$\beta_\pm = \left[2\cos(\sqrt{\mu_0}L) - \cos(\gamma_e \pm q)\right]\left[1 \mp i\tan(q/2)\right] \quad \text{triangular}$$

Observe that for the triangular ladder $\beta_\pm$ are *complex numbers*; for $q = 0$ it is $\beta_\pm = 1$. The non trivial solution of (2.28) requires

$$\begin{vmatrix} \beta_+ & -1 \\ -1 & \beta_- \end{vmatrix} = 0$$

or equivalently

$$\cos(\sqrt{\mu_0}L) = \frac{2}{3}\left[\cos\gamma_e \cos q + \sqrt{\sin^2\gamma_e \sin^2 q + \frac{1}{4}}\right] \qquad \text{rhombic}$$

$$\cos(\sqrt{\mu_0}L) = \frac{1}{2}\left[\cos\gamma_e \cos q + \sqrt{\sin^2\gamma_e \sin^2 q + \cos^2\frac{q}{2}}\right] \quad \text{triangular}$$

(2.29)

The N-S boundary is given by the lower envelope of the lines $\mu_0 = \mu_0(\phi_e)$ obtained from (2.29) for each value of q in the interval $[-\pi, +\pi]$, as shown in Figs. 2.17a and 2.18a. These plots will be analyzed below.

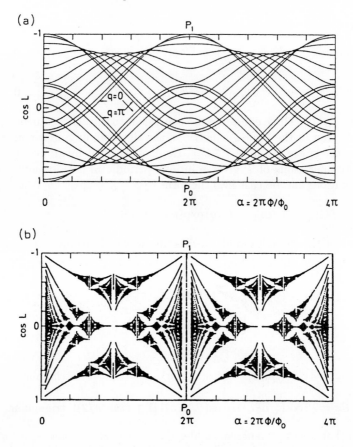

Fig. 2.17. a) Spectrum of eigenvalues for the square ladder. The Floquet index takes on the values $(k\pi/10)$ where k is integer. b) Spectrum of eigenvalues for the square lattice. In both cases the lower part corresponds to symmetric modes and the upper part to antisymmetric modes. Note that both figures have the same symmetry properties relative to the transformations of Table 2.1. In both cases the flux goes up to 2 flux quanta per unit cell (see text). (Taken from Ref. [4]; note that $\alpha = 2\pi\phi/\phi_0$ and L in the figure correspond to ε and $\sqrt{\mu_0}L$ in this article.)

It is possible to obtain an analytical expression for the N-S phase boundary taking the lower value μ_0 corresponding to a given applied flux ϕ_e; this condition gives us the q value characterizing the nucleation mode of the order parameter at the N-S transition, which we call q_0. Since the solutions are

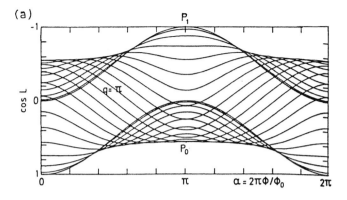

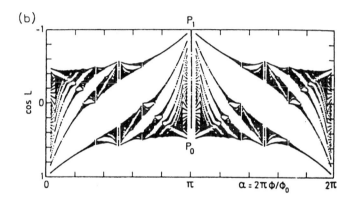

Fig. 2.18. a) Spectrum of eigenvalues for the triangular ladder. The Floquet index q takes on the same values as in Fig. 2.17a). b) Spectrum of eigenvalues for the triangular lattice. Note that both spectra have the same symmetry properties under transformations of α and L. Both spectra can be seen to evolve out of those of Fig. 2.17 in the way described in the text. (Taken from Ref. [4]; as in Fig. 2.17, $\alpha = 2\pi\phi/\phi_0 \equiv \varepsilon$ and $L \equiv \sqrt{\mu_0}L$.)

periodic in ϕ_e, for $0 \leq \phi_e \leq 1$ we have

$$\cos q_0 = \begin{cases} 1 & (0 \leq \phi_e \leq .215) \\ \cos{(\pi\phi_e)}\sqrt{1 + [2\sin{(\pi\phi_e)}]^{-2}} & (.215 < \phi_e < .785) \quad \text{rhombic} \\ -1 & (.785 \leq \phi_e \leq 1) \end{cases}$$

$$\cos q_0 = \begin{cases} 1 & (0 \leq \phi_e \leq .167) \\ \cos{(2\pi\phi_e)} + \frac{1}{4}[1 - \cos{(2\pi\phi_e)}]^{-1} & (.167 < \phi_e < .833) \quad \text{triangular.} \\ 1 & (.833 \leq \phi_e \leq 1) \end{cases}$$

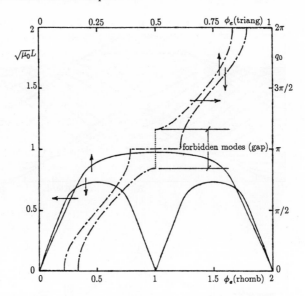

Fig. 2.19. The effect of the additional branch transforming the unit cell of a rhombic ladder (with $\delta = \pi/3$) into the unit cell of a triangular ladder, can be seen both in the N-S phase boundary (solid lines) and in the superconductive nucleation modes at the N-S transition (dotted lines). Note that for the triangular ladder superconductivity can not be nucleated with $q_0 \simeq \pi$, showing a band gap of forbidden modes

Replacing these values in (2.29) we obtain the desired phase boundary. In Fig. 2.19 we plot the phase diagrams for both cases and also the values of q_0 characterizing the superconductive nucleation modes at the transition. *In the rhombic case all possible q-modes are present at the N-S phase boundary, whereas in the triangular case the modes with q_0 close to π are excluded*; this can be understood considering that if a regular rhombic ladder with $q_0 \sim \pi$ is transformed into a triangular one by adding new branches, the resulting triangular ladder has an order parameter varying strongly along the branches, so that energy increases with respect to the rhombic case. This shifts the corresponding eigenvalues μ_0 to higher levels. In Fig. 2.18a we see that the modes $q \sim \pi$ are clearly shifted toward lower temperature and apart from the normal state.

2.3.3   Nodal Order Parameters

The form (2.27) for the nodal order parameter is imposed by the symmetry of the networks, in particular by the translation space group T, and it is enough to know it to predict the phase boundary. But there are also other symmetries included in the point space group C_{2h} and we expect the solutions of Alexander equations to be distributed into two classes: one of them

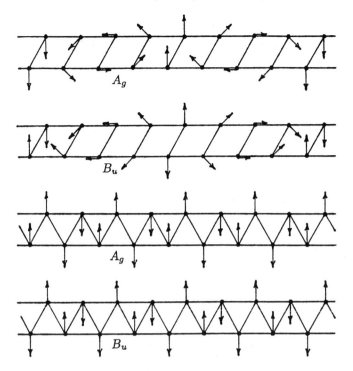

Fig. 2.20. Two modes, symmetric (A_g) and antisymmetric (B_u), are shown for the rhombic ladder ($\phi_e = 0.295$, $q = \pi/4$) and for the triangular ladder ($\phi_e = 0.5$, $q = \pi$)

transforming like the symmetric irreducible representation A_g, the other one like the antisymmetric irreducible representation B_u; this is to say that they must satisfy (Fig. 2.20)

$$\psi_{-a^-} = \pm\psi_{a^+}$$

where the $+$ $(-)$ sign corresponds to the A_g (B_u) mode.

From equations (2.29) it can be seen that the modes $+q$ and $-q$ are degenerate, so the nodal order parameter ψ_{a^σ} will be written as a linear combination of those modes; the adequate linear combinations that transform according to A_g and B_u are given by the 2-uples

$$\psi_0(a) = \begin{bmatrix} \psi_0(a^+) \\ \psi_0(a^-) \end{bmatrix} = \frac{1}{(1 \pm \beta_+)} \left[\begin{bmatrix} 1 \\ \beta_+ \end{bmatrix} \exp(iqa) \pm \begin{bmatrix} \beta_+ \\ 1 \end{bmatrix} \exp(-iqa) \right]$$

normalized to $[1, \pm 1]$ for $a = 0$. As before, the $+$ $(-)$ sign holds for the A_g (B_u) modes.

In equations (2.29), the plus sign corresponds to the A_g modes and the minus sign to the B_u modes; the first (symmetric modes) have been considered with great detail in the paper by Simonin et al. [26] for the square ladder, and are generally stable modes; the second type of modes (antisymmetric modes) often of higher energy, metastable or unstable, appear deep into the superconducting region of the phase diagram. This is the case of the square ladder, but it does not hold in the case of the triangular ladder because the B_u modes having $q \sim \pi$ approach the N-S boundary for $\phi_e \sim 1/2$ (see Fig. 2.18a); this is due to the gain in condensation energy due to the additional transverse branches which tends to stabilize the modes, similarly to the open branch in the "lasso" treated by de Gennes [9]. For this reason, the B_u modes in this case can no longer be ignored when studying the superconducting phase not far from the phase boundary.

For $q = \pm \pi$ the solutions A_g and B_u of the triangular ladder are degenerate over the complete range of values of ϕ_e (compare the superconductive nucleation for both modes in Fig. 2.20). The rhombic ladder (and also the square ladder) shows in its phase diagram (Fig. 2.17a) a region around $\phi_e \sim 1/2$, $\mu_0 \sim (\pi/2L)^2$ where there are no eigenvalues μ_0, where superconductivity only can be nucleated in A_g modes with finite order parameter. These characteristics of each ladder are absent in the other.

2.3.4   Current Distribution near the Phase Transition

The superconducting currents in the branches can be obtained as usual using equation (2.22), except for the undetermined amplitude factor that requires use of the non linear GL terms for its determination. For the rhombic ladder we have

$$j_{0a+} = -J\frac{\sqrt{\mu_0}L}{\sin(\sqrt{\mu_0}L)(1\pm\beta_+)^2}\{\sin(\pi\phi_e + q) + \beta_+^2 \sin(\pi\phi_e - q)\pm$$

$$2\beta_+ \sin(\pi\phi_e)\cos[(2a+1)q]\}$$

$$j_{0a-} = -j_{0a+}$$

$$j'_{0a} = \pm J\frac{\sqrt{\mu_0}L}{\sin(\sqrt{\mu_0}L)}\left(\frac{1\mp\beta_+}{1\pm\beta_+}\right)\sin(2aq)$$

where j_{0a+} and j_{0a-} are the current densities in the branches starting at nodes $a+$ and $a-$ and ending at nodes $(a+1)+$ and $(a+1)-$ (upper and lower branches in Fig. 2.16); the currents in the transversal branches starting at $a+$ are denoted by primes. In turn, for the triangular ladder we obtain

$$j_{0a+} = -J\frac{\sqrt{\mu_0}L}{\sin(\sqrt{\mu_0}L)}\frac{N(a+\frac{1}{2})}{(1 + |\beta_+|^2 \pm 2\operatorname{Re}\{\beta_+\})}$$

$$j_{0a-} = J\frac{\sqrt{\mu_0}L}{\sin(\sqrt{\mu_0}L)}\frac{N(a)}{(1 + |\beta_+|^2 \pm 2\operatorname{Re}\{\beta_+\})}$$

$$j'_{0a} = \pm J\frac{\sqrt{\mu_0}L}{\sin(\sqrt{\mu_0}L)}\frac{(1 - |\beta_+|^2)\sin(2aq)}{(1 + |\beta_+|^2 \pm 2\operatorname{Re}\{\beta_+\})}$$

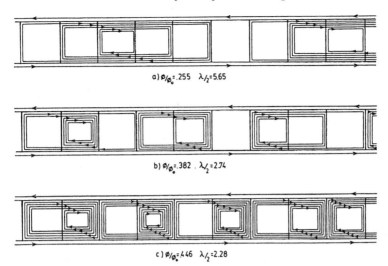

a) ϕ/ϕ_o=.255 $\lambda_{1/2}$=5.65

b) ϕ/ϕ_o=.382 . $\lambda_{1/2}$=2.74

c) ϕ/ϕ_o=.446 $\lambda_{1/2}$=2.28

Fig. 2.21. Vortex structure in the ladder showing the evolution of current distribution as field is increased. The net current in each branch is the sum of all current lines shown. The external flow lines represent the shielding currents. Normalization is arbitrarily defined for each graph and a suitable discretization was applied. (Taken from Ref. [26]; λ in the figure is $2\pi/q$.)

$$j''_{0a} = \pm J \frac{\sqrt{\mu_0}L}{\sin(\sqrt{\mu_0}L)} \frac{(1 - |\beta_+|^2)\sin[(2a + 1)q]}{(1 + |\beta_+|^2 \pm 2\,\mathrm{Re}\{\beta_+\})}$$

where

$$N(a) = \sin(2\pi\phi_e + q) + |\beta_+|^2 \sin(2\pi\phi_e - q) \pm 2\sin(2\pi\phi_e)\,\mathrm{Re}\{\beta_+ e^{-i2aq}\}$$

In these expressions the upper sign corresponds to the A_g modes, and the lower sign to the B_u modes.

In Fig. 2.21 we show the current distribution near the N-S phase boundary for a mode incommensurate with L in the square ladder, quite similar to the rhombic one.

2.3.5   Some Features of the Ladders

In Figs. 2.17a, 2.18a and 2.19 it can be seen that addition of a new branch in the primitive cell of a rhombic ladder transforms it into a triangular ladder and gives rise to various noticeable effects:

a) The period of the phase diagram for the triangular ladder, when considered as a function of the magnetic field, is twice the period of the rhombic ladder because the minimal loops area is diminished by half.

b) In the case of the triangular ladder, the normal state can be found at lower temperatures than in the case of the rhombic ladder, because the augmented connectivity of the network facilitates the "phase frustration" [24] of the complex order parameter in the network, when there are "vortex condensation" modes [26].

c) The starting slopes of the N-S boundary in the (μ_0, ϕ_e) plane are different for the rhombic ladder and for the triangular ladder, because the "densification" of the micronetwork in the triangular ladder increases the superconductive material mass in the unit cell [26], facilitating the nucleation of the almost uniform superconducting phase at low fields.

d) In the rhombic ladder all symmetric modes A_g contribute to the configuration of the N-S boundary, while the antisymmetric modes B_u remain confined to deep regions of the phase diagram.

e) In the triangular ladder not all symmetric modes determine the N-S boundary; the A_g modes with $q \sim \pi$ are excluded because the order parameter varies strongly on the additional branch, thus raising the energy; instead, the B_u modes are stabilized by the extra branch and localized near the phase transition boundary; this behavior is analogous to that of the simple YY network (see paragraph 2.2.1) where the B_u mode comes to be more stable than the A_g mode for $\phi_e \sim 1/2$.

f) For $q \sim 0$ both ladders show similar behavior: the symmetric A_g modes are more stable than the antisymmetric ones B_u.

g) The general distribution of modes in the rhombic ladder phase diagram strongly resembles the characteristic "butterfly" form of the Hofstadter [1,3,17,4] spectrum (Fig. 2.17b) for the regular 2-D square network. In turn, for the triangular ladder it can be seen that the B_u modes appear in the superconducting region with similar distribution to that of the A_g modes (which give the N-S boundary), but displaced by $\Delta\phi_e \sim 1/2$; besides this, the general distribution of modes is clearly similar to the eigenvalue spectrum of the regular 2-D triangular network [8] (Fig. 2.18b).

h) In the low field limit with extended modes $(\gamma_e, q, L \to 0)$ equation 2.29 gives

$$\mu_0 \simeq \frac{2}{3} \frac{(q^2 + \gamma_e^2)}{L^2} \quad \text{rhombic}$$

$$\mu_0 \simeq \frac{5}{8} \frac{(q^2 + \frac{4}{5}\gamma_e^2)}{L^2} \quad \text{triangular} \tag{2.30}$$

This kind of dependence of μ_0 with respect to q and γ_e is typical of Landau levels for an electron confined in a box of width L on the y axis and free to move along the x axis, save for a term of "confinement energy" coming from the different boundary conditions for each problem. The influence of boundary conditions when solving micronetworks is of great importance; the presence of boundaries or limiting surfaces determines the type of solutions.

Due to the above mentioned equivalence between this problem and that of an electron in the tight binding approximation, we see that equations (2.30) give the dispersion relation for the electron in the network geometry, at low fields. The "effective mass", depends on the length of the branches and on the "connectivity" of the unit cell of the ladders.

i) Finally we can mention that, at the transition to the normal state, the extended vortex modes ($q \to 0$) give rise in both ladders (rhombic and triangular), to "solitonic solutions" for the phase of the order parameter; this is to say that the phase remains almost uniform on certain regions of the ladders, and varies strongly in small regions (a few unit cells) in the same way as the magnetization in Bloch walls between ferromagnetic domains. In this limit, taking the node index a as a continuous variable x in Eqs. (2.25) or (2.26), it can be shown that the phase satisfies the Sine-Gordon equation [30]:

$$\frac{d^2\varphi(x)}{dx^2} \propto \sin\left[2\varphi(x)\right] \qquad (q \to 0)$$

2.3.6 Square and Triangular Networks

Of special interest is the case of the square, triangular and other networks. These planar networks can be studied starting from our knowledge of ladders, discussed in the previous section. In the absence of a magnetic field, ladders and planar networks have different translation properties. The presence of the magnetic field and the dependence of the nodal equations (2.21) on the vector potential imposes a breaking of 2D translational symmetry. The best one can do the exploit some translational symmetry is to choose a Cartesian gauge which preserves 1D translations and is compatible with the C_{2h} group. As a consequence it turns out that the square, triangular and hexagonal lattices have the same symmetry group, $T_1(d) \otimes C_{2h}$, where d is the lattice parameter [4].

The modes for the planar structures and for the strips (ladders) are classified according to the Floquet index or wave vector q. For each q value there must be a mode symmetric under inversion (A_g) and an antisymmetric one (B_u). These exhaust all possible irreducible representations.

Using the same gauge for ladders as for planar lattices, it follows that the symmetry properties of the spectra and the transformation properties of the corresponding modes of the lattices can be deduced from those of the associated ladders. In Fig. 2.22 we show the choice of origin for the square lattice and for the corresponding ladders. With the Landau gauge $A_x = -H_e y$, we preserve translational invariance in the x direction. In both cases we chose $\psi_{mn} = \psi_m e^{iqn}$, where m and n are lattice indices for the nodes. For the ladders, ψ_m exists only for two values of m, corresponding to each one of the two longitudinal wires; the corresponding amplitudes are related trough (2.25) and (2.26). For the square lattice the different values of ψ_m are related through the Harper equation [17,8] of the Aubry model. Both spectra (for the square ladder and the corresponding planar lattice) remain

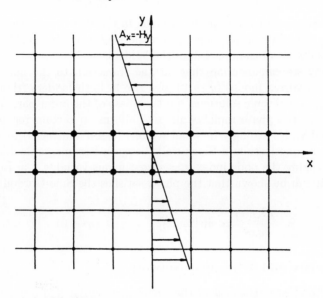

Fig. 2.22. Square lattice and square ladder. Shown is the coordinate system associated with the Landau gauge. (Taken from Ref. [4]; note that $A_x \equiv A_{ex}$ and $H \equiv H_e$)

invariant under the transformations of L and $\varepsilon = 2\pi\phi_e$ indicated in Table 2.1. The associated space symmetries are also indicated:

A similar table can be constructed for the triangular ladder and the corresponding planar lattice.

These symmetry properties allow us to draw a relation between the spectra of the triangular structures and those of the square ones. It is possible to construct a triangular ladder, and similarly a triangular lattice starting from a $\pi/3$ tilted, rhombic structure. This again has the C_{2h} symmetry and its spectrum is the same as for a square network. Adding a new transverse strand produces a triangular network. Fig. 2.17a shows the spectrum for the square ladder, whereas Fig. 2.17b is the well known spectrum for the square lattice, for periodic boundary conditions. The corresponding graphs for the triangular structures are given in Figs. 2.18a and 2.18b. In each case the lower part of the spectrum corresponds to symmetric modes, whereas the upper part belongs to antisymmetric modes. It can be seen that although the spectra for the lattices are much more complex with the characteristic fractal structure, there are patterns which are already contained in the spectra of the ladders. In particular one can see that the density of states in both cases is quite similar, a fact that can be appreciated by looking at the figures at a glazing angle. Also, one can appreciate the inversion symmetry of the spectra by turning the page up side down.

Table 2.1. Transformations of the parameters ε and L that leave the spectra of the square structures (ladder and planar network) invariant. The Floquet index q and the point group mode change as indicated in each case. L has been adimensionalised (multiplied by $\sqrt{\mu_0}$)

Spectrum		$T_1(L)$	C_{2h}	
ε	L	q	A_g	B_u
$\varepsilon + 2\pi$	L	$\pi + q$		
$\varepsilon + 4\pi$	L	$2\pi + q$		
$2\pi - \varepsilon$	L	$\pi - q$	A_g	B_u
$4\pi - \varepsilon$	L	$2\pi - q$		
ε	$\pi - L$	$\pi - q$		
$2\pi - \varepsilon$	$\pi - L$	q	B_u	A_g
$2\pi + \varepsilon$	$\pi - L$	$-q$		
$4\pi - \varepsilon$	$\pi - L$	$q - \pi$		

2.3.7   Incommensurate, Fractal and Disordered Systems

Several types of incommensurate structures have been studied in the literature, including incommensurate, fractal, quasicrystalline and disordered structures.

Franco Nori and Qian Niu [21] have studied lattices with Penrose tiling design and have compared their results with experiments by J.M.Gordon and A. M. Goldman [16]. The peculiarities of the spectrum in this case follow from the fact that different loops in the network have uncommesurate areas, a fact that breaks down the periodicity of the spectrum in the applied flux. Fig. 2.23 shows the normal superconducting boundary for this case and a comparison with experimental results for a Josephson junction array, with the same geometry, which shows similar behavior to the superconductive network.

Another interesting case considered are fractal lattices, like the triangular Sierpinsky gasket, theoretically studied first by S.Alexander [1,2], and experimentally worked out by J.M.Gordon et al. [16]. In Fig. 2.13 we can see the results for the phase boundary. The experiments were performed on a gasket of order four. For this reason the fractal characteristics of the spectrum are only partially present.

Disordered systems were studied by Simonin el al. [28,29]. These authors considered a square lattice on which nodes are removed at random. The branches connected to the missing node are also removed. The fraction of missing nodes is $(1 - p)$. In this case, being a two dimensional system the classical limit for site percolation is $p_c = 0.59$.

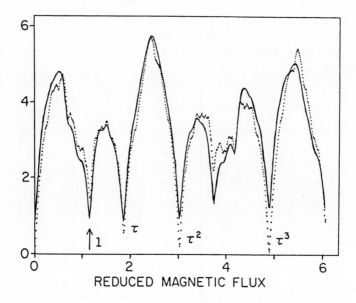

Fig. 2.23. Superconducting-normal phase boundary for the Penrose pattern. The solid line is experimental data for a Josephson-junction array, and the points are theoretically obtained values for a lattice with 301 nodes. The vertical axis represents voltage ($10nV$) for the experimental data, and $T_c(0) - T_c(H)$ (arb. units) for the theory. (Taken from Ref. [22])

For the perfect square lattice ($p = 1$) the critical magnetic field near $H = 0$ goes as $H_e \propto (T - T_c)$. As $1 - p$ increases the lattice opens up and there are a certain number of isolated loops, with dangling branches, for which the critical field varies as for de Gennes "lasso", $H_e \propto (T - T_c)^{1/2}$. The divergence in the slope can be represented by

$$\frac{dH_e}{dT} \propto (p - p_c)^{-k}.$$

Several authors [11,19] have predicted the value of k using scaling laws to be $k = 0.87$. The numerical experiments of Simonin et al. give $k = 0.93 \pm 0.06$ in good agreement with the predictions. Experiments in heterogenous superconductors give $k = 0.6 \pm 0.05$; the discrepancy is due to the fact that these are three dimensional systems. Microfabrication techniques have allowed for real systems to be built which are identical to the model systems. In the work by Gordon and Goldman [16], experimental results are reported for a disordered square lattice, for which $k \sim 1.0$, showing how good numerical experiments are for this case.

2.4 Beyond de Gennes–Alexander Theory: Inside the Phase Diagram

The de Gennes-Alexander theory, as a limiting case of the Ginzburg-Landau theory applied to microcircuits, is valid only on the second order normal/ superconductor phase transition boundary, where the order parameter and the currents are vanishingly small throughout the superconducting material. The solution to the ensuing linearized equations suggests the order parameter and current distribution in the system, giving their relative amplitude. The theory does not determine the absolute amplitude of these physical quantities, which strictly speaking are zero at the phase transition boundary. It can be said that the dGA theory refers more to the stability of the normal phase than to properties of the superconducting phase.

It is well known that the Ginzburg-Landau theory can describe the interplay between the superconducting condensation energy and the kinetic and magnetic energy of the induced currents. This interplay can lead to a change in the type of transition, from second order to first order at the so-called Landau critical points of the phase transition boundary. Since a first order transition implies a finite jump in the order parameter, it is clear that the dGA theory, which deals with infinitesimal or nascent quantities, is not adequate to describe this kind of transitions.

A complete study of the phase transition boundary, of the superconducting region immediately close to it, and of the phase diagram deep inside the superconducting region, up to the limit of validity of the Ginzburg-Landau theory, requires consideration of the complete expression for the free energy and of the ensuing equations, a more complete treatment than that of de Gennes-Alexander.

This problem has been approached by two different roads: one is a perturbative approximation [14,5,6] and the other is a variational method [5,7,18]. As these works go beyond the scope of the present review of the dGA theory, we present here only briefly the main results of the variational approach.

The variational approach allows for a description of the behavior of networks inside the superconducting region of the phase diagram. An important result of general validity is that a superconductive network behaves, in the presence of a magnetic field, similarly to a bulk material with a renormalized G-L constant and three additional Abrikosov-like constants; all these parameters depend on the geometry and topology of the network.

Let us first mention that at the second order phase transition boundary the Ehrenfest-Keesom (EK) relation, analogous to the well known Clapeyron-Clausius relation for first order transition, is satisfied. The EK equation gives the slope of the phase transition line and for superconducting networks it can be written as

$$\frac{dH_e}{dT} = \Omega \frac{B_1}{S_\sigma B_2}$$

where

$$\Omega = \frac{\Phi_0}{2\pi\xi^2(0)T_c}$$

and B_1 and B_2 are coefficients associated to the order parameter and current distribution on the network,

$$B_1 = \sum_b \int_0^{L_b} |\psi_0(s)|^2 \, ds$$

$$B_2 = (2/\phi_e) \sum_b j_{0b} \int_0^{L_b} A_e(s) ds \; ,$$

S_σ being the area of the loop used to measure the flux ϕ_e.

The theory predicts the existence of Landau critical points (LCP). At these points the second order phase transition boundary predicted by the dGA theory goes over into the supercooling boundary for a first order transition. The presence of a LCP depends essentially on the relative weight of the magnetic inductive effects on the network; a LCP is present if the following condition is satisfied [5,6]

$$\frac{S}{2\pi\kappa^2} = \frac{B_3^2}{c^2 B_4^2}$$

Here S is the uniform cross section of the branches and B_3^2 and B_4^2 are coefficient related to the order parameter and the current distribution, similar to B_1 and B_2,

$$B_3^2 = \sum_b \int_0^{L_b} |\psi_0(s)|^4 \, ds$$

$$B_4^2 = \sum_l \Lambda_l \, j_{0l}^2 + 2 \sum_{l,l'>l} M_{ll'} \, j_{0l} \, j_{0l'}$$

The coefficients Λ and M are the self induction and mutual induction coefficients of the loops; j_{0l} are the loop currents obtained from the branch currents j_{0b}. Neumann's theorem ensures that the result is independent of the assumed decomposition of the network into loops. From the expression for B_3^2 it follows that this quantity represents double the superconducting condensation energy, while B_4^2 represents twice the magnetic energy of the system. There is a competition between these two energies. When the condensation energy supersedes the magnetic energy the transition is second order whereas in the converse case the transition is first order. Since $\kappa = \lambda/\xi$, given a value for the coherence length, the inductive effects depend on the ratio between the transverse area of the branches and the penetration area for the field in the material as measured by λ^2.

Finally, for the phase transition boundary the theory predicts a jump in the specific heat, in the coefficient of variation of magnetization with temperature and in the magnetic susceptance, given by the derivatives of the volume average of the free energy density, g:

$$\Delta C_H = T_0 \frac{\partial^2 g}{\partial T^2} = \frac{T_0 \Omega^2}{4\pi \beta'_A (2K^2 - 1)}$$

$$\Delta \nu = \frac{\partial^2 g}{\partial T \partial H} = \frac{\Omega}{4\pi \beta''_A (2K^2 - 1)}$$

$$\Delta \eta_T = \frac{\partial^2 g}{\partial H^2} = \frac{1}{4\pi \beta'''_A (2K^2 - 1)}$$

Here K denotes a kind of "renormalized" Ginzburg-Landau constant,

$$K = \sqrt{\frac{\pi}{S} \frac{B_3}{c B_4}} \kappa.$$

The coefficients β are analogous to Abrikosov's distribution factor β_A. For a micronet $\beta_A = (V/S)(B_3/B_1)^2$ and

$$\beta'_A = \frac{V}{\pi}\left(\frac{c^2 B_4^2}{B_1^2}\right)$$

$$\beta''_A = -\frac{V}{\pi}\left(\frac{c^2 B_4^2}{B_1 S_\sigma B_2}\right)$$

$$\beta'''_A = \frac{V}{\pi}\left(\frac{c^2 B_4^2}{S_\sigma^2 B_2^2}\right),$$

V being the material volume of the lattice, i.e. the product of the cross section times the total length of the micronet.

Acknowledgments

We want to thank Elsevier Publishing Company and Pergamon Press for permission to reproduce previously published figures.

References

1. S.Alexander, Phys. Rev. B **27**, 1541 (1983).
2. S.Alexander, Physica **126B**, 294 (1984).
3. S.Alexander and E.Halevi, J.Physique **44**, 804 (1983).
4. J.I.Castro and A.López, Solid State Commun **82**, 787 (1992).
5. J.I.Castro, Ph. D. Thesis, Instituto Balseiro (UNCuyo-CNEA, Argentina), (1992).
6. J.I.Castro and A.López, Phys. Rev. B, **46**, 2, 1075 (1992).
7. J.I.Castro and A.López, Phys. Rev. B, **52**, 7495 (1995).
8. F.H.Claro and G.H.Wannier, Phys Rev B **19**, 6068 (1979).
9. P.G.de Gennes, C.R.Acad.Sc. Paris, **292**, Série II, 9 (1981).

10. P.G.de Gennes, C.R.Acad.Sc. Paris, **292,** Série II, 279 (1981).

11. G.Deutscher, I.Grave and S. Alexander, Phys. Rev. Lett. **48,** 1497 (1982).

12. H.J.Fink, A.López and R. Maynard, Phys. Rev. B **26,** 5237 (1982).

13. H.J.Fink, A.López and D.Rodrigues, Jap. J. of Appl. Phys. **26,** 1465(1987).

14. H.J.Fink, D.Rodrigues and A.López, Phys. Rev. B **38,** 8767 (1988).

15. V.L.Ginzburg and L. D.Landau, Zh.Eksperim.i Teor. Fiz., **20,** 1064 (1950).

16. J.M.Gordon, A.M.Goldman, J.Maps, D.Costello, R.Tiberio and B.Whitehead, Phys. Rev. Lett. **56,** 2280 (1986).

17. D.R.Hofstadter, Phys. Rev B **14,** 2239 (1976).

18. E.Horane, J.Castro, G.Buscaglia and A.López, Phys. Rev. B, **53,** 14, 9296 (1996).

19. S.John and T.C. Lubensky, Phys. Rev. Lett. **55,** 1014 (1985).

20. L.D.Landau, Phys. Z. Sowjet U. **11,** 545 (1937).

21. F.Nori and Q.Niu, Phys. Rev B **37,** 2364 (1988).

22. F.Nori and Q.Niu, Physica B **152,** 105 (1988).

23. P.Gandit, J.Chaussy, B.Pannetier and R.Rammal, Physica B **152,** 32 (1988).

24. See Coherence in Superconducting Networks, Edited by J.E.Mooij and G.B.J.Schoen, Physica B **152** (1988).

25. R.Rammal, T.C.Lubensky and G.Toulouse, Phys. Rev. B **27,** 2820 (1983).

26. J.Simonin, D.Rodrigues and A.López, Phys. Rev Lett. **49,** 944 (1982).

27. J.Simonin, C.Wiecko and A.López, Phys. Rev B **28,** 2497 (1983).

28. J.Simonin and A.López, Phys. Rev. Lett. **56,** 2649 (1986).

29. J.Simonin, C.Wiecko and A.López, Phys.Rev.B **28,** 2497 (1983).

30. F.C.Frank and J.H.van der Merve, Proc.Roy.Soc. **A198,** 205 (1949).

3 Nodal Sets, Multiplicity and Superconductivity in Non-simply Connected Domains[1,2]

Bernard Helffer[1], Maria Hoffmann-Ostenhof[2], Thomas Hoffmann-Ostenhof[3,4], and Mark Owen[5]

[1] Département de Mathématiques: UMR CNRS 8628, Université Paris-Sud
[2] Institut für Mathematik, Universität Wien
[3] Institut für Theoritische Chemie, Universität Wien
[4] International Erwin Schrödinger Institut for Mathematical Physics
[5] Institute of Statistics, Probability Theory and Actuarial Mathematics, Technische Universität Wien

Abstract. This is a survey on [HHOO] and further developments of the theory [He4]. We explain in detail the origin of the problem in superconductivity as first presented in [BeRu], recall the results of [HHOO] and explain the extension to the Dirichlet case. As illustration of the theory, we detail some semi-classical aspects and give examples where our estimates are sharp.

3.1 Zero Sets in Superconductivity

Following the paper by Berger-Rubinstein [BeRu], who refer to the Little Parks experiment [LiPa] we would like to understand the minima (or more generally the extrema) of the following Ginzburg-Landau functional which plays an important role in superconductivity. We are working in a bounded, connected, regular, open set $\Omega \subset I\!\!R^2$ and, for any $\lambda > 0$ and $\kappa > 0$, we consider $G_{\lambda,\kappa}$ defined, for $u \in H^1(\Omega; \mathbb{C})$ and $A \in H^1_{loc}(I\!\!R^2; I\!\!R^2)$ such that $\operatorname{curl} A \in L^2$, by

$$G_{\lambda,\kappa}(u, A) = \int_\Omega (\lambda(-|u|^2 + \tfrac{1}{2}|u|^4) + |(\nabla - iA)u|^2)dx_1 \cdot dx_2 \\ + \kappa^2\lambda^{-1}\int_{I\!\!R^2}|\operatorname{curl} A - H_e|^2dx_1 \cdot dx_2 \ . \tag{3.1}$$

We note here that u (which is called in this theory the order parameter) is defined only in Ω, but that the magnetic potential A is defined in $I\!\!R^2$. Here H_e is a C_0^∞ function on $I\!\!R^2$ or more generally some function in $L^2(I\!\!R^2)$. Physically H_e represents the exterior magnetic field.

We shall sometimes use the identification between vector fields A and 1-forms ω_A.

[1] Funded by the European Union TMR grant FMRX-CT 96-0001
[2] Supported by the Austrian Science Foundation grantnumber P-12864-MAT

When analyzing the extrema of the GL-functional, it is natural to first analyze the corresponding Euler-Lagrange equations. This is a system of two equations (with a boundary equation) :

$$
\begin{aligned}
(GL)_1 & \ -(\nabla - iA)^2 u + \lambda u(|u|^2 - 1) = 0 \ , \ \forall x \in \Omega \ , \\
(GL)_2 & \ \operatorname{curl}^*(\operatorname{curl} A - H_e) = \lambda \kappa^{-2} \Im[\bar{u} \cdot (\nabla - iA)u)] \cdot 1_\Omega \\
(GL)_3 & \ (\nabla - iA)u \cdot \nu = 0 \ , \ \forall x \in \partial\Omega \ .
\end{aligned}
\tag{3.2}
$$

Here ν is a unit exterior normal to $\partial\Omega$ and 1_Ω is the chracteristic function of Ω. The operator curl^* is defined by $\operatorname{curl}^* f := (\partial_{x_2} f, -\partial_{x_1} f)$.
Moreover, without loss of generality in our problem, we may add the condition

$$
(GL)_4 \qquad \operatorname{div} A = 0 \ \text{in} \ \Omega.
\tag{3.3}
$$

One can also assume if necessary that the vector potential satisfies

$$
A \cdot \nu = 0 \ ,
\tag{3.4}
$$

on the boundary of Ω.
Let A_e be a solution of

$$
\begin{aligned}
\operatorname{curl} A_e &= H_e \\
\operatorname{div} A_e &= 0 \ .
\end{aligned}
\tag{3.5}
$$

Such a vector potential always exists in $H^1_{loc}(\mathbb{R}^2)$.
The first remark is that the corresponding pair $(0, A_e)$ is a solution of the Ginzburg-Landau system. This solution is called the normal solution.

Remark 1. Note also that the normalization of the functional leads to the property that

$$
G_{\lambda,\kappa}(0, A_e) = 0 \ .
\tag{3.6}
$$

Following for example [DGP], one can show that, if Ω is bounded, then the functional $G_{\lambda,\kappa}$ admits a global minimizer which is a solution of the equation. If (u, A) is such a minimizer (or more generally a local minimizer) and if it is different from the normal solution, it is a natural question to ask for the structure of the nodal sets of u. This problem was analyzed in [ElMaQi] in the case of a simply connected domain Ω and with different boundary conditions. The authors show that the solutions have as zero sets points or lines which necessarily end at the boundary.
Inspired by [BeRu], our aim is to analyze the non simply-connected situation. In the next section we shall show how to reduce this non-linear problem to a linear problem. Instead of following closely [BeRu], we will show how to implement some ideas of [HHOO] for an alternative presentation of their results and extend it to the multiple holes case. We refer to [DuHe] (or [He4]) for a more complete analysis of the bifurcation (stability) (in connection with [Du]).

3.2   Bifurcation from Normal Solutions. Preliminaries

Starting from one normal solution, a natural way for finding new solutions is to increase λ from 0 and to see if one can bifurcate for a specific value of λ. One can show by a priori estimates that this is impossible for small λ, if the lowest eigenvalue $\lambda^{(1)}$ of the Neumann realization of $-\Delta_{A_e}$ in Ω is strictly positive :

$$\lambda^{(1)} > 0 \ . \tag{3.7}$$

Here, we denote by $-\Delta_A$, for a magnetic potential A, the magnetic Laplacian :

$$-\Delta_{A_e} = \sum_{j=1}^{2} (\frac{1}{i}\partial_{x_j} - A_j(x))^2 \ . \tag{3.8}$$

It is rather standard that a necessary condition for having a bifurcation is actually that λ becomes an eigenvalue of $-\Delta_{A_e}$ ([BeRu], [DuHe]). We shall consider only what is going on around $\lambda^{(1)}$. Note here that there is an intrinsic degeneracy to the problem related to the existence of an S^1 action. We observe indeed the property :

If (u, A) is a solution of the GL-system,   then $(\exp i\theta\, u, A)$ is a solution, (3.9)

for any $\theta \in \mathbb{R}$.
In order to go further, we add the assumption

$$\lambda^{(1)} \text{ is a simple eigenvalue.} \tag{3.10}$$

We will come back later to the properties of the spectrum of $-\Delta_A$ and in particular to the strict positivity (3.7) and to the multiplicity of the lowest eigenvalue. Let us only say here that (3.10) is rather generic.
Now, one can try to apply the general bifurcation theory due to Crandall-Rabinovitz [CrRa]. Note that, although, the eigenvalue is assumed to be simple, it is not exactly a simple eigenvalue in the sense of Crandall-Rabinowitz who are working with real Banach spaces. Actually, this is only simple modulo this S^1-action which was observed in (3.9). In this context, the theory developped for example [GoSc], (Chapter VII, §3 and Chapter VIII (Proposition 2.2)) is relevant, but special cases involving Schrödinger operators with magnetic field are treated in [Od], [BaPhTa], [Ta] and [Du]. There are actually various ways to remove the degeneracy and we shall present one which seems quite adapted to our peculiar situation by introducing a notion of real solution in Section 3.5.
All the considered operators are (relatively to the wave function or order parameter) suitable realizations of operators of the type

$$u \mapsto -\Delta_A u - \lambda f(|u|^2)u \ ,$$

with $f(0) = 1$. For our model (see $(GL)_1$), we have indeed $f(t) = 1 - t$.
In the case when $||u||_{C^0(\overline{\Omega})}$ is sufficiently small (which will be for example a
consequence of $||u||_{H^2(\overline{\Omega})}$ small), we observe that if u is a non zero solution
of

$$-\Delta_A u - \lambda f(|u|^2)u = 0 , \qquad (3.11)$$

then

$$v := \sqrt{f(|u|^2)}u \qquad (3.12)$$

is a non zero solution of the linear equation :

$$-(f(|u|^2))^{-\frac{1}{2}}\Delta_A(f(|u|^2))^{-\frac{1}{2}}v - \lambda v = 0 , \qquad (3.13)$$

and has the same nodal set as u. Moreover u and v satisfy the same boundary
condition.

The last remark is here that, if $\lambda - \lambda^{(1)}$ and $||u||_{C^0(\overline{\Omega})}$ are small enough and
(3.10) is satisfied, then λ is the lowest eigenvalue (with multiplicity 1) of the
Neumann-realization of the linear operator $\mathcal{L}_{u,A} := -f(|u|^2)^{-\frac{1}{2}}\Delta_A f(|u|^2)^{-\frac{1}{2}}$
and we have to analyze the properties of the groundstate v of this opera-
tor. Note that the assumptions that $\lambda - \lambda^{(1)}$ and $||u||_{C^0(\overline{\Omega})}$ are small enough
will be satisfied when we consider bifurcation starting from the normal so-
lution. The conclusion is that we can apply the linear theory developed for
the Schrödinger operator with magnetic field Δ_A modulo a small (but easy)
extension to the operators $\mathcal{L}_{u,A}$. So before to go further it seems useful to
recall some basics on the Schrödinger operator with magnetic field.

3.3   The Linear Schrödinger Operator with Magnetic Field

Our starting point is the Laplace operator with magnetic field in a C^∞,
relatively compact domain $\Omega \subset I\!R^2$. Here we discuss also the more general
Schrödinger operators of the form $-\Delta_A + V$ where V is a C^∞ potential.
This was not needed for the application to superconductivity but we want
to emphasize how general is our approach. In particular there would be no
use of analyticity properties like for example in [ElMaQi]. More generaly, we
could also look at operators in the form $-g(x)^{-1}\Delta_A g(x)^{-1} + V(x)$ in order
to treat the examples appearing previously but we do not do this here for
simplicity. So we are considering the operator

$$u \mapsto P_{A,V}u := -(\nabla - iA)^2u + Vu . \qquad (3.14)$$

We are interested in the Neumann-realization $P_{A,V}^N$ or in the Dirichlet realization $P_{A,V}^D$. The basic properties of these operators are discussed in [AHS] (See
also [He2]). It is well known that if we denote by $\lambda^{(1),N}(A)$ (resp. $\lambda^{(1),D}(A)$)
the lowest eigenvalue of $P_{A,V}^N$ (resp. $P_{A,V}^D$) , then we always have the property

$$\lambda^{(1),N}(A) \geq \lambda^{(1),N}(0) ;$$
$$\lambda^{(1),D}(A) \geq \lambda^{(1),D}(0) . \qquad (3.15)$$

The natural question is then to decide when we have equality. The first necessary condition is

$$B = \operatorname{curl} A \equiv 0 \quad \forall x \in \Omega. \tag{3.16}$$

This condition is also sufficient in the case when Ω is simply connected. When Ω is not simply connected, we have also to consider, for any closed path γ in Ω, the quantity $\Phi(\gamma) := \frac{1}{2\pi} \int_\gamma (A_1 dx_1 + A_2 dx_2)$ which will be called the (normalized) circulation of A along γ. Of course, when $B = 0$ and Ω is simply connected, this quantity is equal to 0 but it is no more the case when we have holes. In the general context, it can be proved that $\lambda^{(1),N}(A) = \lambda^{(1),N}(0)$ if and only if $B = 0$ and, for any closed path γ in Ω

$$\Phi(\gamma) \in \mathbb{Z} . \tag{3.17}$$

A similar result holds for the Dirichlet case. This was proved in the case of the Dirichlet realization (See Lavine, O'Caroll [LaO], [He1], [He2]) and a similar result is also true in the case of Neumann [HHOO]. We observe that (3.16) and (3.17) are just the conditions under which there exists a gauge transform such that the operator is unitary equivalent to the case without magnetic potential.

In this last case, the lowest eigenvalue is simple and the corresponding ground-state has no zero.

We will discuss later the natural question of determining for which potentials A, such that $\operatorname{curl} A = 0$ in Ω, the eigenvalues $\lambda^{(1),N}(A)$ or $\lambda^{(1),D}(A)$ are maximal.

Note also that, in the case when $V \equiv 0$ and for the Neumann realization, the condition (3.7) is a consequence of the non validity of (3.16) or (3.17). We have indeed $\lambda^{(1),N}(0) = 0$ and strict inequality in the first line of (3.15).

We close this section by discussing briefly the question of multiplicity. It has been observed [AHS] that the property that the lowest eigenvalue is simple when $A = 0$ (which is a consequence of the Krein-Rutman theorem, cf for example [ReSi]) is no more true in general. As we shall discuss later, there is a strong connection between the nodal sets and the multiplicity.

3.4   The Operator K

We now consider the case with zero magnetic field (Assumption (3.16)). We consider the non simply connected situation. The connected components of the complementary set of Ω in $\mathbb{R}^2$ is the union of k holes O_i and of an unbounded connected component. Let us denote by σ_i a closed path[3] which

---

[3] A piecewise smooth mapping $\gamma : [0, 1] \to X$ is called a path in X. The point $\gamma(0)$ is called the initial point and $\gamma(1)$ is called the final point. The image $\Gamma = \gamma([0, 1])$ of the path is called a curve.

parameterizes the boundary Σ_i of the i-th hole and turns once in an anti-clockwise direction. Let us consider the circulations

$$\Phi_i = \frac{1}{2\pi} \oint_{\sigma_i} A \cdot \mathrm{dx}, \qquad (3.18)$$

of A round the i-th hole ($i = 1, \ldots, k$).
Let us assume that

$$2\Phi_i = 2\ell + 1 , \quad \text{with } \ell \in \mathbb{Z} . \qquad (3.19)$$

This case is quite specific in the sense that, in this case,

$$P_{A,V} \text{ and } P_{-A,V} \text{ are gauge equivalent.} \qquad (3.20)$$

One can indeed find a multivalued C^∞ function[4] ϕ in Ω such that $\exp i\phi$ is univalued and such that

$$d\phi = -2(A_1 \, dx_1 + A_2 \, dx_2) . \qquad (3.21)$$

We then have

$$\exp -i\phi \, P_{A,V} \, \exp i\phi = P_{-A,V} . \qquad (3.22)$$

We also observe that, if Γ is the conjugation operator

$$u \mapsto \Gamma u = \overline{u} , \qquad (3.23)$$

then

$$\Gamma \Delta_A = \Delta_{-A} \Gamma , \qquad (3.24)$$

and consequently

$$\Gamma P_{A,V} = P_{-A,V} \Gamma , \qquad (3.25)$$

Combining (3.22) and (3.25), we obtain, for the operator

$$K := \exp -i\phi \, \Gamma , \qquad (3.26)$$

which satisfies

$$K^2 = Id , \qquad (3.27)$$

the following commutation relation

$$K \, \Delta_A = \Delta_A \, K . \qquad (3.28)$$

Let us also observe that the Neumann conditions and the Dirichlet conditions are respected by K.
As a corollary, we get

Lemma 1. *If u is an eigenvector of Δ_A^N (resp. Δ_A^D), then Ku has the same property.*

---

[4] Along a path γ of index 1 around one of the holes contained in Ω, the variation of ϕ is $2n\pi$ ($n \in \mathbb{Z}$).

Let us now explain the possible role of the operator K. Similarly to what is done with the Laplacian and the operator Γ, we observe that one can find an orthonormal basis of "real" eigenvectors u_j, i. e. they satisfy the condition

$$Ku_j = u_j \ . \tag{3.29}$$

In particular, if u is a simple eigenvector, then we can multiply it by a constant so that it becomes a "real" eigenvector.

Remark 2. Here we would like to say that all the properties of $P_{A,V}$ described in this section are also true for the operators $-g(x)^{-1}\Delta_A g(x)^{-1} + V(x)$. Note also that if $Ku = u$, then the function v defined by (3.12) will also satisfy the same condition.

3.5   Bifurcation for Special Spaces

Let us come back to the functional $G_{\lambda,\kappa}$ introduced in (3.1). Following [BeRu], we assume that

$$H_e \equiv 0 \text{ in } \overline{\Omega} \ , \tag{3.30}$$

and that (3.19) is satisfied for $B = H_e$. We would like to show how to take advantage of the existence of the operator K introduced in Section 3.4. We look for solutions of the GL equation in the form (u, A_e) with $Ku = u$. Let us observe that

$$L^2_K(\Omega; \mathbb{C}) := \{u \in L^2(\Omega; \mathbb{C}) \mid Ku = u\} \ , \tag{3.31}$$

is a real Hilbert subspace of $L^2(\Omega; \mathbb{C})$.

We denote by H^m_K the corresponding Sobolev spaces :

$$H^m_K(\Omega; \mathbb{C}) = H^m(\Omega; \mathbb{C}) \cap L^2_K \ . \tag{3.32}$$

We now observe the

Lemma 2. *If $u \in H^1_K$, then the current $\Im(\overline{u} \cdot (\nabla - iA_e)u) = 0$ almost everywhere.*

Proof of Lemma 2:
Let us consider a point where $u \neq 0$. Then we have $u = \rho \exp i\theta$ with $2\theta = \phi$ modulo $2\pi \mathbb{Z}$. Remembering that $A_e = \frac{1}{2}\nabla\phi$, it is easy to get the property. ∎
Once this lemma is proved, one immediately sees that (u, A_e) (with $Ku = u$) is a solution of the GL system if and only if $u \in H^1_K$ and

$$\begin{aligned} -\Delta_{A_e} u - \lambda u(1 - |u|^2) &= 0 \ , \\ (\nabla - A_e)u \cdot \nu &= 0 \ , \text{ on } \partial\Omega \ . \end{aligned} \tag{3.33}$$

We shall call this new system the reduced GL-equation.
But now we can apply Theorem 2.4 by Crandall-Rabinowitz [CrRa]. By assumption (3.10), the kernel of $(-\Delta_{A_e} - \lambda^{(1)})$ is now a one-dimensional real

subspace in L^2_K. Let us denote by u_1 a normalized "real" eigenvector. Note that u_1 is unique up to the multiplication by ± 1 and that $u_1 \in H^m_K$, $\forall m \in I\!N$. Therefore, we have (see [BeRu] for a weaker form in the one hole case) the

Proposition 1. *Under assumptions (3.10), (3.16) and (3.19), there exists a bifurcating family of solutions $(u(\cdot; \alpha), \lambda(\alpha))$ in $H^1_K \times I\!R^+$ with $\alpha \in]-\epsilon_0, +\epsilon_0[$, for the reduced GL-equation such that*

$$u(\alpha) = \alpha u_1 + \alpha^3 u_2(\alpha) \, ,$$
$$< u_1, u_2(\alpha) >_{L^2} = 0 \, , \tag{3.34}$$
$$\|u_2(\alpha)\|_{H^2(\Omega)} = \mathcal{O}(1) \, ;$$

$$\lambda(\alpha) = \lambda^{(1)} + c\alpha^2 + \mathcal{O}(\alpha^4) \, , \tag{3.35}$$

with

$$c = \lambda^{(1)} \cdot \int_\Omega |u_1|^4 \, dx \, . \tag{3.36}$$

Moreover

$$u(-\alpha) = -u(\alpha) \, , \quad \lambda(-\alpha) = \lambda(\alpha) \, . \tag{3.37}$$

Remark 3. Note that the property (3.37) is what remains of the S^1-invariance when one considers only "real" solutions.

Let us give here the formal computations of the main terms. Denote by L_0 the operator $L_0 := -\Delta_{A_e} - \lambda^{(1)}$. Writing $u_2(\alpha) = u_{20} + \mathcal{O}(\alpha)$, we get modulo $\mathcal{O}(\alpha^4)$.

$$(L_0 - c\alpha^2)(\alpha u_1 + \alpha^3 u_{20}) + \lambda^{(1)} \alpha^3 u_1 |u_1|^2 = \mathcal{O}(\alpha^4) \, .$$

Projecting on u_1, we get (3.36). Projecting on the orthogonal space u_1^{perp} and denoting by M_0 the operator equal to the inverse of L_0 on this subspace and to 0 on Ker L_0, we get

$$u_{20} := -\lambda^{(1)} M_0(u_1 |u_1|^2) = -\lambda^{(1)} M_0(u_1 |u_1|^2 - c u_1) \, . \tag{3.38}$$

The second result is that, once we have observed that $u(\alpha)$ is a "real" solution of the reduced GL equation, then one can show that, for small $\alpha \neq 0$, the nodal set of $u(\alpha)$ will behave like the nodal set of u_1 (see Section 3.8 for a precise statement). In particular, if there is only one hole, then the nodal set of $u(\alpha)$ consists exactly of one line joining the interior boundary and the exterior boundary.

3.6   The Toy Model

Before to discuss the general results of [HHOO] in Sections 3.7 and 3.8, we think it is worth to discuss a model which in some sense is a limiting case of a very thin ring like open set (see Berger-Rubinstein-Schatzman [BeRuSc]).

This is the case of the Schrödinger operator on the circle which is actually a rewriting of the Hill's equation. If we consider the operator

$$P_\alpha = -(\partial_\theta - i\alpha)^2$$

on $L^2(S^1)$, then the spectrum is easily seen as given by

$$\sigma(P_\alpha) = \{(n - \alpha)^2 \, , \, n \in \mathbb{Z}\} \, .$$

The minimum is obtained for the n's such that $(n-\alpha)^2$ is minimal. When α is an integer, one obtains $n = \alpha$ and the eigenvalue is simple with corresponding eigenfunction $\theta \mapsto \exp i\alpha\theta$.

This property is still the same for any $\alpha \notin \frac{1}{2} + \mathbb{Z}$. When $\alpha \in \frac{1}{2} + \mathbb{Z}$, the lowest eigenvalue is $\frac{1}{4}$ and with multiplicity 2. For example, when $\alpha = \frac{1}{2}$, the eigenvectors are generated by 1 and $\exp i\theta$ or $\exp i\theta/2 \cos \theta/2$ and $\exp i\theta/2 \sin \theta/2$. The last one vanishes on $\theta = 0$. Of course one gets a continuous family of eigenvectors with zeros by considering $\exp i\theta/2 \sin(\theta - \theta_0)/2$.

It is easy to see that this degeneracy disappears when considering

$$P_{\alpha, \epsilon v} = -(\partial_\theta - i\alpha)^2 + \epsilon v(\theta) \, ,$$

perturbatively as $\epsilon \neq 0$ is small, and if the condition $\int_0^{2\pi} \exp i\theta v(\theta) d\theta \neq 0$. We now also see from the general theory of the Hill's equation (See Eastham [East]) that for any C^∞ potential v on S^1 the lowest eigenvalue $\lambda_1(\alpha)$ of $P_{\alpha,v}$ is periodic and takes its maximum for $\alpha \in \mathbb{Z} + \frac{1}{2}$. Moreover, the multiplicity is one if this last condition is not satisfied.

Our aim is to analyze the zeros of the groundstate as a function of α. We recall that it has been proved that a "real" groundstate vanishes when $\alpha = \frac{1}{2}$ modulo $\mathbb{Z}$. This was the object (in a more general case) of [HHOO].
The case when $\alpha = 0$ modulo $\mathbb{Z}$ is well known because the operator is gauge equivalent to the Schrödinger operator with $\alpha = 0$.

There is a lot of information in this case because it was analyzed in great detail when considering the Schrödinger operator with periodic potential on $L^2(\mathbb{R})$. We shall recall some of these results. The following theorem can be found in [ReSi] (Theorem XIII.89, p. 293).

Theorem 1. *Let $\lambda_1(\alpha)$ the first eigenvalue of $P_{v,\alpha}$. Then we have*

1. $\lambda_1(\alpha)$ is simple for $\alpha \in [0, \frac{1}{2}[$.
2. $\alpha \mapsto \lambda_1(\alpha)$ increases strictly on $[0, \frac{1}{2}[$.

So the only case, where $\lambda_1(\alpha)$ may be degenerate is consequently the case when $\alpha = \frac{1}{2}$, modulo $\mathbb{Z}$.
Let us look in this context at the nodal sets.

Theorem 2.

1. If $\alpha \notin \frac{1}{2} + \mathbb{Z}$, the groundstate does not vanish.
2. If $\alpha \in \frac{1}{2} + \mathbb{Z}$, the multiplicity may be 1 or 2 and the "real" groundstates have a single zero in S^1.

Proof
We observe that $\theta \mapsto v_1(\theta, \alpha) = \exp -i\alpha\theta \, u_1(\theta, \alpha)$ is a solution in $C^\infty(\mathbb{R})$ of $-d^2u/d\theta^2 + v(\theta)u - \lambda_1(\alpha)u = 0$ on $\mathbb{R}$. Here v_1 is considered as a function of θ (here α is fixed and $\alpha \notin \mathbb{Z}$) and has the same zeros as $u_1(\theta, \alpha)$. Let us assume the existence in $[0, 2\pi[$ of a zero of u_1 and consequently of v_1. Let us consider the decomposition : $v_1 = f + ig$, with f and g real.

Two cases have to be considered.

- f and g are linearly dependent. In this case, possibly changing v_1, we can assume that v_1 is real. But the Floquet condition will give $v_1(\theta + 2\pi, \alpha) = \exp -i2\pi\alpha \, v_1(\theta, \alpha)$ and this implies $\sin(2\pi\alpha) = 0$. This corresponds to the case when $\alpha \in \mathbb{Z}/2$. We are in the second case and get the announced result.
- We now consider the case when f and g are linearly independent. Here we get two linearly independent solutions of the same second order equation vanishing at the same point. By the Cauchy uniqueness theorem they have to be proportional. This gives the contradiction in the first case. So u_1 can not vanish as initially assumed.

3.7   Variation of the Ground State Energy

Let us consider the non-simply connected situation. We observe that, if the circulations $\Phi = (\Phi_1, \ldots, \Phi_k)$ of two distinct vector potentials A and A' are equal modulo $\mathbb{Z}^k$ then the corresponding operators $P_{A,V}$ and $P_{A',V}$ are unitarily equivalent under a gauge transformation. The following theorem was given in [HHOO].

Theorem 3. Let $\Omega \subset \mathbb{R}^2$ be a region with smooth boundary, which is homeomorphic to a disk with k holes. For a given smooth potential V, the first eigenvalue $\lambda^{(1)}$ of the magnetic Schrödinger operator $P_{A,V}$, where A satisfies (3.16), depends only on the circulations $\Phi = (\Phi_1, \ldots, \Phi_k)$ of A:

$$\lambda^{(1)} = \lambda_1(\Phi) \,.$$

The function $\lambda_1(\Phi)$ has the following properties (in which $l \in \mathbb{Z}^k$ is arbitrary):

$$\lambda_1(\Phi + l) = \lambda_1(\Phi) \,. \tag{3.39}$$

$$\lambda_1(l/2 + \Phi) = \lambda_1(l/2 - \Phi) \,. \tag{3.40}$$

$$\lambda_1(\Phi) > \lambda_1(0, \ldots, 0) \quad for \ \Phi \notin \mathbb{Z}^k \,. \tag{3.41}$$

For the case $k = 1$, we have in addition to inequality (3.41) that

$$\lambda_1(\Phi) < \lambda_1(1/2) \tag{3.42}$$

for $\Phi \notin 1/2 + \mathbb{Z}$.

Equations (3.39), (3.40) and inequality (3.41) are straightforward. In this context we should also mention the recent very interesting results [HN] by Herbst and Nakamura concerning large magnetic fields. The Neumann boundary conditions on $P_{A,V}$ are motivated by superconductivity but our results are also valid for the case of Dirichlet boundary condition. Dirichlet boundary conditions are related to the Aharonov-Bohm effect for bound states. See [LaO,He1,He2,He3]. This will be discussed in Section 3.11.

Remark 4. The proof of (3.42) uses a connection between the maximality of the first eigenvalue for circulation $1/2$ and the structure of the nodal set of groundstates. In a recent paper [HHON], it is conjectured that for $\Phi \in [0, 1/2)$ (3.42) can be strenghtened to $\lambda_1'(\Phi) \geq 0$.

Using semiclassical arguments, one can show that in general the first eigenvalue is not necessarily maximized for circulation $(1/2, \ldots, 1/2)$, this will be discussed in Section 3.11.

3.8   Nodal Sets

Let us now analyze the structure of the nodal set of the groundstate of $P_{A,V}$ under the assumptions (3.16) and (3.19). For the toy model, one has to think that α was the constant magnetic potential on the circle, and the condition (3.19) corresponds exactly to the second case in Theorem 2. We have seen in Section 3.4, that, without any assumption on the multiplicity, it is enough to analyze the zero sets of "real" groundstates (See (3.29)). But let us start by a precise definition of this set.

Definition 1. The nodal set $\mathcal{N}(u)$ of an eigenfunction u of a magnetic Schrödinger operator on a manifold Ω with smooth boundary is defined in $\overline{\Omega}$ by

$$\mathcal{N}(u) := \overline{\{x \in \Omega : u(x) = 0\}}. \tag{3.43}$$

Useful information on nodal sets of real valued eigenfunctions of non-magnetic Schrödinger equations in two dimensions is given in [Bers,Che,HOMN] and in Proposition 4.1 in [HHOO]. In particular such nodal sets consist of the finite union of smoothly immersed circles and lines. "Generically", the nodal set of every complex eigenfunction of a magnetic Schrödinger operator consists of isolated points of intersection of the lines of zeros of the real and imaginary parts of the function. See [ElMaQi].

The local properties of the nodal sets of eigenfunctions of the operator $P_{A,V}$ are the same as the local properties of complex solutions of non-magnetic Schrödinger equations. More precisely, since we may find at every point a local gauge $e^{i\phi}$ satisfying : $\nabla\phi = A$, we may multiply any eigenfunction of $P_{A,V}$ by a local gauge so that the product solves a non-magnetic Schrödinger equation. The nodal set is invariant under local gauge transformations.

We shall see in what follows that although the local properties of nodal sets of eigenfunctions of our magnetic Schrödinger operator are the same as the properties of a non-magnetic Schrödinger operator, the global properties differ in the case where $\Phi = (1/2, \ldots, 1/2)$. In particular, in the non-magnetic case we see that (since a real eigenfunction must change sign at the nodal set) an even number of nodal lines (or perhaps no nodal lines) of an eigenfunction emerges from each boundary component of the region. In Theorem 4 we show that for $\Phi = (1/2, \ldots, 1/2)$, an odd number of nodal lines of the groundstate emerge from each component. The following definition was introduced in [HHOO].

Definition 2. We say that a (nodal) set $\mathcal{N}$ **slits** $\overline{\Omega}$ if it is the union of a collection of piecewise smooth, immersed lines such that

1. each line starts and finishes at the boundary $\partial\Omega$ and leaves the boundary transversally;
2. internal intersections between lines are transversal;
3. the complement $\Omega \setminus \mathcal{N}$ is connected;
4. an odd number of nodal lines leaves each interior boundary component.

We shall say that a collection of paths slits $\overline{\Omega}$ if the union of the images of the paths slits $\overline{\Omega}$.

See Figure 3.1 for some examples of regions which are slit.

Note that part 3 of Definition 2 is the reason why a nodal set which slits $\overline{\Omega}$ contains no immersed circles, and also implies that each line of a slitting set links together a unique pair $\{\Sigma_i, \Sigma_j\}$ of distinct (i.e. $i \neq j$) boundary components. Note also that for the single hole case, a set which slits $\overline{\Omega}$ consists of one line which joins the outer boundary of Ω to the inner boundary. The main theorem is then

Theorem 4. *Let Ω be a region with smooth boundary, which is homeomorphic to a disk with k holes. Let V be a smooth potential and let A be a smooth magnetic vector potential satisfying equation (3.16), such that the value of the circulations around each hole lie in $1/2 + \mathbb{Z}$ (that is $\Phi = (1/2, \ldots, 1/2)$, modulo $\mathbb{Z}^k$).*

1. *If the first eigenvalue of $P_{A,V}$ is simple then the nodal set of the corresponding eigenfunction slits $\overline{\Omega}$. Otherwise there exists an orthonormal*

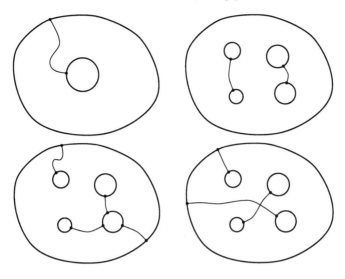

Fig. 3.1. Examples of some sets which slit $\overline{\Omega}$

"real" basis $\{u_1, \ldots, u_m\}$ of the groundstate eigenspace such that the nodal set of any non-zero combination $\sum_{i=1}^{m} a_i u_i$, with $a_i \bar{a}_j \in \mathbb{R}$ for each $1 \le i, j \le m$, slits $\overline{\Omega}$.

2. *For $k = 1, 2$ with groundstate multiplicity two, the nodal sets of two linearly independent groundstates do not intersect. It follows that the nodal set of a combination $a_1 u_1 + a_2 u_2$ is empty whenever $a_1 \bar{a}_2 \notin \mathbb{R}$.*

Remark 5.

1. For the cases $k \ge 3$, we expect that there could be intersection of nodal sets of two independent groundstates, and correspondingly that the nodal set of a combination $a_1 u_1 + a_2 u_2$ will not in general be empty when $a_1 \bar{a}_2 \notin \mathbb{R}$.
2. If we assume that $P_{A,V}$ has Dirichlet boundary conditions then Theorem 4 holds with suitable changes to the proofs.

The proof given in [HHOO] is inspired by one step in the proof of [BeRu] who treat only the one hole case. But it appears more convenient to lift the problem on some two-fold covering manifold where one can gauge away the magnetic potential and analyze real solutions of a real operator with some additional anti-symmetry. This will be explained in Section 3.9.

Let us compare with the point of view developed in [BeRu].
In the one hole case, we can consider for any slitting path γ, the selfadjoint realization of the Schrödinger operator $-\Delta + V$ in $\Omega(\gamma) := \Omega \setminus \{\gamma\}$ with Neumann boundary condition on $\partial\Omega$ and Dirichlet condition on γ. We denote the lowest eigenvalue of this operator by $\mu(\gamma)$ and the corresponding strictly

positive, normalized, eigenvector by ψ_γ. Let $\chi = \chi_\gamma$ be a solution of

$$d\chi = A_1\, dx_1 + A_2\, dx_2\,, \quad \text{in } \Omega(\gamma)\,.$$

Such a solution exists because $\Omega(\gamma)$ is simply connected, without any assumption on the circulation of A. Considering the function $u_\gamma = \exp i\chi\,\psi_\gamma$ (considered by extension as defined on $L^2(\Omega)$) (which belongs to $H^1(\Omega)$), we get by applying the minimax lemma,

$$\lambda^N_{A,V} \leq \mu(\gamma). \tag{3.44}$$

We consequently have proved, without any special assumption on the multiplicity and without any restrictive condition on the circulation, the first part of the

Proposition 2. *Let V be C^∞ and A a C^∞ potential on $\overline{\Omega}$ such that $B = 0$ in Ω. Then*

$$\lambda^N_{A,V} \leq \inf_{\gamma \in \mathcal{L}^1(\Omega)} \mu(\gamma)\,, \tag{3.45}$$

where $\mathcal{L}^1(\Omega)$ is the set of all piecewise C^1 curves γ such that $\Omega(\gamma)$ is simply connected.
If (3.19) is satisfied, then we have actually equality in (3.45)

$$\lambda^N_{A,V} = \inf_{\gamma \in \mathcal{L}^1(\Omega)} \mu(\gamma)\,, \tag{3.46}$$

Remark 6. The quantity appearing in the r. h. s of (3.45) was introduced in [BeRu] in the case $V = 0$. The condition of non degeneracy which was introduced by these authors and used in their argument was actually the condition:
(C) : The infimum $\inf_\gamma \mu(\gamma)$ is realized for a unique slitting path γ.
We emphasize that we shall not use this assumption. Moreover we can prove, as a result of our analysis that either this condition is satisfied (and the lowest eigenvalue is then simple) or there is an infinite family of paths realizing the minimum.

We assume now condition (3.19) and would like to find some γ such that $\mu(\gamma) \leq \lambda^N_{A,V}$. This would give as a consequence the equality (3.46). A solution γ which gives the solution is actually the nodal line of one "real" ground state.

One can interpret all this theory as the analysis of an excited eigenvalue of a Schrödinger operator defined on the two-fold covering of Ω. This will be detailed in Section 3.9.

On the other hand, according to the arguments developed in Sections 3.3 and 3.5, we obtain the following theorem

Theorem 5. *Under the assumptions of Proposition 1, there exists $\epsilon_1 > 0$ such that, for all α such that $0 < |\alpha| \leq \epsilon_1$, the zero set of the "real" bifurcating state $u(\alpha)$ slits $\overline{\Omega}$.*

Although we have not verified the mathematical details there are strong reasons to think that this nodal set tends as $|\alpha| \to 0$ to the nodal set of u_1.

3.9   A Twofold Riemannian Covering Manifold

In this section we keep the assumption of Section 3.8 and explain some elements of the proof of the characterization of the nodal sets. The proofs of the results of [HHOO] use a twofold Riemannian covering manifold $\widetilde{\Omega}$ of the domain Ω. For the case of more than one hole, there exists more than one twofold Riemannian covering manifold of Ω. We shall take a particular choice of covering manifold on which the circulation of the lifted magnetic potential $\widetilde{A}$ along any closed curve is an integer. Before the precise definition, we introduce some basic notation. For further details see for example [GHL].

Definition 3. Let $\widetilde{\Omega}$ be a covering manifold of Ω, and let Π be the associated covering map. We denote the lifts of various quantities as follows:
  For a set $\mathcal{N}$ define $\widetilde{\mathcal{N}} = \{x \in \widetilde{\Omega} : \Pi(x) \in \mathcal{N}\}$. For a function $f : \Omega \to \mathbb{C}$, define $\widetilde{f} : \widetilde{\Omega} \to \mathbb{C}$ by $\widetilde{f} = f \circ \Pi$. For a path $\sigma : [0,1] \to \Omega$ and a point $x \in \widetilde{\Omega}$ such that $\Pi(x) = \sigma(0)$ let $\widetilde{\sigma} : [0,1] \to \widetilde{\Omega}$ denote the unique lifted path such that $\widetilde{\sigma}(0) = x$ and $\Pi \circ \widetilde{\sigma} = \sigma$.
  We endow the covering manifold with the metric obtained by lifting the flat Euclidean metric of Ω to $\widetilde{\Omega}$. This is the unique metric which makes Π a local isometry, and therefore a Riemannian covering map. Let $\widetilde{\Delta}$ denote the Laplace-Beltrami operator on $L^2(\widetilde{\Omega})$ induced by the lifted metric on Ω, and let $w_{\widetilde{A}}$ be the 1-form on $\widetilde{\Omega}$ obtained by lifting the 1-form w_A associated with the smooth vector potential A defined on Ω.

Let $\widetilde{\Omega}_\infty$ be the universal covering manifold of Ω and let Π_∞ be the associated covering map. Note that due to (3.19) if two points $x_\infty, y_\infty \in \widetilde{\Omega}_\infty$ satisfy $\Pi_\infty(x_\infty) = \Pi_\infty(y_\infty)$ then for any path σ joining x_∞ to y_∞, the integral

$$\frac{1}{2\pi} \int_{\Pi_\infty \circ \sigma} A \cdot dx \tag{3.47}$$

lies either in $1/2 + \mathbb{Z}$ or in $\mathbb{Z}$ and is independent of the path σ because $\operatorname{curl} A = 0$ and because the universal covering manifold is simply connected. We therefore construct the twofold covering manifold (as a quotient of the universal covering manifold) as follows:

Definition 4.

1. We define the twofold covering manifold $\widetilde{\Omega}$ by identifying points x_∞, y_∞ in $\widetilde{\Omega}_\infty$ according to the equivalence relation $x_\infty \sim y_\infty$ if and only if

$$\Pi_\infty(x_\infty) = \Pi_\infty(y_\infty) \tag{3.48}$$

and for each path σ in $\widetilde{\Omega}_\infty$ joining x_∞ to y_∞ we have

$$\frac{1}{2\pi} \int_{\Pi_\infty \circ \sigma} A \cdot dx \in \mathbb{Z}. \tag{3.49}$$

The covering map $\Pi : \widetilde{\Omega} \to \Omega$ is defined by $\Pi(x) = \Pi_\infty(x_\infty)$, where $x = [x_\infty]$ is the equivalence class (under $\sim$) containing x_∞.

2. On our twofold covering manifold we define the symmetry map $G : \widetilde{\Omega} \to \widetilde{\Omega}$ by setting Gx to be the other point in $\widetilde{\Omega}$ which lies above $\Pi(x) \in \Omega$. Note that $\Pi^{-1}(\Pi(x)) = \{x, Gx\}$.

3. We say that a function $f : \widetilde{\Omega} \to \mathbb{C}$ is symmetric (resp. antisymmetric) if $f(Gx) = f(x)$ (if $f(Gx) = -f(x)$) for all $x \in \widetilde{\Omega}$.

Note that the identity map and G form a group $\mathbf{G} = \{I, G\}$, with the composition $G^2 = I$, which acts freely on $\widetilde{\Omega}$. The quotient of $\widetilde{\Omega}$ by $\mathbf{G}$ is the original manifold Ω. Using equation (3.49), we have

$$\frac{1}{2\pi} \oint_\sigma \widetilde{A} \cdot d\widetilde{x} = \frac{1}{2\pi} \oint_{\Pi \circ \sigma} A \cdot dx \in \mathbb{Z}, \tag{3.50}$$

for any closed path σ in $\widetilde{\Omega}$. Hence there exists a smooth, multivalued function θ on $\widetilde{\Omega}$ such that $\exp i\theta$ is singlevalued and

$$\nabla \theta = \widetilde{A}. \tag{3.51}$$

The crucial lemma in [HHOO] is

Lemma 3.

1. The operator $\mathbf{L} : L^2(\Omega) \to L^2(\widetilde{\Omega})$ defined by

$$\mathbf{L}u = \frac{1}{\sqrt{2}} e^{i\theta} \widetilde{u} \tag{3.52}$$

   is a isometry onto the antisymmetric functions in $L^2(\widetilde{\Omega})$.

2. $\mathbf{L}$ maps the "real" functions (that is such that $Ku = u$) onto the real antisymmetric functions on $\widetilde{\Omega}$.

As a consequence, we get that $\mathbf{L}$ maps eigenfunctions of $P_{A,V}$ onto antisymmetric eigenfunctions of the Schrödinger operator

$$\widetilde{P}_{0,V} = -\widetilde{\Delta} + \widetilde{V} \tag{3.53}$$

acting on $L^2(\widetilde{\Omega})$ with Neumann boundary conditions. The "real" eigenvectors become real eigenvectors of $\widetilde{P}_{0,V}$.

In particular $\lambda^{(1)}$ is the lowest eigenvalue of $\widetilde{P}_{0,V}$ restricted to the antisymmetric space. Any real corresponding antisymmetric groundstate has to vanish (this is the initial argument for the proof of Theorem 4) and can not consequently correspond to the lowest eigenvalue of $\widetilde{P}_{0,V}$. Actually, we know that the groundstate corresponds to a single eigenvalue and is symmetric. This does not mean that $\lambda^{(1)}$ is necessarily the second eigenvalue of $\widetilde{P}_{0,V}$. For any $\ell \in I\!\!N$ ($\ell > 0$) and in the one hole case, it is indeed possible to construct potentials V such that the first ℓ eigenvalues correspond to symmetric eigenstates [Ni]. Another consequence for the Ginzburg-Landau functional is that

$$G_\lambda(u, A_e) = \int_{\widetilde{\Omega}} \left(\lambda(-|\mathbf{L}u|^2 + \frac{1}{2}|\mathbf{L}u|^4) + |\widetilde{\nabla}\mathbf{L}u|^2 \right) d\widetilde{x} . \qquad (3.54)$$

In particular, to look for minima for G_λ with "real" u and $A = A_e$ leads to minimize the functional

$$v \mapsto \int_{\widetilde{\Omega}} \left(\lambda(-|v|^2 + \frac{1}{2}|v|^4) + |\widetilde{\nabla}v|^2 \right) d\widetilde{x} ,$$

reduced to the real, antisymmetric functions on $L^2(\widetilde{\Omega})$.

3.10 About the Multiplicity

We have seen at many places that assumptions on the multiplicity of the groundstate play an important role. As we shall see in this section, this problem of multiplicity is deeply connected with the structure of nodal sets of groundstates. It has been proven in [HHOO] the following theorem.

Theorem 6. *If (3.19) is satisfied then the multiplicity m of the first eigenvalue of $P_{A,V}$ (Dirichlet or Neumann realization) satisfies*

$$m \leq \begin{array}{ll} 2 & , k = 1, 2; \\ k & , \text{ if } k \text{ is odd, } k \geq 3; \\ k - 1 & , \text{ if } k \text{ even, } k \geq 4 . \end{array} \qquad (3.55)$$

Here we make some remarks connected to the above theorem.

Remark 7.

1. The above bound on the multiplicity of the first eigenvalue is sharp in the case of one hole (see Example 5.3 in [HHOO]), two holes (see Section 3.11), three holes (see Section 3.12), but it is not expected to be sharp for a large number of holes. It would be interesting to know an asymptotic result about the growth of the maximum multiplicity with the number of holes.

2. The proof is obtained by taking advantage of topological obstructions to nodal sets caused by the holes. These obstructions prevent the existence of high dimensional groundstate eigenspaces. Our type of method was first discovered in [Che] and has since then been taken up and used by others, e.g. [Na,HOHON,HOMN]. See also [Col] for explicit constructions of examples with high multiplicity.
3. This result bears similarities to bounds on multiplicities of higher eigenvalues of non-magnetic Schrödinger operators on surfaces with boundary. Some related literature on this topic is given in [Col,Na,HOHON,HOMN].
4. It has been shown in [BCC] that no upper bound on the multiplicity exists when one adds a general magnetic field, even on the sphere.
5. In [HHON] it is conjectured for the single hole case but for arbitrary circulation Φ that the groundstate eigenvalue is always simple unless Φ is a half integer.

About the proof:
Let us explain the case of one hole for the Dirichlet case. The proof is by contradiction. If there are three linearly independent ground states, then one can always be reduced to the case when there are three linearly independent states u_1, u_2 and u_3 such that

$$Ku_j = u_j \, , \quad \text{for } j = 1, \, 2, \, 3 \, .$$

Then we show, that, for any x and y on $\partial\Omega$, one can find a non zero ground state u ($u = \sum_j \lambda_j u_j$ with $\lambda_j \in I\!R$) such that $Ku = u$ and $(\frac{\partial}{\partial\nu}u)(x) = (\frac{\partial}{\partial\nu}u)(y) = 0$ where $\frac{\partial}{\partial\nu}$ denotes the normal derivative. This means that at x and y nodal lines hit the boundary (see below) and if $x \neq y$ and belong to the same connected component of $\partial\Omega$, then this is in contradiction with the property of a ground state satisfying $Ku = u$.
We have indeed, as a consequence of the Hopf boundary point lemma, the following property

Lemma 4. *If $x \in \partial\Omega$, and u is a groundstate for Neumann such that $u(x) = 0$, then $x \in \mathcal{N}(u)$. If u is a groundstate for the Dirichlet case and $x \in \partial\Omega$ such that $\partial u/\partial n = 0$ then $x \in \mathcal{N}(u)$.*

3.11 About the Maximality at Circulation $\frac{1}{2}$

If we introduce a semi-classical parameter like the analysis of the Aharonov-Bohm effect in [He1], one can naturally ask the following question :
Is there some limiting curve for the nodal set as $h \to 0$, for ground states of $P_{A, \frac{v}{h^2}}$ when curl $A = 0$ and (3.19) is satisfied. One suspects that the Agmon distance will play an important role in the picture but this problem remains open.
We recall that in the case of one hole, we have mentioned in Section 3.7 that

the lowest eigenvalue of $P_{A, \frac{V}{\hbar^2}}$ is maximal, as a function of the circulation Φ around the hole, when $\Phi = \frac{1}{2}$ modulo $\mathbb{Z}$. We consider the Dirichlet situation but one can similarly treat the Neumann situation.

We would like to show that we can not hope for a similar result in the many hole case. More precisely, we shall consider the 2 holes case and construct, using semiclassical analysis, an example where $\lambda^{(1)}(\frac{1}{2}, \frac{1}{2}) < \lambda^{(1)}(\frac{1}{4}, \frac{1}{4})$.

The comparison of $\lambda^{(1)}(\frac{1}{2}, \frac{1}{2})$ and $\lambda^{(1)}(\frac{1}{4}, \frac{1}{4})$ is just an application of the semi-classical analysis of the tunneling effect presented in [He1] (see also [HeSj1,HeSj2]). But let us be more precise in the assumptions. We consider the operator $-\Delta_A + \frac{V(x)}{\hbar^2}$ in Ω where Ω has two holes O_1 and O_2. We assume that the magnetic field $B = \operatorname{curl} A$ vanishes in Ω and keep the two circulations Φ_1 and Φ_2 as parameters, together with h. We assume also that the electric potential V has a non degenerate unique minimum at x_0 in Ω and assume for normalization that $V(x_0) = 0$.

In this case it is well established that, in the semiclasssical limit (i.e. as $h \to 0$), the lowest eigenvalue is simple and given modulo $\mathcal{O}(1)$ by the lowest eigenvalue of some harmonic oscillator attached to x_0 :

$$-\Delta_y + h^{-2}\frac{1}{2}(\operatorname{Hess} V(x_0)y, y) . \tag{3.56}$$

We denote by $\lambda_\Omega^{(1)}(h; \Phi_1, \Phi_2)$ this lowest eigenvalue.

Moreover the normalized groundstate u_1 is known to be decaying at least as $\mathcal{O}_\epsilon(\exp -\frac{(1-\epsilon)d_V(x, x_0)}{h})$ for any $\epsilon > 0$, where d_V is the Agmon distance associated with the Agmon metric defined as $V \cdot dx^2$ on $\overline{\Omega}$. This result was proven in [HeSj2].

We now assume that V creates a big barrier not only near $\partial \Omega$ but also between the two holes O_1 and O_2. The point is that there is a minimal path (for the length associated to the Agmon distance) starting from x_0, turning once around $O_1 \cup O_2$ and coming back to x_0 whose length is strictly less than the length of any closed path starting from x_0 and going once around only O_1 or O_2 or two-times the distance from x_0 to $\partial \Omega$. If γ is this minimal path (supposed to be unique and non degenerate), then one can find an one-hole open set Ω' such that $\Im \gamma \subset \Omega' \subset \Omega$ such that, for some $\epsilon_0 > 0$,

$$\lambda_\Omega^{(1)}(h; \Phi_1, \Phi_2) = \lambda_{\Omega'}^{(1)}(h, \Phi_1 + \Phi_2) + \mathcal{O}(\exp -\frac{S + \epsilon_0}{h}) ,$$

where S is the length of this minimal path. The hole O' contains O_1 and O_2 and the circulation along γ is $\Phi_1 + \Phi_2$. The analysis of $\lambda_{\widetilde{\Omega}}(h, \Phi)$ was performed in [He1] (Theorem 4.1.1, with $t = h$). It is shown that, for some $\epsilon_1 > 0$ and some $\alpha_0 > 0$,

$$\lambda_{\Omega'}^{(1)}(h, \Phi) = \lambda_{\Omega'}^{(1)}(h, 0) + h^{\frac{1}{2}}(1 - \cos \Phi)a(h) \exp -\frac{S}{h} + \mathcal{O}(\exp -\frac{S + \epsilon_1}{h}) ,$$

with $a(h) \sim \alpha_0 > 0$.

This shows that this is when $\Phi = \Phi_1 + \Phi_2$ is near $\frac{1}{2}$ (modulo $\mathbb{Z}$) that $\lambda_{\Omega'}^{(1)}(h, \Phi)$ and consequently $\lambda_{\Omega}^{(1)}(h; \Phi_1, \Phi_2)$ is maximal in the semiclassical limit.

An indirect consequence is that in this case the nodal set is a line between the two holes. The maximality of the eigenvalue for flux $(\frac{1}{2}, \frac{1}{2})$ can be proven in the case when $\Omega \setminus \mathcal{N}$ is simply connected, where $\mathcal{N}$ is the nodal set of some "real" groundstate. When discussing all the possibilities of slitting of two holes the only possibility is indeed the case when Ω is slit by a line between two holes.

Moreover, we have a good WKB approximation of the ground state in the neighborhood of γ (see [He1]) and we know from this approximation that the nodal line when Φ is an integer can not meet γ in the semiclassical limit. This is another way to show that there is necessarily in this case a nodal line between the two holes in the case when $(\Phi_1, \Phi_2) = (\frac{1}{2}, \frac{1}{2})$.

It is easy to construct a case when a groundstate has as nodal set two lines joining the boundary of the hole to the exterior boundary. This is for example the case when the multiplicity is 2 because one can always in this case construct, by considering a real linear combination of two linearly independent groundstates u_1 and u_2 satisfying $K u_j = u_j$ $(j = 1, 2)$, a groundstate with a nodal line arriving at the external boundary (and consequently with two!).

In particular we have shown that the two possible situations of nodal sets predicted by the topology in the case of two holes are effectively observed.

Remark 8. The semi-classical behavior of the nodal sets is a nice problem. In the case considered before where a potential V creates a single well, the already developed semi-classical analysis gives some information. If the minimal non trivial path going from x_0 to x_0 (this can be interpreted in the covering manifold (See Section 3.9)) corresponds to an integer flux, then as already mentioned there are no nodal lines crossing this path. On the contrary, in the half integer case, one can show that the nodal line crosses as $h \to 0$ the curve γ at the point which is at the Agmon distance $S/2$. One suspects that the nodal set will asymptotically contain a line defined (say in the case of one hole) (under suitable assumptions on V) as the projection of the set in the covering $\widetilde{\Omega}$ given by

$$d_{\widetilde{V}}(x, \widetilde{x}_0) = d_{\widetilde{V}}(x, G\widetilde{x}_0) \,,$$

where $\Pi(\widetilde{x}_0) = x_0$.

Remark 9. One can also present, in the case of two holes, examples with $V \equiv 0$ giving rise to the two nodal sets situation. The case of three holes will be detailed in the next section.

3.12   The Case of 3 Holes

Here we want to construct a domain Ω with 3 holes such that the multiplicity of the first eigenvalue is 3. Thereby we assume Dirichlet boundary conditions.

Pick a disk with radius 1 centered at the origin and remove 3 small disks of radius ρ such that their centers are in polar coordinates (r, ω) given by $(r, \pi/2)$ $(r, 7\pi/6)$ $(r, 11\pi/6)$. Note that it has the symmetry of an equilateral triangle.

We denote the resulting domain by $\Omega(r, \rho)$. As usually we assume that the circulation around each hole is 1/2. Clearly $r \leq 1 - \rho$. Define $\lambda_1(\Omega(r, \rho))$ by

$$\lambda_1(\Omega(r, \rho)) := \lambda_1(r, \rho) = \inf_S \inf\{ \int |\nabla \phi|^2 dx : \phi \in C_0^\infty, ||\phi|| = 1\} \quad (3.57)$$

according to [HHOO] where S runs over all slitting sets. We will always assume that ρ is sufficiently small. In fact it is not difficult to show that we could even let ρ tend to 0 and we would get still a well defined operator since the associated quadratic form stays semibounded.

We have two extreme cases.

(i) $r \to 1 - \rho$, $(\rho << 1)$.

For $r = 1 - \rho$ the disks touch the boundary and we can gauge the field away. Furthermore

$$\lim_{\rho \to 0} \lambda_1(1 - \rho, \rho) = \lambda_1(1, 0) \quad (3.58)$$

the lowest eigenvalue of the unit disk with Dirichlet boundary conditions. Here we used results on the dependence of Dirichlet eigenvalues with respect to domain variations [Stoll]. To be more precise we can go to the double covering $\widetilde{\Omega}(r, \rho)$. The spectrum of the lifted Laplacian depends continuously on r and ρ as well as the eigenprojections (in the appropriate sense).

(ii) If r and ρ tend to zero then $\Omega(r, \rho) \to \Omega(0, 0)$ and $\Omega(0, 0)$ is just the punctured unit disk. The slitting set hence consists of a nodal arc γ connecting the boundary with the center of the disk.

Clearly

$$\lambda_1(0, 0) > \lambda_1(1, 0). \quad (3.59)$$

(The additional Dirichlet condition at γ raises the eigenvalue, because the capacity of γ is strictly greater than zero.)

We first show that for ρ small and $r = 1 - \rho - \epsilon$, ϵ small, the slitting set must consist of 3 nodal arcs which connect the holes to the boundary of the unit disk such that we have the full symmetry of the equilateral triangle. To see this suppose that we have not the full symmetry, then also the nodal set rotated by $2\pi/3$ or reflected by one of the symmetry axis will be a minimizing slitting set. Hence we would have at least multiplicity 2 for $\lambda_1(1 - \rho - \epsilon, \rho)$. But then there is a function u in the corresponding eigenspace, such that $u(0) = 0$. Hence a nodal arc must pass through the origin and hit eventually two holes or a hole and the boundary. Again this cannot happen since for $\epsilon \to 0$, we would not obtain the groundstate energy.

Now consider the case (ii), namely r small and ρ even smaller. There we must have at least multiplicity 2. To see this suppose that we have multiplicity 1. Then the slitting set S must have the total symmetry, in particular it must be invariant under a rotation by $2\pi/3$, $4\pi/3$. Again this cannot happen since for $r \to 0$ and, say, $\rho = r/10$ we would end up with a slitting set which consists of three nodal arcs connecting the origin to the boundary in such a way that the symmetry of the equilateral triangle is respected, which clearly cannot be the groundstate.

We now fix ρ sufficiently small and consider $\lambda_1(r, \rho)$ as a function of r. For r small we have $m(\lambda_1(r, \rho)) = 2$ and for $r = 1 - \rho - \epsilon$, $m(\lambda_1(r, \rho)) = 1$ for ϵ small. Our previous analysis shows that we have groundstates in two orthogonal symmetry subspaces, one with the symmetry of the equilateral triangle which corresponds to multiplicity 1 and the nonsymmetric one which corresponds to multiplicity (at least) 2. These eigenvalues depend continuously on r and hence there must be for some r_0 a crossing proving our assertion.

As mentioned before it would be interesting to construct examples with high multiplicity of the groundstate eigenvalue for more holes. In particular the question arises what the asymptotics of the multiplicity is as a function of the number of holes.

References

[AHS] Y. Avron, I. Herbst, B. Simon : Schrödinger operators with magnetic fields. I- General interactions. Duke Math. J. 45, p. 847-883 (1978).

[BaPhTa] P. Bauman, D. Phillips, Q. Tang : Stable nucleation for the Ginzburg-Landau system with an applied magnetic field. IMA Preprint Series 1416 (1996).

[BeRu] J. Berger, J. Rubinstein : On the zero set of the wave function in superconductivity. Commun. in Math. Physics 202, p. 621-628 (1999).

[BeRuSc] J. Berger, J. Rubinstein, M. Schatzman : Multiply connected mesoscopic superconducting structures. Proceedings of the International Conference on Calculus of Variations, A. Ioffe et al. Eds., Chapman and Hall, 21-40, 1999.

[Bers] L. Bers : Local behaviour of solution of general linear elliptic equations. Comm. Pure Appl. Math., 8 (1955), p. 473-476.

[BCC] G. Besson, B. Colbois, and G. Courtois, *Sur la multiplicité de la première valeur propre de l'opérateur de Schrödinger avec champ magnétique sur la sphère S^2*, Trans. Amer. Math. Soc. **350** (1998), 331–345.

[Che] S. Y. Cheng : Eigenfunctions and nodal sets. Commentarii Math. Helv. , 51 (1976), p. 43-55.

[Col] Y. Colin de Verdière, *Multiplicités des valeurs propres. Laplaciens discrets et laplaciens continus*, Rend. Mat. Appl. **VII** (1993), 433–460.

[CrRa] M.G. Crandall, P.H. Rabinowitz : Bifurcation from simple eigenvalues. Journal of Functional Analysis 8, p. 321-340 (1971).

[Du] M. Dutour : Bifurcation vers l'état d'Abrikosov et diagramme de phase. Thèse Orsay, Décembre 1999.

[DuHe] M. Dutour, B. Helffer : On bifurcations for normal solutions to superconducting solutions. Preprint May 2000 (submitted).

[DGP] Q. Du, M.D. Gunzburger and J.S. Peterson : Analysis and approximation of the Ginzburg-Landau model of superconductivity. Siam Rev. 34 (1992), p. 431-448.

[East] M. S. P. Eastham : The spectral theory of periodic differential equations. Scottish Acad. Press, Edinburgh-London (1973).

[ElMaQi] C. M. Elliott, H. Matano, T. Qi : Zeros of complex Ginzburg-Landau order parameter with applications to superconductivity, Eur. J. Appl. Math. 5, p. 431-448, 1994.

[GHL] S. Gallot, D. Hulin, and J. Lafontaine, *Riemannian geometry*, 2nd ed., Universitext, Springer-Verlag, 1990.

[GoSc] M. Golubitsky, D.G. Schaeffer, Vol 1. *Singularity and groups in bifurcation theory*, Appl. Math. Sciences 51, Springer Verlag.

[He1] B. Helffer : Effet d'Aharonov-Bohm pour un état borné. Comm. in Math. Phys. 119 (1988), p. 315-329.

[He2] B. Helffer, *Semi-classical analysis for the Schrödinger operator and applications*, Lecture notes in mathematics, vol. 1336, Springer-Verlag.

[He3] B. Helffer : On spectral theory for Schrödinger operators with magnetic fields. Advanced studies in Pure Mathematics 23, p. 113-141 (1994).

[He4] B. Helffer : Some of the results were presented in a talk in Matrei (Austria) in the meeting: Geometrical aspects of spectral theory. (July 99).

[HeSj1] B. Helffer, J. Sjöstrand : Multiple wells in the semi-classical limit I- Comm. in PDE 9(4), p. 337-408 (1984).

[HeSj2] B. Helffer, J. Sjöstrand : Effet tunnel pour l'équation de Schrödinger avec champ magnétique. Annales de l'ENS de Pise 14 (1987), p. 625-657.

[HHOO] B. Helffer, M. and T. Hoffmann-Ostenhof and M. Owen : Nodal sets for the groundstate of the Schrödinger operator with zero magnetic field in a non simply connected domain. Commun. in Math. Physics. 202, p. 629-649 (1999).

[HHON] B. Helffer, M. and T. Hoffmann-Ostenhof, N. Nadirashvili : In preparation.

[HN] I. Herbst and S. Nakamura, *Schrödinger operators with strong magnetic fields: Quasi-periodicity of spectral orbits and topology*, Amer. Math. Soc. Trans, **189**, (1999), p. 105-123.

[HOHON] M. Hoffmann-Ostenhof, T. Hoffmann-Ostenhof, and N. Nadirashvili, *On the multiplicity of eigenvalues of the Laplacian on surfaces*, Ann. Global Anal. Geom.,**17**,(1999), p. 43-48

[HOMN] T. Hoffmann-Ostenhof, P. Michor, and N. Nadirashvili, *Bounds on the multiplicity of eigenvalues for fixed membranes*, Preprint 1998, to appear in Geometrical and Functional Analysis.

[LaO] R. Lavine, M. O'Caroll : Ground state properties and lower bounds on energy levels of a particle in a uniform magnetic field and external potential, J. Math. Phys. 18 (1977), p. 1908-1912.

[LiPa] W.A. Little and R.D. Parks, *Observation of quantum periodicity in the transition temperature of a superconducting cylinder*, Phys. Rev. Lett. **9** (1962), p.9-12.

[Na] N. S. Nadirashvili : *Multiple eigenvalues of the Laplace operator*, Math. USSR, Sb. **61** (1988), p. 225-238.

[Ni] Chang-Shou Lin, Wei-Ming Ni : *A Counterexample to the nodal Domain conjecture and a related semilinear equation*. Proc. Amer. Math. Soc. **102**, 1988, p. 271-277

[Od] F. Odeh : Existence and bifurcation theorems for the Ginzburg-Landau equations. J. Math. Phys. 8:12, p. 2351-2357.

[ReSi] M. Reed, B. Simon : Methods of modern mathematical Physics, IV. Analysis of operators. Academic Press. New York (1978).

[Stoll] P. Stollmann: A convergence theorem for Dirichlet forms with applications to boundary value problems with varying domains. Math. Z. **219** (1995), p. 275-287.

[Ta] P. Takac : Bifurcations and vortex formation in the Ginzburg-Landau equations. Preprint 1999.

4 Connectivity and Flux Confinement Phenomena in Nanostructured Superconductors

Victor V. Moshchalkov, Vital Bruyndoncx, and Lieve Van Look

Laboratorium voor Vaste-Stoffysica en Magnetisme, Katholieke Universiteit Leuven, Celestijnenlaan 200 D, B-3001 Leuven, Belgium

Abstract. We have analyzed the effect of connectivity on flux confinement by studying the normal/superconducting phase boundaries $T_c(H)$ in structures with different topologies. Three different types of nanostructured superconductors are considered: individual nanoplaquettes (loops, dots), one- and two-dimensional clusters of plaquettes, and finally huge arrays of nanoplaquettes (e.g. antidot lattices which cover a macroscopically large area). In all these structures connectivity plays an essential role in the topology dependent superconducting critical temperature $T_c(H)$, which can be strongly increased by optimizing the flux confinement through the modification of the topology and the connectivity of the sample. This concept implies that, for nanostructured superconductors, $T_c(H)$ is determined not only by the type of material used, but also by the 'topological portrait' of the superconducting condensate, which is imposed via nanostructuring.

4.1 Introduction

4.1.1 Ginzburg–Landau theory

The superconducting state is usually described by introducing the complex order parameter $\Psi = |\Psi|\, e^{i\delta}$ which plays the same role for the charged condensate as the wave function for a charged particle in quantum mechanics. The square of the modulus, $|\Psi|^2$, is proportional to the superfluid density n_s. In the presence of an applied magnetic field $\boldsymbol{H}$, the order parameter pattern $\Psi(x,y)$ or 'topological portrait' (here the plane $(x,y) \perp \boldsymbol{H}$) determines the state the superconductor is in, depending both on the temperature T, and the magnetic field H.

The order parameter Ψ can be found by minimizing the Ginzburg-Landau (GL) functional for the free energy density $\mathcal{F}_s$ [1]

$$\mathcal{F}_s = \mathcal{F}_n + \alpha|\Psi|^2 + \frac{\beta}{2}|\Psi|^4 + \frac{1}{2\,m^\star}\left|(-i\hbar\boldsymbol{\nabla} - 2\,e\boldsymbol{A})\,\Psi\right|^2 + \frac{\mu_M\, h^2}{2} \qquad (4.1)$$

where the vector potential $\boldsymbol{A}$ is related to the local magnetic field $\boldsymbol{h}$, through $\mu_M\, \boldsymbol{h} = \boldsymbol{\nabla} \times \boldsymbol{A}$. Here $\mathcal{F}_n$ is the free energy density in the normal state, α is a temperature dependent coefficient, which is positive in the normal and negative in the superconducting state, and the coefficient β is a positive material dependent constant. The GL free energy functional (Eq. (4.1)) is minimized with respect to Ψ and $\boldsymbol{A}$ to obtain the two celebrated GL coupled differential equations [1–3]:

$$\frac{1}{2\,m^\star}(-i\hbar\boldsymbol{\nabla} - 2\,e\boldsymbol{A})^2\Psi + \beta|\Psi|^2\Psi = -\alpha\Psi \tag{4.2}$$

for the order parameter Ψ, and

$$\boldsymbol{j} = \boldsymbol{\nabla}\times\boldsymbol{h} = \frac{e}{m^\star}\left[\Psi^\star(-i\hbar\boldsymbol{\nabla} - 2\,e\boldsymbol{A})\Psi + \Psi(i\hbar\boldsymbol{\nabla} - 2e\boldsymbol{A})\Psi^\star\right], \tag{4.3}$$

which is the second GL equation, for the supercurrent density $\boldsymbol{j}$. This expression can be rewritten, with δ the phase of the order parameter, in a more transparent form:

$$\boldsymbol{j} = \frac{2\,e}{m^\star}|\Psi|^2(\hbar\boldsymbol{\nabla}\delta - 2\,e\boldsymbol{A}) \equiv 2\,e|\Psi|^2\boldsymbol{v}_s \tag{4.4}$$

where $\boldsymbol{v}_s$ is identified as the superfluid velocity.

The GL equations form the standard theoretical framework to analyze the superconducting state. In the general case, the two GL equations are coupled nonlinear differential equations which have to be calculated in a self-consistent way. At the onset of superconductivity, i.e. close to the super-conducting/normal phase boundary, the order parameter Ψ is so small that the nonlinear term $\beta|\Psi|^2\Psi$ in Eq. (4.2) can be omitted. Moreover, the super-currents $\boldsymbol{j}$ are very weak, so the microscopic field $\boldsymbol{h}$ is equal to the applied field $\boldsymbol{h} = \boldsymbol{H}$. We can therefore assume $\mu_M\,\boldsymbol{H} = \boldsymbol{\nabla}\times\boldsymbol{A}$, for the vector potential $\boldsymbol{A}$ close to the phase boundary. The linearized first GL equation for the superconducting order parameter Ψ [1] becomes:

$$\mathcal{H}\,\Psi = \frac{1}{2m^\star}(-i\hbar\boldsymbol{\nabla} - 2\,e\boldsymbol{A})^2\Psi = -\alpha\Psi \tag{4.5}$$

which is formally identical to the Schrödinger equation for a free particle with charge $2\,e$ in a uniform magnetic field $\boldsymbol{H}$, with $E = -\alpha = |\alpha|$ (without potential term $U(\boldsymbol{r})$). The mass $m^\star$ is usually taken as two times the free electron mass. The parameter $-\alpha$ in the linearized GL equation (4.5) thus plays the role of the energy E in the Schrödinger equation:

$$E = -\alpha = \frac{\hbar^2}{2m^\star\,\xi^2(T)} = \frac{\hbar^2}{2m^\star\,\xi^2(0)}\left(1 - \frac{T}{T_{c0}}\right) \tag{4.6}$$

Here $\xi(T)$ is the temperature dependent coherence length, T is the actual temperature and T_{c0} is the critical temperature at zero magnetic field.

4.1.2   Fluxoid Quantization

The quasiclassical Bohr-Sommerfeld quantization rule for a Cooper pair of charge $2\,e$ in a uniform magnetic field can be applied to the impulse $\boldsymbol{p}$ [2]:

$$\oint \boldsymbol{p}\cdot\boldsymbol{dl} = \oint (2\,e\boldsymbol{A} + m^\star\boldsymbol{v}_s)\cdot\boldsymbol{dl} = Lh \tag{4.7}$$

It is easy to see that the above requirement is necessary in order to have a single-valued Ψ everywhere in space. With the use of Eq. (4.4), inserted in Eq. (4.7), this gives $\oint \nabla \delta \cdot dl = L \, 2 \pi$, i.e. the phase winds over integer L times 2π, after integration along a closed contour. For this reason, L is often called the *phase winding number*, specific to the area enclosed by each contour. The same condition can be written in a slightly different form:

$$\Phi' = \Phi + \frac{1}{2e} \oint m^\star v_s \cdot dl = L \, \Phi_0 \qquad (4.8)$$

where the *fluxoid* Φ' is quantized in units of the superconducting flux quantum $\Phi_0 = h/2e$, for all closed paths within the superconductor. The fluxoid is equal to the sum of the magnetic flux $\Phi = \oint A \cdot dl$ through the contour, and an additional term due to the supercurrents j. For this reason, L is also called the *fluxoid quantum number*, counting the number of flux quanta Φ_0, penetrating the enclosed area.

If v_s is constant along a circular contour of radius r, integration of Eq. (4.8) gives

$$m^\star v_s = \frac{h}{2 \pi r} \left(L - \frac{\Phi}{\Phi_0} \right) = \frac{h}{2 \pi r} \left(L - \frac{\mu_M H \pi r^2}{\Phi_0} \right) \qquad (4.9)$$

It follows that the velocity v_s is divergent at the origin $r = 0$ for any $L \neq 0$. Since this would imply an infinite energy, the divergence is cut off at a certain critical radius, at which the superfluid velocity exceeds the critical depairing velocity. This cut-off creates a 'normal core' with suppressed superconductivity $|\Psi| = 0$, making the superconductor multiply connected. This is a necessary condition in order to have finite supercurrents flowing in the system. The only exception for this 'self-destruction' of superconductivity is the $L = 0$ case, where the divergence at $r \to 0$ in Eq. (4.9) is lifted and the Meissner state builds up (for a more formal proof, see the text below), with no normal spots created spontaneously.

4.1.3 Type-I versus Type-II Superconductivity

Let us now consider a bulk system, without taking into account any sample interfaces. Throughout the whole chapter, we will not discuss effects arising from demagnetizing factors, since we are mostly discussing the superconducting state at or very close to the nucleation field. Depending on the type of bulk material used, the 'identity' of the superconductor can be of two types, namely Type-I and Type-II.

In *Type-I materials*, the superconducting state is the perfectly diamagnetic Meissner phase with a singly connected $|\Psi(x, y)|$ pattern over the complete sample area, without any zeros of $|\Psi|$ present. So, *the singly connected phase always corresponds to the Meissner state*. The definition 'singly connected' indeed implies that any closed contour inside the sample does not enclose a normal state area, i.e. a node (or area) where $|\Psi| = 0$.

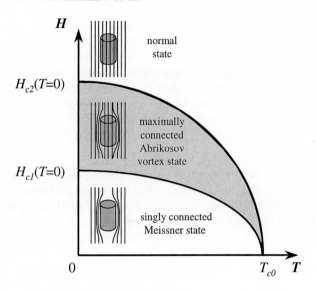

Fig. 4.1. Phase diagram of a *bulk* Type-II superconductor. At $H_{c2}(T)$ the Abrikosov vortex lattice is formed. Below $H_{c1}(T)$ the sample is in the perfectly diamagnetic Meissner state. The surface critical field $H_{c3}(T)$ is not shown

Remarkably different behavior for the connectivity is typical for *Type-II superconductors* (see the $H(T)$ phase diagram in Fig. 4.1): the Meissner phase, as a singly connected state, is still possible, but, *as the field increases, a spontaneous transition from singly connected to the multiply connected Abrikosov vortex state takes place at the lower critical field* $H_{c1}(T)$. The Abrikosov vortex lattice (AVL) creates the maximum possible number of normal spots ($|\Psi| = 0$) in a superconductor, i.e. 'maximum connectivity', since the lattice is composed of singly quantized flux lines, each carrying a flux quantum $\Phi_0 = h/2e$. The latter are called 'vortices' or 'fluxons', and are not to be confused with 'fluxoids'. It is possible to trap a fluxoid $L \neq 0$ in a certain area, while there are no vortices in the material, as we will see later when the Little-Parks experiment is discussed. The area between the $H_{c1}(T)$ and the $H_{c2}(T)$ curves can then be interpreted as a transition regime, from minimum (Meissner state), to maximum connectivity (AVL). At fields higher than the upper critical field $H_{c2}(T)$, a bulk Type-II superconductor undergoes the transition to the normal state.

For bulk samples, the distinction of superconductors into Type-I and Type-II materials can be made by using the expression [1,2]

$$H_{c2}(T) = \kappa \sqrt{2}\, H_c(T) \qquad (4.10)$$

with $H_c(T)$ the thermodynamic critical field, and $\kappa = \lambda(T)/\xi(T)$ the temperature independent GL parameter ($\lambda(T)$ is the penetration depth). When

$H_{c2} < H_c$ (or $\kappa < 1/\sqrt{2}$) the superconductor is of the first Type, while $H_{c2} > H_c$ (or $\kappa > 1/\sqrt{2}$) corresponds to a Type-II superconductor. We remind here that in the first case, the superconducting state is necessarily singly connected, i.e. the fully diamagnetic Meissner state. In a Type-I superconductor, the upper critical field $H_{c2}(T)$ plays the role of a 'supercooling' field, up to which the sample might remain normal as it is cooled down. At this point, a discontinuous and irreversible jump to the Meissner state takes place.

4.1.4   Surface Nucleation of Superconductivity

If the sample has finite dimensions, superconductivity can exist even for fields above the upper critical field $H_{c2}(T)$. It was shown by Saint-James and de Gennes [4] that, for a semi-infinite superconductor with a plane superconductor/vacuum (or superconductor/insulator) interface, the nucleation field can be as high as $1.69\,H_{c2}$. Since superconductivity initially develops near the surface, the nucleation field was called the *surface critical field* (or third critical field), and was labeled H_{c3}. To distinguish between $H_{c3} = 1.69\,H_{c2}$ for a plane interface, and the nucleation field of the specific sample geometries we are studying in this chapter, we will rather use the notation $H_{c3}^*(T)$ for the latter. It should be noted that in the case of finite samples, $H_{c3}^*(T)$ is not a single-valued function of temperature, since the $T_c(H)$ phase boundary in general shows an oscillatory component. $H_{c3}^*(T)$ denotes simply the 'rotated' $T_c(H)$ dependence measured in our experiments, and not a function in the mathematical sense.

In 'finite' samples (with interfaces), the $|\Psi|$ behavior in the area between the nucleation field $H_{c3}^*(T)$ and the upper critical field $H_{c2}(T)$ is less understood. For example, for an infinitely long cylinder, subjected to a magnetic field H along its axis, a surface superconducting sheath with thickness $\xi(T)$ is created at $T_c(H)$ (or $H_{c3}^*(T)$). This gives rise to the creation of a 'giant vortex state' (GVS). At some point in the crossover regime between $H_{c3}^*(T)$ and H_{c1}, the GVS decays into individual Φ_0-vortices. This may occur as a multi-stage process [5]. For example, there are different possibilities for the decay of a $6\,\Phi_0$ GVS into a linear combination $\sum c_L \Psi_L$, namely $6\,\Phi_0 = 3\,\Phi_0 + 3\,\Phi_0 = 2\,\Phi_0 + 4\,\Phi_0 = 1\,\Phi_0 + 2\,\Phi_0 + 3\,\Phi_0 = 1\,\Phi_0 + 1\,\Phi_0 + 4\,\Phi_0 = \ldots = 1\,\Phi_0 + 1\,\Phi_0 + 1\,\Phi_0 + 1\,\Phi_0 + 1\,\Phi_0 + 1\,\Phi_0$. Here, multiple-quanta vortices $L\,\Phi_0$-vortices with different L can be realized, coexisting with each other, and/or with single Φ_0-vortices. The evaluation of the free energy (Eq. (4.1)) is a decisive factor in determining which configurations of different vortices have a lower energy. The systematic analysis of all these vortex phases has not been performed yet. The dotted line in Fig. 4.2 shows schematically where in the phase diagram the Abrikosov state with maximal connectivity is established, as the sample is cooled down. The location of this line is expected to be strongly dependent on the chosen sample geometry.

Palacios has calculated the magnetic response of a Type-II superconducting disk of infinite length, i.e. a cylinder [6], and found *nonmonotonous*
behavior of the magnitude of the magnetization jumps below $H_{c2}(T)$, which
indicates the existence of a single-fluxoid 'vortex glass'. Between $H_{c2}(T)$ and
the nucleation field $H_{c3}^*(T)$ the 'giant vortex state' is established, accompanied by monotonous jumps in the magnetization, each time a fluxoid is added
to or removed from the sample. Schweigert *et al.* [7] have investigated the role
of the disk thickness τ on the magnetization of thin disks in the field regime
below $H_{c2}(T)$. They found that in very thin disks a distribution of several
single fluxoids (which they call 'multivortex' state) is favorable, while in the
limit of thick disks only the giant vortex state survives below $H_{c2}(T)$.

If we define the connectivity C in a more quantitative way as *the number
of nodes $|\Psi| = 0$ in the area enclosed by the outer sample boundary*, then the
transition takes place between the ultimate connectivity of the Abrikosov
mixed state to the GVS, with connectivity C = 1.

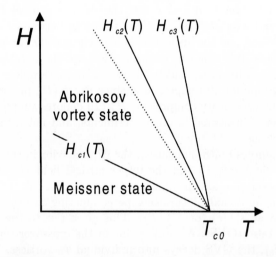

Fig. 4.2. Phase diagram of a finite sample, close to T_{c0}, where the GL equations
can be used. Somewhere on the way between the nucleation field $H_{c3}^*(T)$ and the
first critical field $H_{c1}(T)$, the giant vortex state (GVS) decays into a collection of
Φ_0-vortices. This is shown schematically as a *dotted line*, which position depends
on the chosen sample geometry. A multi-stage GVS decay process, giving several
crossovers ($L\Phi_0 \rightarrow (L-1)\Phi_0 \rightarrow (L-2)\Phi_0 \rightarrow \ldots \rightarrow \Phi_0$), might be also possible [5]

So far, the relation between connectivity and superconductivity has been
considered for homogeneous bulk superconductors, where connectivity can be
varied by a spontaneous formation of 'normal spots' or vortex cores, which
breaks the symmetry of $|\Psi|$ in the (x, y) plane. This spontaneous process,

however, can be controlled by using nanostructuring. It is possible, for example, to create lateral $|\Psi|$ patterns which are not singly connected as such, even in the absence of an applied magnetic field. This can be obtained by making openings into the sample, so that it automatically implies $\mathsf{C} > 0$. In such situations, our original definition of connectivity needs to be reexamined. The singly connected Meissner state ($\mathsf{C} = 0$) is in principle not possible in a multiply connected geometry. The general definition of 'connectivity' C can be given as follows: C *is the total number of normal spots or areas with* $|\Psi|=0$, *in the whole sample. Around each of those spots* $|\Psi|=0$, *a contour can be found along which* $|\Psi|$ *has a finite value.* According to $j \propto |\Psi|^2 v_s$ (Eq. (4.4)), this implies that there is a supercurrent j, flowing around such a spot ($|\Psi|=0$) or area (e.g. antidot).

Knowing in advance the type of spontaneously formed 'topological portrait' of $|\Psi(x,y)|$, needed to stabilize specific vortex configurations, we can prefabricate, for example, microholes ('antidots') exactly at those locations where superconductivity exhibits a 'suicidal' behavior, resulting in the appearance of normal cores. In this case, we can perfectly match the required boundary conditions for the distribution of zeros of $|\Psi|$ imposed by a specific topological portrait. The optimum distribution of zeros for a certain fixed applied field H is not a priori valid for other fields. Therefore, by varying H, the commensurability effects between the $|\Psi|$ distribution and the prefabricated underlying structure imposed by the microholes in the material can be tuned. It may also turn out that the application of an external field H results in the spontaneous formation of extra normal spots (or 'interstitial' vortices), in addition to the pattern of artificially created antidots [8,9].

If now, in a finite size sample, the nucleation field exceeds H_{c2}, say $H_{c3}^*(T) = \eta\, H_{c2}(T)$ with $\eta > 1$, then Eq. (4.10) becomes

$$H_{c3}^*(T) = \eta\, H_{c2}(T) = \eta\, \kappa\, \sqrt{2}\, H_c(T) \qquad (4.11)$$

Obviously, the crossover between the regimes $H_{c3}^*(T) < H_c(T)$ and $H_{c3}^*(T) > H_c(T)$ must occur at a lower κ value: $\kappa = \kappa_c = (1/\sqrt{2}\,\eta)$. Therefore, *it is not necessary to fabricate a sample from a Type-II superconductor ($\kappa > 1/\sqrt{2}$) in order to initially have 'surface nucleation', with the bulk of the sample still normal.* The nucleation mechanism for superconductivity will be initiated at the sample boundary, even in a Type-I material, if η can be made sufficiently large, as for very small samples. Hence, the multiply connected 'vortex' state can exist as well in 'nanostructured' superconductors, where the presence of superconductor/vacuum (or superconductor/insulator) interfaces is crucial. If this requirement can not be fulfilled, i.e. when $\kappa < \kappa_c$, the singly connected Meissner state ($\mathsf{C} = 0$) is the only thermodynamically stable state. Even then, for strong Type-I samples, quantum oscillations can be observed in the supercooling field, as it was measured for In microcylinders by Michael and McLachlan [10].

It should be noted that, when studying thin films with thickness $\tau \ll \lambda(T)$, with the magnetic field applied perpendicular to the film plane, the penetration depth effectively increases to $\lambda_{eff} = \lambda^2/\tau$ [1]. This implies an enhancement of κ as well, and thus has a supplementary effect on the crossover between Type-I and Type-II behavior.

4.1.5   Solution of the Linear GL Equation in Cylindrical Coordinates

Before moving to real sample geometries, we give the solution of the linearized GL equation (4.5) in cylindrical coordinates, with $\mu_M H = \nabla \times A$. This general derivation can also be found in Ref. [11]. Using a cylindrical system of coordinates (r, φ, z), and the London gauge $\nabla \cdot A = 0$ for the vector potential $A = (\mu_M H r/2) e_\varphi$, the Hamiltonian from 4.5 reads

$$\mathcal{H} \Psi = -\frac{\hbar^2}{2\, m^\star} \left\{ \frac{1}{r} \frac{\partial}{\partial r} \left(r \frac{\partial \Psi}{\partial r} \right) + \frac{1}{r^2} \frac{\partial^2 \Psi}{\partial \varphi^2} + \frac{\partial^2 \Psi}{\partial z^2} \right\} \qquad (4.12)$$
$$- \frac{i e \mu_M H \hbar}{m^\star} \frac{\partial \Psi}{\partial \varphi} + \frac{(e \mu_M H r)^2}{2\, m^\star} \Psi$$

The z-dependence will be neglected, and Ψ can be split in a radial r and an angle φ dependent part: $\Psi(r, \varphi) = |\Psi(r)| \exp (i\delta(\varphi))$, leading to $\delta = \pm L\varphi$ for the imaginary part, when inserting this into Eq. (4.12). The number L should be an integer, since the order parameter Ψ should be a single-valued function $\Psi(r, \varphi + 2\pi) = \Psi(r, \varphi)$, or $\oint \nabla \delta \cdot dl = L\, 2\,\pi$, after a turn along the circle circumference. The r-dependent part of Eq. (4.12) becomes

$$\frac{1}{r} \frac{\partial}{\partial r} \left(r \frac{\partial |\Psi|}{\partial r} \right) + \left\{ \frac{2\, m^\star}{\hbar^2} \left(|\alpha| \mp L \frac{\hbar\omega}{2} \right) - \frac{L^2}{r^2} - \gamma^2 r^2 \right\} |\Psi| = 0 \qquad (4.13)$$

where $\gamma = e\mu_M H/\hbar$ and the cyclotron frequency ω is given by

$$\omega = \frac{2\, e\, \mu_M H}{m^\star} \qquad (4.14)$$

With the substitutions $y = \gamma r^2 = R^2$ and $|\Psi| = r^{-1} \left(\gamma r^2 \right)^{\frac{L+1}{2}} \exp\left(-\frac{\gamma r^2}{2} \right) g$, Eq. (4.13) transforms into

$$y \frac{\partial^2 g}{\partial y^2} + (L + 1 - y) \frac{\partial g}{\partial y} + n\, g = 0 \qquad (4.15)$$

where the number n is defined via

$$|\alpha| \mp L \frac{\hbar\omega}{2} = \frac{\hbar\omega}{2} (2\, n + L + 1) \qquad (4.16)$$

or

$$E = |\alpha| = \frac{\hbar\omega}{2} (2\, n \pm L + L + 1) \qquad (4.17)$$

Eq. (4.15) is the confluent hypergeometric differential equation. The most general solution for g can be presented as a linear combination of the two types of Kummer functions $M(-n, L+1, y)$ and $U(-n, L+1, y)$, where the Kummer function of first kind is given by the series

$$M(a, c, y) = 1 + \frac{a}{c}y + \frac{a(a+1)}{c(c+1)}\frac{y^2}{2!} + \frac{a(a+1)(a+2)}{c(c+1)(c+2)}\frac{y^3}{3!} + \cdots \quad (4.18)$$

and is defined for all complex parameters a, c, and y, provided that $c \neq 0, -1, -2, \dots$. In other words, the number $L = 0, 1, \dots$, i.e. L is a positive integer.

The second type of Kummer function U can be expressed in terms of M [12]:

$$U(a, c, y) = \frac{\pi}{\sin(\pi c)} \left\{ \frac{M(a, c, y)}{\Gamma(1 + a - c)\,\Gamma(c)} - \quad (4.19) \right.$$
$$\left. y^{1-c} \frac{M(1 + a - c, 2 - c, y)}{\Gamma(a)\,\Gamma(2 - c)} \right\}$$

with Γ the Gamma function.

The order parameter finally becomes:

$$\Psi_L(r, \varphi) = e^{\pm iL\varphi}\, r^L \gamma^{\frac{L+1}{2}} \exp\left(-\frac{\gamma r^2}{2}\right) \quad (4.20)$$
$$\times \left\{ c_1 M(-n, L+1, \gamma r^2) + c_2 U(-n, L+1, \gamma r^2) \right\}$$

where $a = -n$, $c = L+1$, $y = \gamma r^2$. The parameters c_1 and c_2 are, in general, complex numbers, which have to be found from the boundary conditions. Introducing the dimensionless radius $R = \sqrt{\gamma}\, r$, the superconducting order parameter can be written in the form

$$\Psi_L(R, \varphi) = e^{\pm iL\varphi}\, R^L \exp\left(-\frac{R^2}{2}\right) \quad (4.21)$$
$$\times \left\{ c_1 M(-n, L+1, R^2) + c_2 U(-n, L+1, R^2) \right\}$$

where the r-independent prefactor $\sqrt{\gamma}$ has been omitted.

Keeping in mind the main focus of the chapter, the interplay between connectivity and superconducting critical parameters (in our case mostly $T_c(H)$), we shall start below from the nucleation process in bulk superconductors. After that, we move on to 'single plaquette' structures. In particular, we study dots and loops, which intrinsically have a different connectivity C. By varying the size of the opening in the loop, we shall investigate the crossover between the dot and the loop, as deduced from the shape of $T_c(H)$. Then, we move on to other mesoscopic samples built up from several 'nanoplaquettes' in a one- or two-dimensional fashion and containing several (0, 1, 2, 3 or 4) prefabricated microholes to confine the flux. We shall investigate the competition between simple zeros located at the sites of the microholes and the 'degenerate' zero corresponding to the giant vortex state, imposed by the outer sample boundary. Finally, we discuss the $T_c(H)$ behavior of films containing huge arrays of antidots confining the zeros of $|\Psi|$. In dense arrays, we focus mainly on the possibility of a surprisingly large enhancement of $H_{c3}^*(T)$. This becomes possible through modification of the sample's connectivity by laterally patterning the sample with antidots.

4.2   Connectivity Effects in Bulk Superconductors

For bulk samples, Ψ should be finite everywhere in space. If we impose in Eq. (4.20) that $\Psi(r \rightarrow \infty, \varphi)$ is finite, this automatically implies that n should be 0 *or a positive integer number*, for which the notation $n = N$ will be used further on. The discrete set of eigenvalues $E = |\alpha|$ of 4.5, is

$$E_N = |\alpha_N| = \hbar\omega \left(N + \frac{1}{2} \right) \tag{4.22}$$

where the energy contribution $m^\star v_{s,z}^2/2$ due to the motion of Cooper pairs along the field direction is omitted, since the highest possible nucleation field $H_{c3}^*(T)$ is the present field of interest. It corresponds to the lowest energy, and therefore to $v_{s,z} = 0$. The minus sign in Eq. (4.17) has been chosen in order to find the ground state. These Landau levels (Eq. (4.22)) correspond to the quantized motion for a particle of charge $2\,e$, in the plane perpendicular to the applied magnetic field $\boldsymbol{H}$. The particle moves in circular orbits, with the cyclotron frequency [13] (see Eq. (4.14)). In other words, the lowest Landau level $E = |\alpha| = \hbar\omega/2$, with $N = 0$, has to be selected from the eigenenergies of Eq. (4.22), in order to find the highest possible bulk nucleation field, which is the upper critical field H_{c2}:

$$\mu_M H_{c2}(T) = \frac{\hbar}{2\,e\,\xi^2(0)} \left(1 - \frac{T}{T_{c0}} \right) = \frac{\Phi_0}{2\,\pi\,\xi^2(T)} \tag{4.23}$$

Note that, in a bulk superconductor, H_{c2} has a linear dependence on T, or alternatively, $T_c(H)$ is linear.

When there are no sample boundaries, and the lowest energy simply corresponds to the bulk Landau level for $N=0$ (Eq. (4.22)), then the two confluent hypergeometrical functions become $M(N = 0, L + 1, R^2) = 1$ and $U(N = 0, L + 1, R^2) = 1$ for all L [12]. In that case, the solution for Ψ can be written as (Eq. (4.21)):

$$\Psi_L(R, \varphi) = e^{-iL\varphi} R^L \exp\left(-\frac{R^2}{2} \right) \tag{4.24}$$

where the constant prefactor $c_1 + c_2 = 1$ can be chosen, since we are dealing with a linear differential equation. The set of functions (Eq. (4.24)) for L

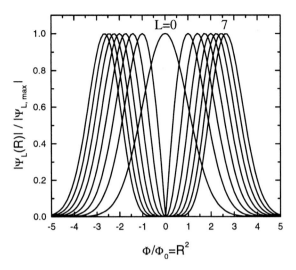

Fig. 4.3. Radial dependence of the functions $|\Psi_L|$ for a bulk superconductor (Eq. (4.24)). All functions are normalized to the same maximum value at $R_L = \sqrt{L}$

values ranging from 0 to 7 are shown in Fig. 4.3, where they are normalized to their maximum value (for the 3D pattern of $|\Psi_L|$, see also Fig. 4.4). The functions $|\Psi_L|$ have their maxima at $R^2 = R_L^2 = L$, i.e., the area enclosed by the circle with a radius corresponding to the maximum in $|\Psi_L|$ is always penetrated by an integer number L of flux quanta: $R^2 = \Phi/\Phi_0 = L$.

While Abrikosov solved 4.5, using a cartesian system of coordinates (x, y, z), the use of cylindrical coordinates has certain advantages. The bulk eigenenergies (Eq. (4.22)) only depend on the main quantum number N, and not on the winding number L. Therefore, for the ground state $(N = 0)$, Ψ can be represented as an expansion over all Ψ_L [14]

$$\Psi = \sum_{L=0}^{\infty} c_L \Psi_L = \sum_L c_L e^{-iL\varphi} R^L \exp\left(-\frac{R^2}{2}\right) \qquad (4.25)$$

If the L values used in the expansion are sufficiently far from each other, then Ψ consists of a series of concentric nonoverlapping rings, as is shown in Fig. 4.4. The phases φ_L of the complex coefficients $c_L = |c_L| \exp(-i\varphi_L)$ are chosen in a random fashion, and the moduli $|c_L|$ are taken such that $|\Psi_L(max)|=1$. When the separation between the different L values in the summation is large, as for $L = 16\,m^2$, with m=0, 1, 2, and 3, then the relative phases φ_L are unimportant (Fig. 4.4). For a closer packed set of solutions $L = 4\,m^2$, on the contrary, the overlap of the Ψ_L functions causes interference, resulting in the creation of single Φ_0-vortices, i.e. increasing the sample's connectivity to $C = L_{max} = 4\,m^2 = 36$. This is demonstrated in

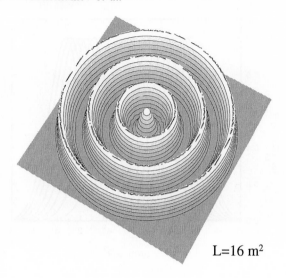

L=16 m²

Fig. 4.4. $|\Psi|$ patterns for a bulk superconductor where $\Psi = c_0\,\Psi_0 + c_{16}\,\Psi_{16} + c_{64}\,\Psi_{64} + c_{144}\,\Psi_{144}$, and Ψ_L is given by Eq. (4.24). The phases of the complex coefficients c_L are chosen randomly. Interference between the Ψ_L solutions is negligible

Fig. 4.5, which shows the pattern calculated in a similar way as Fig. 4.4 with random phases φ_L.

Let us now formulate the expansion in Eq. (4.25) somewhat differently. First, we define the complex quantity $Z = R\,e^{-i\varphi}$. Then Ψ can be rewritten

$$\Psi = \left(\sum_{L=0}^{\infty} c_L\,Z^L\right)\exp\left(-\frac{R^2}{2}\right) = f(Z)\exp\left(-\frac{R^2}{2}\right) \qquad (4.26)$$

Since the coefficients $\{c_L\}$ can be chosen arbitrarily, an important observation can be made immediately: *any* function $f(Z)$ analytical at $Z = 0$, may be used to construct a solution Ψ of the linearized GL equation, in the bulk case. This statement is easy to prove, when expanding $f(Z)$ in a Taylor series around the origin $Z = 0$

$$f(Z) = f(0) + \frac{1}{1!}f'(0)Z + \frac{1}{2!}f''(0)Z^2 + \dots \qquad (4.27)$$

$$= \sum_{L=0}^{\infty} \frac{f^{(L)}(0)}{L!}\,Z^L = \sum_{L=0}^{\infty} c_L\,Z^L$$

In order to find the Ψ distribution below, but close to the H_{c2} line, Abrikosov used a perturbative method by setting the vector potential $\boldsymbol{A} = \boldsymbol{A}_0 + \boldsymbol{A}_1$, where $\boldsymbol{A}_0$ is the unperturbed $\mu_M\,\boldsymbol{H} = \boldsymbol{\nabla} \times \boldsymbol{A}_0$, and the perturbation $\boldsymbol{A}_1 \ll \boldsymbol{A}_0$ [3,15]. He found that the sample free energy density $\langle \mathcal{F}_s \rangle$,

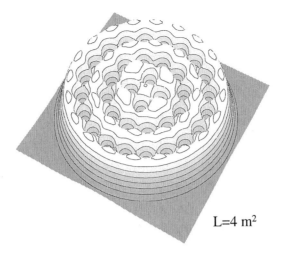

L=4 m²

Fig. 4.5. $|\Psi|$ patterns for a bulk superconductor where $\Psi = c_0\,\Psi_0 + c_4\,\Psi_4 + c_{16}\,\Psi_{16} + c_{36}\,\Psi_{36}$, and Ψ_L is given by Eq. (4.24). The phases of the complex coefficients c_L are chosen randomly. Interference between the Ψ_L solutions gives rise to a $|\Psi|$ pattern with single Φ_0-vortices

averaged over the sample, becomes

$$\langle \mathcal{F}_s \rangle = \mathcal{F}_n + \frac{1}{2\,\mu_M}\left(B^2 - \frac{(\mu_M\,H - B)^2}{1 + \beta_A\,(2\kappa^2 - 1)}\right) \qquad (4.28)$$

where the induction $B = \mu_M\,\langle h \rangle$ is the macroscopic average of local fields h. The notation $\langle\,\rangle$ stands for the average over the sample space. The Abrikosov parameter β_A in Eq. (4.28), characterizing the flatness of $|\Psi|$, is defined as

$$\beta_A = \frac{\langle |\Psi|^4 \rangle}{\langle |\Psi|^2 \rangle^2} \geq 1 \qquad (4.29)$$

It is clear that the free energy can be reduced by decreasing β_A, which physically corresponds to a $|\Psi|$ distribution as flat as possible. While Abrikosov, in his original paper [15], obtained a value $\beta_A = 1.18$, corresponding to a square vortex lattice (Fig. 4.6 (**a**)), it was shown later on that this situation corresponds to a saddle point of the free energy. Kleiner *et al.* [16] calculated the lowest value $\beta_A = 1.16$, for which an equilateral triangular vortex lattice is created (Fig. 4.6 (**b**)). The dark spots in Fig. 4.6 correspond to the vortices $|\Psi| = 0$.

Here, we will follow the Abrikosov method in order to find a $|\Psi|$ function, which gives β_A as small as possible. It is obvious that the nonoverlapping functions Ψ_L in Fig. 4.4 give rise to a very high value for β_A. The problem is now reduced to obtaining an ideal set of $\{c_L\}$ complex numbers which

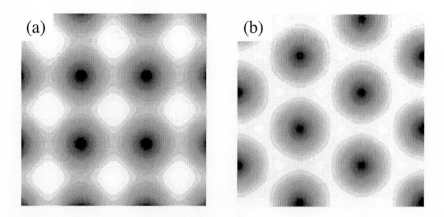

Fig. 4.6. Abrikosov vortex lattices, (a) square (β_A=1.18), (b) triangular (β_A=1.16). The dark spots indicate the vortex positions

minimizes β_A. From Eq. (4.26), it can been seen that the function $f(Z)$ has to be built in such a way that its modulus eventually becomes

$$|f(Z)| \to \exp\left(\frac{|Z|^2}{2}\right) \qquad (4.30)$$

In any finite size sample, a finite number of terms L in Eq. (4.26) has to be taken. The maximum value for $L = L_{max}$ determines the 'vorticity', i.e. the total number of fluxoids in the sample. Indeed, the summation may be presented as a product

$$c_0 + c_1 Z + c_2 Z^2 + \ldots + c_{L_{max}} Z^{L_{max}} = (Z - Z_1)(Z - Z_2) \ldots (Z - Z_{L_{max}}) \quad (4.31)$$

where $\{Z_L\}$ are the poles, corresponding to the points where $|\Psi| = 0$. From this point of view, the β_A minimization problem can be reformulated in terms of the *optimization of the positions of the poles Z_L in the (x, y) plane*. Eqs. (4.26) and (4.31) provide a simple mathematical formulation of the relation between connectivity and superconductivity. When the $|\Psi| = 0$ vortices are all located at different positions ($Z_i \neq Z_j, \forall i, j$), the connectivity is equal to $\mathsf{C} = L_{max}$. The vorticity L_{max} is determined by the amplitude of the applied magnetic field $\boldsymbol{H}$. In very low fields $L_{max} = 0$, and the Meissner state is realized. Then, the sample is singly connected ($\mathsf{C} = 0$). As $\boldsymbol{H}$ increases, more and more Φ_0-vortices penetrate the sample, each of them adding 1 to the connectivity C. If, instead of forming a new Φ_0-vortex at a different spot each time, multiple-quanta vortices are formed, then $\mathsf{C} < L_{max}$. Ultimately, $\mathsf{C} = 1$ corresponds to the giant vortex state, with only one 'big' vortex.

By taking different terms in the summation (Eq. (4.25)), it is possible to impose $|\Psi|$ patterns possessing a specific symmetry. In Fig. 4.7, for example,

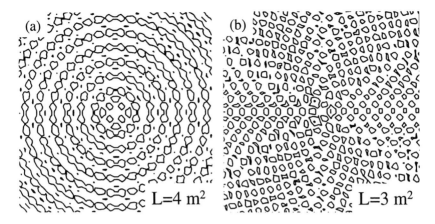

Fig. 4.7. Contour $|\Psi|$ plots for (a) $L = 4\,m^2$, (b) $L = 3\,m^2$, with m a positive integer. The phases and the moduli of the complex coefficients c_L are taken such that the $|\Psi|$ distribution is as flat as possible, i.e. with minimal β_A. Note that a k-fold symmetry ($L = k\,m^2$) is reflected in the $|\Psi|$ patterns

$|\Psi|$ contours are shown for $L = 4\,m^2$ (a), and $L = 3\,m^2$ (b), with m an integer. The resulting $|\Psi|$ patterns (together with the $|\Psi|$ nodes) obey the C_4 and C_3 symmetries, respectively. For the calculation of the Ψ distribution in Fig. 4.7, the phases φ_L were varied, in order to obtain the Abrikosov parameter β_A as low as possible. For these representations β_A lies in the range β_A=1.17-1.21, i.e. slightly above the value for the triangular vortex array in bulk. The $|\Psi|$ configurations calculated from a summation over $L = k\,m^2$ reflect a k-fold symmetry around the origin [14]. Therefore, these states might be useful for the description of the vortex matter in a superconducting square ($k = 4$ in Fig. 4.7 (a)) and triangle ($k = 3$ in Fig. 4.7 (b)), provided that these specimens do not have any other confining potentials ('pinning centres'), besides the sample boundaries themselves.

4.3 Nanostructured Superconductors: Single Plaquettes

4.3.1 Boundary Conditions

At this point, we will return to the importance of sample boundaries. In finite superconducting samples the order parameter Ψ obeys the boundary condition for a superconductor/insulator interface [1]:

$$(-i\hbar\boldsymbol{\nabla} - 2\,e\boldsymbol{A})\Psi|_{\perp} = 0 \qquad (4.32)$$

This 'Neumann' boundary condition is quite different from the normal 'Dirichlet' boundary condition in the quantum mechanical problem 'particle in a

box', where the density $\Psi\Psi^*$ is zero at the boundary. De Gennes has generalized the boundary condition (Eq. (4.32)) to:

$$(-i\hbar\boldsymbol{\nabla} - 2\,e\boldsymbol{A})\Psi|_\perp = \frac{-i\hbar}{b}\,\Psi \qquad (4.33)$$

for a metal-superconductor interface with no perpendicular current. The
quantity b (real number) is called the extrapolation length, since it mea-
sures the distance outside the boundary where Ψ goes to zero, if the slope at
the interface is maintained. A superconductor-vacuum interface, for exam-
ple, has $b \to \infty$, i.e. the slope $\boldsymbol{\nabla}\Psi|_\perp$ goes the zero, on the condition that a
vector potential $\boldsymbol{A}$, with no perpendicular component can be chosen. While
a positive extrapolation length b generally reduces both T_{c0} and $H_{c3}^*(T)$,
compared to the usual case $b \to \infty$ [17], the opposite is true for negative
b [18]. For this reason, these systems ($b < 0$) are promising from the point of
view of 'enhancing the critical parameters' of superconductors [19,20]. When
the order parameter $|\Psi|$ is enhanced at the interface (as for negative b), a
Type-I superconductor can display an interface delocalization or 'wetting'
transition [21,22].

The boundary condition (Eq. (4.32)) will restrict the possible values for n
in Eqs. (4.17) and (4.20). For cylindrically symmetric samples, the boundary
conditions are considerably simplified, since for the chosen gauge for the
vector potential there is no component of $\boldsymbol{A}$ perpendicular to the interface,
in such a case. Hence, the boundary condition (Eq. (4.32)) can be simply
written as:

$$\left.\frac{\partial|\Psi(r)|}{\partial r}\right|_{r=r_o} = 0\,, \qquad (4.34)$$

with a superconductor/vacuum interface at a radius r_o. Or, in dimensionless
units $R = \sqrt{\gamma}\,r$, with $R_o = \sqrt{\gamma}\,r_o$,

$$\left.\frac{\partial\,|\Psi(R)|}{\partial R}\right|_{R=R_o} = 0 \qquad (4.35)$$

As a result of the relation between E and α (Eq. (4.6)), we have to follow a
simple rule to deduce the $T_c(H)$ phase boundary from $E(H)$: after solving the
Schrödinger equation with proper boundary conditions (Eq. (4.32)), we take
the lowest energy $E_{LLL}(H)$ ('lowest Landau level') which gives the highest
T in Eq. (4.6), coinciding with the phase boundary $T_c(H)$ for the nucleation
of the superconducting state. Hence, we will again choose the minus sign in
Eq. (4.17).

4.3.2 'Mesoscopic' Superconductors

When the sample size becomes 'small', the surface-to-volume ratio simultaneously increases and the boundary conditions (Eq. (4.32)) gain importance.

Which length scale are we interested in? The answer to this question can be given by referring again to the analogy between Schrödinger equation for normal electrons and the linearized GL equation (4.5) for the superconducting order parameter. Dingle [23] treated the Schrödinger equation in connection with the analysis of quantum oscillations in small metallic cylinders. Though he took Dirichlet boundary conditions, his conclusions are expected to stay valid for 'superconducting' Neumann boundary conditions. According to Dingle, the border between small and large samples can be found from the estimate $\mu_M H \times r_o \approx 0.5 \, \mathrm{mT \, cm}$, based on the coincidence between the Larmor radius r_L and the sample radius r_o. If the product of field H times disk radius r_o is smaller than $0.5 \, \mathrm{mT \, cm}$ (or $r_o < r_L$), then boundary conditions essentially modify the solution of 4.5. It is evident that fields $\mu_M H < 0.5 \, \mathrm{mT \, cm}/r_o$ are not extremely low. Indeed, for a sample with dimensions, say 1 mm, the field below which boundary conditions should be taken into account, is 5 mT. In practice, for a superconductor, the 'confinement' effects due to the sample boundary will become important in the 'mesoscopic regime'. This means that the sample size or the characteristic plaquette size, used for lateral nanostructuring, should be comparable to the magnetic penetration depth $\lambda(T)$ and the superconducting coherence length $\xi(T)$.

4.3.3 Nucleation of Superconductivity in a Loop

The nucleation of superconductivity in mesoscopic samples has received a renewed interest after the development of nanofabrication techniques, like electron beam lithography. In the framework of the Ginzburg-Landau (GL) theory, the coherence length $\xi(T)$ sets the length scale for spatial variations of the modulus of the superconducting order parameter $|\Psi|$. The pioneering work on mesoscopic superconductors was carried out already in 1962 by Little and Parks [24,25], who measured the shift of the critical temperature $T_c(H)$ of a (multiply connected) thin-walled Sn microcylinder (a thin-wire 'loop') in an axial magnetic field H. The $T_c(H)$ phase boundary of a loop (made of 1D wires) shows a periodic behavior (see Fig. 4.8, with the magnetic period corresponding to the penetration of a superconducting flux quantum $\Phi_0 = h/2e$. The Little-Parks oscillations in $T_c(H)$ are a straightforward consequence of the fluxoid quantization constraint (Eq. (4.8)) predicted by F. London [26]. This condition can be easily understood by integrating the second GL equation (Eq. (4.3)) along a closed contour

$$\frac{m^\star}{4 e^2} \oint \frac{\boldsymbol{j} \cdot \boldsymbol{dl}}{|\Psi|^2} + \Phi = L \Phi_0 \qquad L = \cdots, -2, -1, 0, 1, 2, \cdots \qquad (4.36)$$

where Stokes theorem $\oint \boldsymbol{A} \cdot \boldsymbol{dl} = \Phi$ was used, with Φ the magnetic flux threading the area inside the contour. In other words, when a non-integer magnetic flux Φ/Φ_0 is applied, a supercurrent j has to be generated in order to fulfill Eq. (4.36). Since, for a cylindrical geometry, the different L states are eigenfunctions of the angular (or orbital) momentum operator as well [27], L is often called the angular momentum quantum number.

When the loop is build up of infinitely thin strips, j and $|\Psi|$ are spatially constant, and from Eqs. (4.9) and (4.36) the following relation is obtained:

$$\frac{T_{c0} - T}{T_{c0}} = \frac{\xi^2(0)}{r_o^2} \left(L - \frac{\Phi}{\Phi_0} \right)^2 \qquad (4.37)$$

The phase boundary $T_c(\Phi)$ is formed when L is the integer number nearest to the applied flux $\Phi/\Phi_0 = \mu_M H \pi r_o^2/\Phi_0$. This is shown as the lower curve in Fig. 4.8. For an infinitely thin loop, $T_c(\Phi)$ is periodic in Φ, and has a parabolic behavior within each period (Eq. (4.37)). Comparing the $T_c(H)$ linear curve for a bulk superconductor with the set of parabolic curves for the ring we can clearly see that connectivity plays an important role in defining the shape of the $T_c(H)$ boundaries.

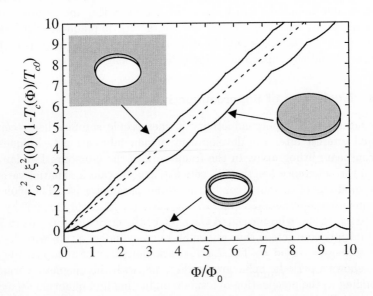

Fig. 4.8. Calculated phase boundaries (i.e., the $H_{c3}^*(T)$ curves) for a circular loop made of infinitely thin wires (Eq. (4.37)), a disk (or cylinder) (Eq. (4.42)), and an antidot (Eq. (4.45)) in normalized units of temperature and magnetic flux. Superconductivity will always nucleate initially near the disk/insulator boundary (the disk has the highest H_{c3}^*). The *dashed line* gives the $H_{c3}(T) = 1.69\, H_{c2}(T)$ curve for a semi-infinite slab [4]

Although, for $L \neq 0$, we are dealing with fluxoids, there are no real 'vortices' (i.e. spontaneously formed $|\Psi|=0$ nodes), present in the loop. The fluxoids rather are 'coreless' vortices, where the usual Abrikosov vortex core is replaced by the sample opening.

4.3.4 Nucleation of Superconductivity in a Disk

In 1965, Saint-James calculated the $T_c(H)$ (or the nucleation magnetic field) for a singly connected cylinder [28] (a mesoscopic 'disk'). The solution of the Hamiltonian (Eq. (4.12)) in cylindrical coordinates has the following form [11]:

$$\Psi(r, \varphi) = e^{\pm iL\varphi}\, r^L\, \gamma^{(L+1)/2} \exp\left(-\frac{\gamma r^2}{2}\right) M(-N, L+1, \gamma r^2) \qquad (4.38)$$

This can be deduced from the more general solution Eq. (4.20), since the second Kummer function U is divergent at the origin: $U(-N, L+1, r^2 \to 0) \to \infty$, thus the second coefficient c_2 should vanish ($c_2 = 0$).

With Eqs. (4.6) and (4.17) this can be rewritten as:

$$\frac{r_o^2}{\xi^2(T_c)} = \frac{r_o^2}{\xi^2(0)}\left(1 - \frac{T_c(H)}{T_{c0}}\right) = 4\left(n + \frac{1}{2}\right)\frac{\Phi}{\Phi_0} = \epsilon(H_{c3}^*)\frac{\Phi}{\Phi_0}, \qquad (4.39)$$

where $\Phi = \mu_M H\pi r_o^2$ is arbitrarily defined, but r_o is most naturally taken as the sample radius. Eq. (4.39) is valid in general, also when the two confluent hypergeometric functions in Eq. (4.20) have to be taken into account. The resulting phase boundary is shown as the solid cusplike line in Fig. 4.8, just below the straight dashed line. The latter represents the surface critical field solution for a semi-infinite slab: $H_{c3} = 1.69\, H_{c2}$ [4]. When the flux (or the radius of the disk) $\to \infty$, the critical field of the disk approaches this value asymptotically.

It is very important to note that in the general form (Eqs. (4.17) and (4.38)) *there are no limitations on the parameter n: it is not necessarily an integer number.* The only argument, which is usually given in favor of taking integer n=N, is a possibility to get a cut off in the summation (Eq. (4.18)). Indeed, if we insert an integer N into the summation, then by adding 1 to N in each new term we shall eventually come to the situation where $-N + N = 0$ and all subsequent terms in the summation will be equal to zero. Thus by the cutoff we just use a finite number of terms in the summation (Eq. (4.18)) and of course M is finite in this case. But we should keep in mind that any converging row also gives a finite solution for M. Therefore, not only positive integer N in Eq. (4.17), but also non-integer and even negative n values are possible. In finite size samples the n value, which we further denote as $n(L, R_o)$, has to be found from the boundary condition at $R = R_o$ (Eq. (4.35)), where R_o is the dimensionless disk radius.

Since we are looking for the lowest possible energy state, we should take the minus sign in the argument of the exponent $\exp(-iL\varphi)$ in the solution given by Eq. (4.38). In this case $-L$ and $+L$ in Eq. (4.17) cancel and *for any L* the energy levels become:

$$E_n = |\alpha_n| = \hbar\omega\left(n + \frac{1}{2}\right) \qquad (4.40)$$

This result coincides again with the well-known Landau quantization but now n is *any real number, including negative real number*, which is to be calculated from Eq. (4.35). Using the expression [12]

$$\frac{d\,M(a,c,y)}{dy} = \frac{a}{c}\,M(a+1,c+1,y) \qquad (4.41)$$

for the derivative of the Kummer function, we can find the $n(L, R_o)$ value, which obeys the boundary condition (Eq. (4.35)), from the equation:

$$(L - R_o^2)\,M(-n, L+1, R_o^2) - \frac{2\,n\,R_o^2}{L+1}M(-n+1, L+2, R_o^2) = 0 \qquad (4.42)$$

The remarkable thing about the $n(L, R_o)$ values in Eq. (4.40), found from the solutions of Eq. (4.42), is that they are negative, which immediately gives the energy E in Eq. (4.41) lower than $\hbar\omega/2$. *As a result of the confinement with the 'superconducting' boundary conditions, the energy levels in finite samples lie below the classical value $\hbar\omega/2$ (found for $n = N = 0$) for infinite samples* [28,29]. The whole energy level scheme (Fig. 4.8), found by Saint-James [28], can be reconstructed by calculating $E = |\alpha|$ vs. Φ/Φ_0 for different L values. In Fig. 4.9, the same cusplike phase boundary is shown as in Fig. 4.8, with the $x-$ and $y-$axis interchanged. Note that the values on $x-$axis are decreasing, so that it corresponds to a temperature axis. The dotted line in Fig. 4.9 gives the bulk upper critical field H_{c2}, which is clearly lower than $H_{c3}^*(T)$, for all temperatures. The Meissner state solution $L = 0$ (dashed line) extrapolates to $H_{c2}(T)$ for high flux values. The 3D plots on the right show the $|\Psi_L|$ profiles for the states $L = 0, 1, 2$. While the $L = 0$ Meissner state ($\mathsf{C} = 0$) demonstrates a maximum at the sample center, the solutions $L \neq 0$ all have a node $|\Psi| = 0$ in the origin ($\mathsf{C} = 1$). The surface superconducting states for $L \neq 0$ concentrate more and more near the sample boundary, as L grows along the phase boundary line. It is worth mentioning that β_A, for certain choices for L and H, can be smaller than the well-known value $\beta_A = 1.16$, which is the minimal value found in bulk, for a triangular arrangement of Abrikosov vortices. For example, for a mesoscopic disk, values in the range 1-1.16 were found for $L = 0, 1, 2$, below $\Phi/\Phi_0 \approx 7.5$ (see Fig. 2 in [30]). The compression of the flux trapped inside the GVS (for $L \geq 1$) leads to the *paramagnetic Meissner effect*, in contrast to the normal diamagnetic Meissner effect for $L = 0$ [30].

The $T_c(H)$ phase boundary (or $H_{c3}^*(T)$) of the disk shows (solid line in Fig. 4.9), just like for the usual Little-Parks effect in a multiply connected sample (i.e. a loop), an oscillatory behavior. Here as well, fluxoid quantization (Eq. (4.36)) is responsible for the oscillations of T_c versus H. In the case of the disk, the oscillation period of $T_c(H)$ is not constant, but slightly decreases as H increases. Since $|\Psi_L|$ is getting more and more localized near the sample boundary, the 'core' of the central vortex enlarges as Φ (or L) grows (see 3D $|\Psi|$ plots in Fig. 4.9). A giant vortex state is formed: a 'normal' core carries L

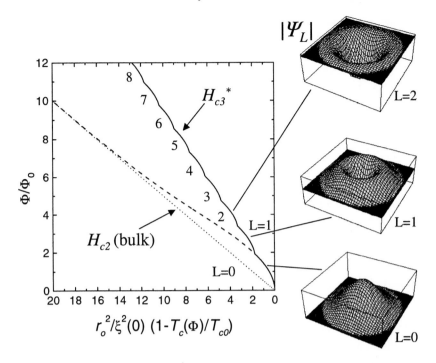

Fig. 4.9. Cusplike phase diagram of a disk (cylinder) (*solid line*). The *dotted* line presents the upper critical field $H_{c2}(T)$ of a bulk superconductor of the same material (same $\xi(0)$). The *dashed* line is the Meissner state solution of 4.5 for $L = 0$. The *3D plots* on the right show the $|\Psi_L|$ distribution for $L = 0$ (Meissner), and $L = 1, 2$ (GVS)

flux quanta, and the 'effective' loop radius increases, resulting in a decrease of the magnetic oscillation period. The enhancement factor $\eta = H_{c3}^*/H_{c2}$ for a disk is always larger than 1.69, the enhancement for a semi-infinite slab. The linear component of the cusplike $H_{c3}^*(T)$ line is $1.69\,H_{c2}$, which is in good agreement with the calculations of H_{c3}^* in the $L \to \infty$ limit [1]. An experimental verification of these predictions was carried out by several groups [10,29,31].

4.3.5   Nucleation of Superconductivity Around an Antidot

For *a single antidot* embedded in an infinite plain film, a similar analysis as for the disk was carried out by Bezryadin and Pannetier [32]. The order parameter Ψ can now be written as

$$\Psi(r,\varphi) = e^{\pm iL\varphi}\, r^L\, \gamma^{(L+1)/2} \exp\left(-\frac{\gamma r^2}{2}\right) U(-N, L+1, \gamma r^2) \qquad (4.43)$$

since now, the first Kummer function M diverges at $r \to \infty$: $M(-N, L + 1, r^2 \to \infty) \to \infty$. In the general solution Eq. (4.20), the first coefficient $c_1 = 0$.

Eq. (4.39) remains valid, and r_o is now taken as the radius of the antidot. The energy levels are again obtained from Eq. (4.40), with negative real numbers n. With the derivative of the function U [12]:

$$\frac{dU(a, c, y)}{dy} = -a\, U(a + 1, c + 1, y) \qquad (4.44)$$

the boundary condition (Eq. (4.32)) translates into:

$$(L - R_o^2)\, U(-n, L + 1, R_o^2) + 2n\, R_o^2\, U(-n + 1, L + 2, R_o^2) = 0\,, \qquad (4.45)$$

when Eq. (4.43) is inserted in Eq. (4.32). The numerical values for n have to be inserted into Eq. (4.39), in order to obtain the $T_c(\Phi)$. The phase boundary $T_c(H)$ is shown in Fig. 4.8 (solid curve above the dashed line). Again, a non-periodic $T_c(\Phi)$ behavior is observed, but this time the period of the oscillatory component increases as the magnetic field grows. This can be understood as a stronger localization of the surface states, as Φ is increasing, through which the effective loop area becomes smaller. Consequently the magnetic period grows with increasing Φ. In the limit $\Phi \to \infty$ (or $L \to \infty$), the $H_{c3}^*(T)$ line will, similar to the case of the disk, tend to $1.69\, H_{c2}$. For a single antidot, the predicted enhancement $\eta(T) = H_{c3}^*(T)/H_{c2}$ will always remain smaller than 1.69, the semi-infinite slab solution. In the other limit, namely $\Phi \to 0$ (vanishingly small antidot), the initial slope of $T_c(H)$ is identical to the case of the upper critical field $H_{c2}(T)$ in a bulk superconductor.

4.3.6   Nucleation of Superconductivity in a Wedge

The nucleation of superconductivity is sensitive not only to the imposed connectivity, but also to the precise sample shape. For example, the presence of sharp corners can enhance $H_{c3}^*(T)$ [33]. The main idea here is that sharp corners reduce the superfluid velocities v_s at these specific locations, thus leading to an increased $T_c(H)$ (or $H_{c3}^*(T)$).

In this section, the nucleation field will be discussed for infinite wedge geometry, with a corner angle Γ. Since the onset of superconductivity occurs near the sample interface, a higher nucleation field is expected for samples with a larger surface-to-volume ratio. The magnetic field is directed parallel to the two intersecting superconductor/vacuum interfaces. In polar coordinates, the superconductor occupies that region of space for which $|\varphi| \leq \Gamma/2$, and $H \parallel e_z$. This problem has been addressed by several authors independently [33–38], all leading to similar conclusions. In the general case of $\Gamma < \pi$, the problem was solved either numerically [38] on a finite space grid, or by using a variational approach with trial functions [34–36]. To very good accuracy, the nucleation field for a wedge can be fitted to [38]:

$$\frac{H_{c3}^*(T)}{H_{c2}(T)} = \frac{\sqrt{3}}{\Gamma}\left(1 + 0.14804\,\Gamma^2 + \frac{0.746\,\Gamma^2}{\Gamma^2 + 1.8794}\right) \tag{4.46}$$

This function is shown as a solid line in Fig. 4.10. The ratio H_{c3}^*/H_{c2} is a monotonically decreasing function of Γ, which tends to the value 1.69, as Γ increases to π (presented as a dashed line). For very small opening angles $\Gamma \ll 1$, an analytical treatment becomes possible. In such case, Ψ varies only weakly with Γ. A rigorous proof of the validity of this assumption was given by Klimin *et al.* [39]. They have used an 'adiabatic approach', and calculated the contribution of the non-adiabaticy operator on the lowest energy eigenvalue of 4.5. For wedges with angles $\Gamma \lesssim 0.1\pi$, this approximation turns out to be very good.

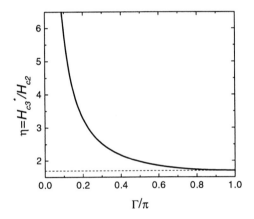

Fig. 4.10. Nucleation field enhancement factor $\eta = H_{c3}^*(T)/H_{c2}(T)$ (*solid line*) for a wedge as a function of the opening angle Γ. As $\Gamma \to \infty$, the well-known ratio $\eta = H_{c2}(T)/H_{c2}(T) = 1.69$ for a semi-infinite slab is recovered (*dashed line*)

In the low angle limit, the solutions of the linear GL equation are known, and can be expressed in terms of Legendre polynomials [39]. The ground state was shown to have a 'corner' vortex, i.e. a *confined circulating superconducting current* appears in the wedge in the vicinity of the corner. This vortex can exist only in a wedge with a sufficiently small angle $\Gamma \lesssim 0.2\pi$.

4.4 Dimensional Crossover in a Loop of Finite Width

The goal of this section is to study the phase boundary $T_c(H)$ of loops made of finite width wires (or disks with an opening in the middle, increasing in size). In a Type-II material, superconductivity is expected to be enhanced, with respect to the bulk $(H_{c3}^*(T) > H_{c2}(T))$, both at the external and the internal sample surfaces, which have a different concavity (see Fig. 4.8). As for a film in a parallel field, a dimensional crossover can be anticipated, since the loops may be simply considered as a film, which is bent in such a way that its ends are joined together. We calculate the phase boundary $T_c(H)$ as the ground state solution of the linearized first GL equation with two superconductor/vacuum interfaces. This calculation has been suggested already by several authors [32,40,41], but was carried out only recently by Bruyndoncx *et al.* [42]. Calculations for a similar geometry, in the framework of the full nonlinear GL equations, were reported by Baelus *et al.* [43]. We will show that the $T_c(\Phi)$ of the loops, for low applied magnetic flux (corresponding to the parabolic regime), can be described within a simple London picture, where the modulus of the order parameter $|\Psi|$ is spatially constant. As the flux increases, the background depression of $T_c(\Phi)$ is changed from parabolic to quasi-linear, which indicates the formation of a GVS, where only a surface sheath close to the sample's outer interface is in the superconducting state. Moreover, the oscillation period of $T_c(\Phi)$ of the loops becomes identical to the one for the full disk, as soon as the transition to quasi-linear behavior has taken place.

Inserting the general solution of 4.5 (Eq. (4.20)) into Eq. (4.34) gives:

$$c_1 \left[(L - R_{o,i}^2) \, M\left(-n, L+1, R_{o,i}^2\right) - \frac{2\,n\,R_{o,i}^2}{L+1} \, M\left(-n+1, L+2, R_{o,i}^2\right) \right]$$

$$+ c_2 \left[(L - R_{o,i}^2) \, U\left(-n, L+1, R_{o,i}^2\right) + 2\,n\,R_{o,i}^2 \, U\left(-n+1, L+2, R_{o,i}^2\right) \right]$$

$$= 0 \tag{4.47}$$

which has to be solved numerically for each integer value of L, resulting in a set of values $n(L, R_o^2)$. The outer R_o, and the inner loop radius R_i are scaled as in Section 4.1.5. Since the boundary conditions have to be fulfilled at the inner radius r_i and at the outer radius r_o, a system of equations is formed by Eq. (4.47), from which n, and c_2 have to be calculated, for fixed L and H. The value $c_1 = 1$ can be chosen.

Fig. 4.11 shows the Landau level scheme (dashed lines) calculated from Eqs. (4.20), (4.39), and (4.47), for a loop with $r_i/r_o = R_i/R_o = 0.5$. The applied magnetic flux $\Phi = \mu_M H \pi r_o^2 = R_o^2$ is defined with respect to the outer sample area. The $T_c(H)$ boundary is composed of Ψ solutions with a different phase winding number L and is drawn as a solid cusplike line in Fig. 4.11. At $\Phi \approx 0$, the state with $L = 0$ is formed at $T_c(\Phi)$ and one by one, consecutive flux quanta L enter the loop as the magnetic field increases. For low magnetic flux, the background depression of T_c is *parabolic*, whereas at higher flux, $T_c(\Phi)$ becomes *quasi-linear*, just like for the case of a filled disk. The crossover point from parabolic to quasi-linear appears at about $\Phi \approx 14 \, \Phi_0$, indicated by the arrow in Fig. 4.11.

For a plain film in a parallel field a parabolic dependence of $T_c(H)$ is obtained [2,31]:

$$\frac{1}{\xi^2(T)} = \frac{1}{\xi^2(0)} \left(1 - \frac{T_c(H)}{T_{c0}}\right) = \frac{\pi^2 \tau^2 \mu_M^2 H^2}{3 \Phi_0^2} \qquad (4.48)$$

with τ the film thickness. Alternatively, the nucleation field $H_{c3}^*(T)$, for a thin strip in a perpendicular field, is determined from the same equation (Eq. (4.48)), with τ replaced by the strip width w.

Comparing the lowest energy eigenvalue $E = |\alpha|$ for a thin strip in a parallel field, with the bulk case (Eq. (4.22) with $N = 0$) gives, at a fixed temperature T:

$$H_{c3}^*(T) = \frac{\sqrt{12}\,\xi(T)}{w} H_{c2}(T) = \eta\, H_{c2}(T) \qquad (4.49)$$

This clearly implies that $H_{c3}^* > H_{c2}$, which will be a general rule of thumb: the smaller size of the sample (or more exact: the sample area perpendicular to the field H), the higher the nucleation field H_{c3}^* is.

If now, the plane film is bent such that its ends are joined together, a closed loop is formed. Within the London limit of a spatially constant $|\Psi|$, $T_c(\Phi)$ can be determined, up to order u^2, by [44]

$$\frac{T_{c0} - T}{T_{c0}} = \frac{\xi^2(0)}{r_m^2} \left[\left(L - \frac{\Phi_m}{\Phi_0}\right)^2 + \frac{4}{3} u^2 \left(\frac{\Phi_m}{\Phi_0}\right)^2\right] \qquad (4.50)$$

where we introduced the loop aspect ratio $u = (r_o - r_i)/(r_o + r_i)$. The first term on the right hand side in Eq. (4.50) represents the oscillatory contribution, like for rings with vanishing strip width $w = 0$ (i.e. the Little-Parks effect, discussed in Section 4.3.3, where the flux needs to be redefined as $\Phi \equiv \Phi_m = \mu_M H \pi r_m^2$, with the mean radius $r_m = (r_o + r_i)/2$. The second term is a parabolic background reduction of $T_c(\Phi)$, due to the finite width of the strips. It is identical to the reduction of T_c of an infinitely long strip of width w, expressed by Eq. (4.48). Hence, Eq. (4.50) can describe only the low field part of the phase diagram, approximately up to the arrow in Fig. 4.11. Above this range, the London limit assumption of a spatially constant $|\Psi|$ does not longer hold, and a more sophisticated calculation is needed.

The solid and dotted straight lines in Fig. 4.11 are the bulk upper critical field $H_{c2}(T)$ and the surface critical field $H_{c3}(T)$ for a semi-infinite slab, respectively. In these units the slopes of the curves (see Eq. (4.39)) are $\epsilon = 2$ for H_{c2} (substitute $n = 0$ in Eq. (4.39)) and $\epsilon = 2/1.69$ for H_{c3}. The ratio $\eta = \epsilon(H_{c2})/\epsilon(H_{c3}) = 1.69$ corresponds then to the enhancement factor $H_{c3}(T)/H_{c2}(T)$ at a constant temperature. For the loops we are considering here (and for the disk and the antidot as well), $\eta = \epsilon(H_{c2})/\epsilon(H_{c3}^*)$ varies as a function of the magnetic field.

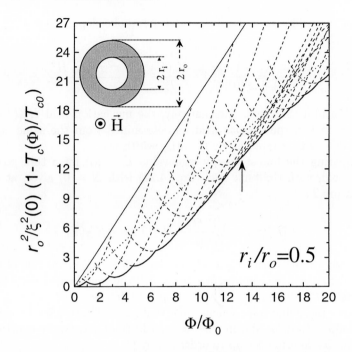

Fig. 4.11. Calculated energy level scheme (*dashed lines*) for a superconducting loop with the ratio of inner to outer radius $r_i/r_o = 0.5$. The *solid* and *dotted* straight lines correspond to $H_{c2}(T)$ and $H_{c3}(T)$, respectively. The *solid* cusplike curve presents the lowest Landau level, corresponding to the $T_c(\Phi)$ phase boundary. The *arrow* indicates the point where a dimensional transition from parabolic to quasi-linear behavior takes place

The energy levels below the H_{c2} line (solid straight line in Fig. 4.11) could be found by fixing a certain L, and finding the real numbers $n < 0$ numerically after inserting the general solution (Eq. (4.20)) into the boundary condition (Eq. (4.32)). Note that the lowest Landau level always has a lower energy $|\alpha(\Phi)|$ than for a semi-infinite superconducting slab, which implies $H_{c3}^*(T) > H_{c3}(T) = 1.69\,H_{c2}(T)$.

For a thin film of thickness τ in a parallel field H, a dimensional crossover is found at $\tau = 1.84\,\xi(T)$. For low fields (high ξ) $T_c(H)$ is parabolic, and for higher fields vortices start penetrating the film and consequently $T_c(H)$ becomes linear [45,46]. In Fig. 4.11 the small arrow indicates the point on the phase diagram $T_c(\Phi)$ where $w = 1.84\,\xi(T)$. For the loops as well, the dimensional transition shows up approximately at this point, although the vortices do not penetrate the sample area in the quasi-linear regime. Instead, the middle loop opening contains a coreless 'giant vortex' with an integer number of flux quanta $L\Phi_0$. Let us now come back to our original definition of connectivity from Section 4.1.4. For the disk, the $L = 0$ state corresponds to the singly connected $C = 0$ Meissner state. For higher flux Φ, the GVS states with $L \neq 0$ all have $C = 1$. By making an opening into the disk, the obtained loop is always in the $C = 1$ multiply connected state, for all fields. Changing the diameter of the opening (different aspect ratio) simply does not influence the connectivity C.

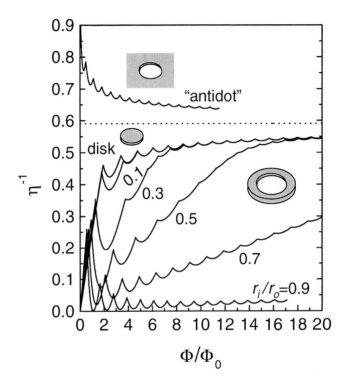

Fig. 4.12. Inverse enhancement factor $\eta^{-1} = \epsilon(H_{c3}^*)/\epsilon(H_{c2})$ (see Eq. (4.39)) for loops with different aspect ratios, compared to the case of a disk and an antidot. The *horizontal dashed* line at $\eta^{-1} = 0.59 = 1/1.69$ corresponds to $H_{c3}(T)/H_{c2}(T) = 1.69$ for a plane superconductor/vacuum boundary [4]

In order to compare the flux periodicity of $T_c(\Phi)$, we have plotted, in Fig. 4.12, the lowest energy levels of Fig. 4.11 as $\eta^{-1} = \epsilon(H_{c3}^*)/\epsilon(H_{c2})$, for loops with a different r_i/r_o. In this representation, the dotted horizontal line at $\eta^{-1} = 0.59$ corresponds to the surface critical field line $H_{c3}(T)$. The nucleation field of a disk $H_{c3}^*(T) > 1.69\, H_{c2}(T)$, and for a circular antidot in an infinite film [32,40] $H_{c3}^*(T) < 1.69\, H_{c2}(T)$. In fact, the curves shown in Fig. 4.12 give the slope of the phase boundaries (η^{-1}), shown in Fig. 4.8, divided by two. As Φ grows (the radius goes to infinity) the $H_{c3}^*(T)$ of both the disk and the antidot approaches the $H_{c3}(T) = 1.69\, H_{c2}(T)$ dotted line.

For all the loops we study here, the presence of the outer sample interface automatically implies that $H_{c3}^*(T) > H_{c3}(T)$ is enhanced ($\eta > 1.69$), with respect to the case of a flat superconductor-vacuum interface. For loops with a small r_i/r_o, the $T_c(\Phi)$ boundary very rapidly collapses with the $T_c(\Phi)$ of the disk (η becomes the same). The presence of the opening in the sample is not relevant for the giant vortex formation in the high flux regime. On the contrary, in the low flux regime, the surface sheaths along the two interfaces overlap, giving rise to a different periodicity of $T_c(\Phi)$ and to a parabolic background. This regime can be described within the London limit [44], since superconductivity nucleates almost uniformly within the sample. The aspect ratio of the loop 'tunes' the flux range, for which the parabolic and the quasi-linear regime in $T_c(\Phi)$ can be seen. The larger the opening, the broader the flux range where the parabolic $T_c(\Phi)$ behavior is found. This demonstrates that the $T_c(H)$ boundaries are sensitive not only to the imposed connectivity, but also to the exact sample geometries.

Summarizing this section, we have analyzed the linearized GL equation (4.5) for loops of different wire width, with Neumann boundary conditions at both the outer and the inner loop radius. The critical fields $H_{c3}^*(T)$ are always above the $H_{c3}(T) = 1.69\,H_{c2}(T)$. The ratio $H_{c3}^*(T)/H_{c2}(T)$ is enhanced most strongly when the sample's surface-to-volume ratio is the largest. The $T_c(\Phi)$ behavior can be split in two regimes: for low flux, the background of T_c is parabolic and the Little-Parks $T_c(\Phi)$ oscillations are perfectly periodic. In the high flux regime, the period of the $T_c(\Phi)$ oscillations is decreasing with Φ and the background T_c reduction is quasi-linear. The crossover between the two regimes, at a certain applied flux Φ, is similar to the dimensional transition in thin films subjected to a parallel field. As soon the quasi-linear regime is reached, a 'coreless' giant vortex state is created, where only a sheath close to the sample's outer interface is superconducting.

4.5   Superconducting Micronetworks: Increasing Connectivity by Adding Loops in One Dimension

In this section, we consider one-dimensional multiloop Al structures, with different connectivities [47]. Fig. 4.13 shows a Atomic Force Microscopy (AFM) image of the double and triple loop structures. The loops in the two structures have the same dimensions, thus leading to the same overall magnetic field period $\mu_M\,\Delta H = 1.24\,\mathrm{mT}$. The strips forming the structures are $w = 0.13\,\mu\mathrm{m}$ wide and the film thickness is $\tau = 34\,\mathrm{nm}$.

Fig. 4.14 shows the measured phase boundaries of the double and the triple loop. The curve for the triple loop has been shifted along the temperature axis for clarity. Further on, we will analyze the 'fine structure' in $T_c(\Phi)$, which is clearly different for the two types of samples. The background reduction of $T_c(\Phi)$ shows a parabolic behavior, which is due to the finite width of the strips. When inserting the strip width w into Eq. (4.48), a fit to the

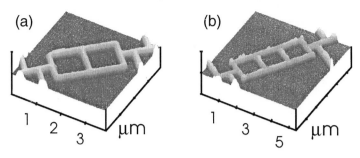

Fig. 4.13. AFM micrographs of the multiloop structures: (**a**) the double loop, and (**b**) the triple loop

parabolic envelope provides us an accurate method to determine $\xi(0)$. In order to allow for a direct comparison between the experiment and the existing 1D models for micronetworks, the background (Eq. (4.48)) is subtracted from the experimental $T_c(H)$ data, and we obtain curves as shown in the right panels of Figs. 4.15 and 4.16.

In the temperature interval where the $T_c(H)$ boundary was measured, the coherence length $\xi(T)$ is considerably larger than the width w of the strips. This makes it possible to use the one-dimensional models for the calculation of $T_c(H)$. The basic idea is to consider $|\Psi| =$ constant across the strips forming the network and to allow a variation of $|\Psi|$ only along the strips. In the simplest approach $|\Psi|$ is assumed to be spatially constant (London limit) [48,49], in contrast to the de Gennes-Alexander (dGA) approach [50–52], where $|\Psi|$ is allowed to vary along the strips. In the latter approach one imposes:

$$\sum_n \left(i\frac{\partial}{\partial x} + \frac{2\pi}{\Phi_0} A_\parallel(x) \right) \Psi(x) = 0 \tag{4.51}$$

at the points (or nodes) where the current paths join. The summation is taken over all strips connected to the junction point. Here, x is the coordinate defining the position on the strips, and $A_\parallel$ is the component of the vector potential along x. Eq. (4.51) is often called the generalized first Kirchhoff law, ensuring current conservation [50]. The second Kirchhoff law for voltages in normal circuits is now replaced by the fluxoid quantization requirement (Eq. (4.8)), which should be fulfilled for each closed contour in the superconducting network (around each loop). It is evident that connectivity plays an important role for both Kirchhoff's laws.

In Figs. 4.15 and 4.16 the $T_c(H)$ boundaries of the double loop and the triple loop are shown [47]. The dashed lines are the phase boundaries calculated in the London limit, while the solid lines give the results from the dGA approach. Attaching contacts modifies the confinement topology, so that the amplitude of the local Little-Parks oscillations is reduced [51,53,54]. This effect has been observed experimentally in a single loop [55], where at low

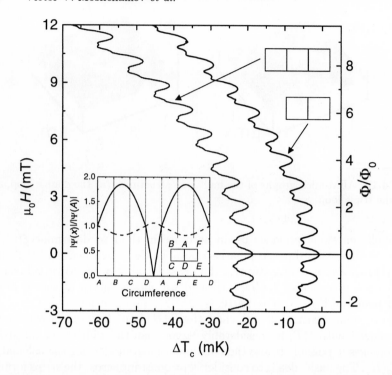

Fig. 4.14. Measured $T_c(H)$ phase boundaries of the double and the triple loop. The curve for the triple loop has been shifted along the temperature axis for clarity. The inset shows the calculated variation of $|\Psi|$ along the circumference of the double loop, at the phase boundary ($\Phi/\Phi_0=0.36$). The *dashed line* is the solution with $|\Psi|$ nearly spatially constant (flux regime I), while the *solid line* is the state with a node in the center of the strip connecting points A and D (flux regime II)

fields H a reduced oscillation amplitude was observed, compared to the case of an isolated single loop. Due to the 'coupling' between the loop and the attached measuring leads, 'nonlocal' Little-Parks oscillations could be seen when voltage probes were put on a strip segment adjacent to the loop. The strength of the coupling is governed by the temperature dependent coherence length $\xi(T)$. In the present case, we have added two side arms of length half of the side of a single cell. The dash-dotted line in Figs. 4.15 and 4.16 gives the result of the dGA calculation where the presence of the leads has been included. The values for $\xi(0)$ obtained from the fits agree within a few percent with the $\xi(0)$ values found independently from the parabolic background of $T_c(\Phi)$ (see Eq. (4.48)).

To facilitate the discussion we divide the flux period in two intervals: flux regime I for $\Phi/\Phi_0 < \phi_c$ or $\Phi/\Phi_0 > (1-\phi_c)$ and flux regime II for $\phi_c < \Phi/\Phi_0 < (1 - \phi_c)$. In the flux regime I the phase boundaries, predicted by the different

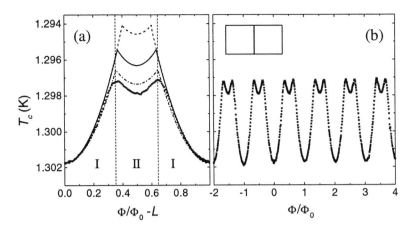

Fig. 4.15. Experimental $T_c(\Phi)$ data for the *double loop* with the parabolic background (Eq. (4.48)) subtracted. The *dots* are the experimental data points, while the *lines* correspond to the different theoretical results as explained in the text. (a) Single period of $T_c(\Phi)$, (b) A few periods of the experimental $T_c(\Phi)$ curve

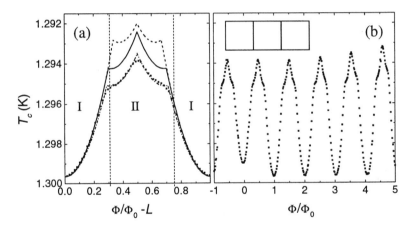

Fig. 4.16. Experimental $T_c(\Phi)$ data for the *triple loop* with the parabolic background (Eq. (4.48)) subtracted. The *dots* are the experimental data points, while the *lines* correspond to the different theoretical results as explained in the text. (a) Single period of $T_c(\Phi)$, (b) A few periods of the experimental $T_c(\Phi)$ curve

models, are nearly identical. Near $\Phi/\Phi_0 = 1/2$ (flux regime II), however, clear differences are found between the dGA approach and the London limit. Both the crossover point ϕ_c between regimes I and II and the amplitude of the T_c oscillations are most accurately predicted by the dGA result. Using the dGA approach, the spatial modulation of $|\Psi|$ and the supercurrents for different values at the $T_c(\Phi)$ boundary have been calculated (see inset of Fig. 4.14). In the flux regime I, $|\Psi|$ varies only slightly and therefore the results of the London limit and the dGA models nearly coincide. The elementary loops have an equal fluxoid quantum number (and consequently an equal supercurrent orientation) for both the double and the triple loop geometry. For the double loop this leads to a cancellation of the supercurrent in the middle strip, while for the triple loop the fluxoid quantization condition (Eq. (4.8)) results in a different value for the supercurrent in the inner and the outer loops. As a result, the inner strips of the triple loop structure carry a finite current.

In the flux regime II, qualitatively different states are obtained from the London limit and the dGA approach: the states calculated within the dGA approach have strongly modulated $|\Psi|$ along the strips. This is most severe for the double loop: Ψ shows a node ($|\Psi| = 0$) in the center of the common strip, the phase φ having a discontinuity of π at this point. This node is a one-dimensional analog of the core of an Abrikosov vortex, where the order parameter also vanishes and the phase shows a discontinuity. In the inset of Fig. 4.14 the spatial variation of $|\Psi|$ along the strips is shown for Φ/Φ_0=0.36 close to the crossover point ϕ_c. The dashed curve gives a quasi-constant $|\Psi|$ in flux regime I. The strongly modulated solution, which goes through zero in the center, is indicated by the solid line. Although there exists a finite phase difference across the junction points of the middle strip, no supercurrent can flow through the strip due to the presence of the node (Eq. (4.4)). This node is predicted to persist when moving below the phase boundary into the superconducting state [56,57]. Already in 1964 Parks [58] anticipated that, in a double loop, 'a part of the middle link will revert to the normal phase', and that 'this in effect will convert the double loop to a single loop', giving an intuitive explanation for the local maximum in $T_c(\Phi)$ at Φ/Φ_0=1/2. This behavior is illustrated schematically in Fig. 4.17. Instead of allowing for the entrance of a fluxoid in one of the two cells, the system chooses to break superconductivity in the center, which is indicated by the dark cross. This makes the sample effectively equivalent to a loop with double area, with the broken strip 'dangling' at the outside of the loop. The obvious energy cost for the breaking of superconductivity causes a uniform shift of $T_c(\Phi)$, for the fluxoid state in flux regime II (see Fig. 4.15 (a)), to lower temperatures, as compared to the $T_c(\Phi)$ of a single loop of double area. This is a remarkable example of a 'spontaneous change of connectivity'. In flux regime I, there is a finite $|\Psi|$ around both cells, thus C = 2. Since superconductivity is destroyed at the sample center in flux regime II, the connectivity effectively becomes C = 1.

Such a modulation of $|\Psi|$ is obviously excluded in the London limit, where the loop currents have an opposite orientation and add up in the central strip, thus giving rise to a rather high kinetic energy. Hence, this model predicts C = 2, also in flux regime II. An extra argument in favor of the presence of the node is given by the much better agreement for the crossover point ϕ_c

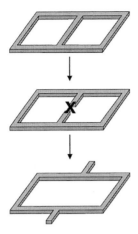

Fig. 4.17. Schematic illustration of the breaking of superconductivity in the center, in flux regime II. The node $|\Psi| = 0$ in the sample center, indicated by the dark cross, prevents the flow of a supercurrent through the middle strip. Therefore, the sample becomes effectively equivalent to a single loop of double area, as shown at the bottom. The broken middle strip are like 'dangling' arms, only reducing the oscillation amplitude of $T_c(H)$. The imposed connectivity $\mathsf{C} = 2$ spontaneously lowers to $\mathsf{C} = 1$ due to the destruction of Ψ in the sample center

when the presence of the leads is taken into account in the calculations (see dash-dotted line in Fig. 4.15).

A similar example, where a change in connectivity can take place, has been recently analyzed theoretically for a *single loop* in Refs. [59–63]: under certain conditions superconductivity spontaneously breaks at some spot along the perimeter of the loop, so that the *superconducting area changes from multiple* (closed loop, $\mathsf{C} = 1$) to *single connectivity* ('open' loop, $\mathsf{C} = 0$). The highest supercurrent (and therefore the strongest reduction of $T_c(\Phi)$) is realized for half integer flux when $\Phi/\Phi_0 - L = 1/2$. In this situation it may turn out, however, that somewhere in the loop the order parameter Ψ is spontaneously suppressed and a sort of 'normal core' is created at a certain location along the loop circumference. The energy of this normal state core, below the $T_c(\Phi)$ line, is, of course, higher than the energy corresponding to a superconducting state everywhere in the loop, but, at the expense of that, the circular supercurrent is interrupted, thus effectively opening the ring for entrance and removal of flux. While Horane *et al.* [59] predicted the existence of the singly connected state for loops made of '1D' strips, Berger and Rubinstein [62] showed that the temperature region where the singly connected state $\mathsf{C} = 0$ exists, can be enhanced by proper tuning the nonuniform 'strip width' profiles along the loop.

For the triple loop (Fig. 4.16 **(a)**) the modulation of $|\Psi|$ is still considerable in flux regime II, but it does not show any nodes. Therefore the supercurrent orientations can be found from the fluxoid quantum numbers $\{L_i\}$, obtained from integrating the phase gradients along each individual loop. When passing through the crossover point between flux regime I and regime II, only the supercurrent in the middle loop is reversed, while increasing the flux above $\Phi/\Phi_0 = 1/2$ implies a reversal of the supercurrent in all loops. The connectivity is $C = 3$, independent of the applied magnetic field.

Surprisingly, the behavior of a microladder with a linear arrangement of m loops appears to be *qualitatively different for even and for odd* m in the sense that m determines the presence or absence of nodes in the common strips. For an infinitely long microladder, $|\Psi|$ was found to be spatially constant below a certain $\Phi < \Phi_c$ [64], which is analogous to the states in flux regime I. For fluxes $\Phi > \Phi_c$ modulated $|\Psi|$ states, with an incommensurate fluxoid pattern, were found. At $\Phi/\Phi_0 = 1/2$, nodes appear at the center of every second common (transverse) strip.

A variety of other structures (micronets) of different topology (coupled rings, bola's, a yin-yang, infinite microladders, bridge circuits, like a Wheatstone bridge, wires with dangling branches, etc.) formed by 1D wires, have been analyzed in a series of publications [50,51,54,56,57,64–70] [71–73] using the approach, initiated in 1981 by de Gennes [51] and further developed by Alexander [52] and Fink *et al.* [50]. For all these structures very pronounced effects of the micronet topology on $T_c(\Phi)$ and critical current have been predicted.

4.6 Perforated Microsquares

In this section, we deal with microstructures built up from several individual nanoplaquettes, assembled in the form of (2×2) 2D structures. The elementary plaquettes, used for this nanoassembly, are filled and open squares.

We will present the measured phase boundaries $T_c(H)$ of three different topologies, which are shown in Fig. 4.18. The three Al structures studied are a filled microsquare, and two squares with 2 and 4 square antidots respectively. Similar structures were studied in Refs. [74,75], where the 4-antidot structure was proposed as a basic cell for a memory based on flux logic. In those papers, several stable vortex configurations were detected at low magnetic fields.

The main idea here is to discuss the influence of connectivity C, varied through bringing in antidots inside a microsquare, on the crossover from the 'network' behavior at low fields to a GVS [30,76] at high fields, and whether eventually the two configurations (vortices pinned by the antidots and the giant vortex state) can coexist. We will mainly focus on the high magnetic field regime.

The square dot has a side $a = 2.04\,\mu\text{m}$ and is taken as a reference sample (a); the square of side $a = 2.04\,\mu\text{m}$ is perforated with four $0.46 \times 0.46\,\mu\text{m}^2$ square antidots (b); and the square with side $a = 2.14\,\mu\text{m}$ has two $0.52 \times 0.52\,\mu\text{m}^2$ antidots, placed along a diagonal (c). The magnetic field was applied

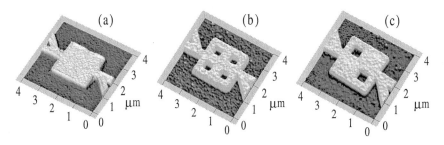

Fig. 4.18. AFM images of the three structures: (a) full, (b) 4-antidot, and (c) 2-antidot microsquares

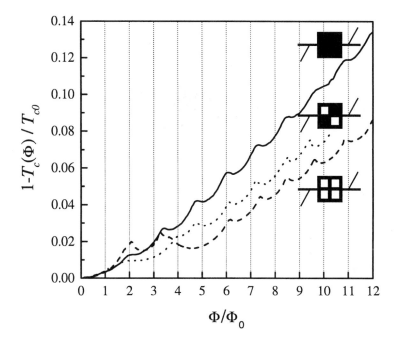

Fig. 4.19. Experimental $T_c(\Phi)$ phase boundaries for a dot (*solid*), 4-antidot (*dashed*), and a 2-antidot microsquare (*dotted* line). For $\Phi/\Phi_0 > 5$ the peaks in $T_c(\Phi)$ appear at the same flux Φ/Φ_0 in all the structures. The flux Φ is defined with respect to the outer sample area, thus including the perforated area of the antidots

perpendicular to the structures. More experimental details can be found in Ref. [77].

In Fig. 4.19 we present the experimental phase boundary $T_c(\Phi)$ of the three structures. For the reference full square, we observe pseudoperiodic oscillations in $T_c(\Phi)$ superimposed with an almost linear background, where the period of the oscillations slightly decreases with increasing field, in agreement with previous studies [28,29,31,78]. These observations are characteristic for the presence of the giant vortex state, as was discussed in Section 4.3.4. For the perforated microstructures, *two different magnetic field regimes* can be distinguished. At *high magnetic flux*, the oscillations in $T_c(\Phi)$ are pseudoperiodic as well. This is similar to the 'single object' regime, as it was found by Bezryadin and Pannetier [40], but here, at $T_c(\Phi)$, a surface superconducting sheath develops near the sample's *outer* boundary only. The comparison of the $T_c(\Phi)$ data obtained on the perforated Al microstructures with that of the reference microsquare without antidots confirms the presence of a giant vortex state in the three structures in the high magnetic flux regime. For the *low flux* part of the phase diagram, distinct features appear (i.e., below $\sim$ $5\,\Phi_0$): for the 2-antidot sample we observe the same number of peaks compared to the full square, but with a considerable shift of the positions of the first peaks. For the 4-antidot structure extra peaks can be clearly seen below $\sim 5\,\Phi_0$.

The $T_c(\Phi)$ curve measured for the full square is quite similar to the result obtained from a calculation [28,30] for a mesoscopic disk in the presence of a magnetic field (see Section 4.3.4). The series of peaks in the $T_c(\Phi)$ curve correspond to transitions between states with different angular momenta $L \to L + 1$ of Ψ as successive flux quanta, $\Phi = L\Phi_0$, enter the superconductor.

It is important to note that here we defined the flux as $\Phi = \mu_M H S_{eff}$, with S_{eff} the effective area of the whole microsquare. It is close to the exact outer sample area S, and was introduced in order to fit the peak positions to the calculated $T_c(\Phi)$ for a circular dot (disk). Doing so, we obtain an effective area of $3.9\,\mu m^2$, close to the actual size of the structure, $4.2\,\mu m^2$. The introduction of this 'effective area' is obviously not needed if the $T_c(\Phi)$ is compared with a calculation performed for a square [79]. From the parabolic shape of the $L = 0$ state, we find the coherence length, $\xi(0) = 92\,nm$. It should be noted that this value might be an underestimation because of the aforementioned coupling to the measuring leads.

In contrast to the experimental result presented in Ref. [29], which was obtained for a substantially larger, but circular dot (disk), the field period for the full square can be matched to the theoretical predictions in the whole measured field interval. When a sufficiently high magnetic field is applied to the sample, a superconducting edge state is formed, where superconductivity only nucleates within a surface layer of thickness w_H. The remaining area acts like a normal core of radius $R_{eff} \approx R - w_H$, and carries L flux quanta in its interior. Due to the expansion of the normal core with increasing H, the sample can then be seen as a loop of variable radius. For this reason, the $T_c(\Phi)$ of the dot shows *nonperiodic* Little-Parks-like oscillations. In comparison to the loop, which has an initial parabolic background on $T_c(\Phi)$ at low fields, the background for the disk is quasi-linear already for $\Phi/\Phi_0 \gtrsim 2$, because of the additional energy cost (i.e. extra reduction of T_c) for suppressing superconductivity in the sample interior. The creation of a giant vortex is accompanied with a crossover, similar as for loops of finite strip width (Section 4.4), which appears at a certain flux above which $T_c(\Phi)$ becomes quasi-linear.

All the structures have peaks in $T_c(\Phi)$ at the same Φ values in the high flux regime. How can we understand this striking coincidence of the peak positions at high fields? For this, we have to look how the order parameter Ψ nucleates along a curved superconductor/insulator boundary. It is seen that the calculated $T_c(\Phi)$ curves for a disk and for a circular antidot (see also Ref. [40]) in an infinite film, (both of radius r_o) (see Fig. 4.8), lie on the other side of the straight dashed line $H_{c3} = 1.69\,H_{c2}$, for a plain normal/insulator interface [4]. Since the disk has a larger $H_{c3}^*(T)$ than the antidot, Ψ is expected to grow initially at the outer sample boundary, as the temperature drops below $T_c(\Phi)$. At slightly lower temperatures, surface superconductivity should as well nucleate around the antidots. In the mean time, however, Ψ has reached already a finite value over the whole width of the strips. The resistively measured $T_c(\Phi)$ curves, probably because of this substantially different $H_{c3}^*(T)$ for a disk and an antidot, only show peaks related to the switching of the angular momentum L, associated with a closed contour along the outer sample boundary. At the $T_c(\Phi)$ boundary, in the high magnetic field regime, there is no such closed superconducting path around each single antidot, and therefore the fluxoid quantization condition does not need to be fulfilled for a closed contour encircling each single antidot. In terms of connectivity, it seems that, at low fields the 4-antidot square is in a state with $C = 4$. As soon as the quasi-linear regime is built up, the peak positions of all three samples are situated at the same Φ values, which suggests the formation of a $C = 1$ giant vortex, which 'ignores' the presence of the antidots inside the microstructures.

The background depression of T_c is different for the three structures studied (Fig. 4.19). The larger the perforated area (in other words the smaller the area exposed to the perpendicular magnetic field), the less $T_c(\Phi)$ is pushed to lower temperatures. Another clear example of a similar behavior is given in Ref. [31], where the $T_c(\Phi)$ of the (square) dot is shown to be lower than the $T_c(\Phi)$ of the loop, when exposed to a perpendicular magnetic field.

Summarizing this section, we have presented the experimental superconducting/normal phase boundaries $T_c(\Phi)$ of Al mesoscopic structures with different connectivity: a full square and two perforated mesoscopic squares. Comparing the results with the behavior of a full square microstructure, we were able to distinguish two regimes as a function of the flux: for *low magnetic flux* the 4-antidot structure behaves like a network consisting of quasi-one-dimensional strips, giving rise to extra peaks in $T_c(\Phi)$ in comparison to the full square. In the 2-antidot structure the peak positions are only shifted compared to the full square. As soon as each antidot confines one flux quantum, a giant vortex develops, resulting in pseudoperiodic oscillations in the

$T_c(\Phi)$ and a quasi-linear background on $T_c(\Phi)$ at high magnetic fields. In this regime, the peak positions coincide for all three structures studied when the phase boundaries are plotted in flux quanta units (where flux is referred to the total sample area). For *high magnetic flux*, the presence of the antidots apparently does not change the phase winding number L for a closed contour around the outer perimeter of the whole square, and strongly indicates the formation of an effective $C = 1$ surface state, localized at the sample's outer boundary.

4.7   Superconducting Films with an Antidot Lattice

As shown in Section 4.3, the enhancement of the critical field beyond the bulk critical field H_{c2} depends strongly on the curvature of the superconducting/normal (S/N) interface. This was illustrated by the two complementary examples studied, namely the disk (convex interface) and the antidot (concave interface). For both systems, the ratio $\eta = H_{c3}^*(T)/H_{c2}(T)$ shows an oscillatory behavior as a function of field and tends to the value $\eta=1.69$, the enhancement factor for a semi-infinite slab. The major difference between the disk and the antidot is, however, the side from which the nucleation field $H_{c3}^*(T)$ approaches the semi-infinite slab solution $H_{c3}^*(T) = 1.69H_{c2}(T)$. For the case of the single antidot, where the S/N interface is concave, the enhancement η is expected to be smaller than 1.69 for all fields.

Knowing the flux confinement by an individual nanoplaquette - in this case a unit cell containing an antidot - the fundamental problem arises: to what extent can this approach be applied when we compose huge arrays (several mm^2) of such nanoplaquettes, like for example an antidot lattice (AL). This problem is of crucial importance for thin film applications, since repeating a certain nanoplaquette laterally over a *macroscopic* area creates a *bulk* nanostructured superconductor, which can be used for numerous applications.

For bulk samples, the surface nucleation of superconductivity plays only a minor role because the thickness of the surface superconductivity sheath, given by the temperature dependent coherence length $\xi(T)$, is negligibly small in comparison with the sample size. This is shown schematically in Fig. 4.20 (**a**), where the dark area represents the surface superconducting state near the sample boundary. The 'bulk' of the sample makes the transition to the superconducting state when the temperature is decreased below $H_{c2}(T)$ (for a Type-II superconductor). In superconducting films with an antidot array, however, supplementary superconductor/vacuum interfaces are introduced. Their presence increases not only the surface-to-volume ratio, but also the connectivity C of the sample. According to a conventional scenario, a 'ring' of surface superconductivity forms at $H_{c3}^*(T)$ around each antidot and, if the antidots are sufficiently closely spaced, almost the entire sample becomes superconducting at a field well above H_{c2} (see Fig. 4.20 (**b**)).

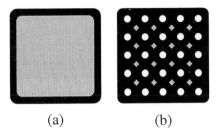

(a)   (b)

Fig. 4.20. Schematic illustration of the nucleation of superconductivity in huge arrays of antidots placed in a perpendicular magnetic field. In a conventional scenario, the surface superconductivity 'rings' of width $\xi(T)$ are overlapping for the antidot lattice (**b**), thus providing the nucleation of superconductivity across the whole sample at a field above the bulk upper critical field $H_{c2}(T)$. In a reference non-patterned film (**a**) the surface superconductivity sheath at the edges of the sample cannot make the bulk of the material superconducting

Therefore, in this case, the nucleation field $H_{c3}^{*}(T)$ is expected to play the role of the bulk critical field. This makes it possible to enhance the critical field above the bulk value $H_{c2}(T)$ up to $H_{c3}^{*}(T)$ in laterally nanostructured films in a perpendicular magnetic field. At first sight it seems, however, impossible to expect an enhancement factor $\eta = H_{c3}^{*}/H_{c2}$ higher than 1.69 (see Fig. 4.8). As we show below, this expectation turns out to be incorrect and much higher enhancement factors - up to 3.6 - can be achieved in films with dense AL's.

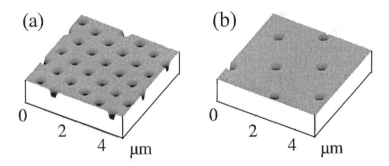

Fig. 4.21. AFM picture of the Pb films with a square antidot lattice with antidot size a=0.4 μm. The lattice period d is 1 μm in sample B (**a**) and 2 μm in sample D (**b**)

We shall analyze the $T_c(H)$ behavior in Pb films with a square AL made by electron-beam lithography on a photoresist layer and standard lift-off processing [8]. The AFM pictures of two of these samples are shown in Fig. 4.21. The antidots have the form of squares with rounded corners. All films (A-E) consist of a 50 nm thick Pb layer covered by a 20 nm protective Ge layer. Besides the reference non-perforated film E, all other samples (A-D) were laterally nanopatterned with a square AL. Samples A, B and C have the same AL period $d=1\,\mu$m, but an antidot size $a=0.2\,\mu$m (sample A), $a=0.4\,\mu$m (sample B) and $a=0.7\,\mu$m (sample C).

The influence of the antidot spacing on the surface superconductivity is analyzed by comparing $H^*_{c3}(T)$ of films B and D (see Fig. 4.21) having the same antidot size ($a=0.4\,\mu$m) but different periods $d=1\,\mu$m (dense AL) and $d=2\,\mu$m (sparse AL).

The residual resistivity at 8 K of the reference Pb film gives the mean free path $\ell=26$ nm. The upper critical field $H_{c2}(T)$ of this film shows the expected linear variation with temperature (Eq. (4.23)), which fits to a value $\xi(0)=38$ nm. This is quite close to the estimated dirty limit coherence length $\xi(0) = 0.865(\xi_o\,\ell)^{1/2} \approx 40$ nm, calculated from the known BCS coherence length of Pb ($\xi_o=83$ nm) [1]. The superconducting transition temperatures T_{c0} for all samples A-E range between 7.164 K-7.186 K. Therefore, nanostructuring has changed T_{c0} of samples A-D only very slightly, compared to $T_{c0}=7.2$ K for bulk Pb. The penetration depth $\lambda(0) \approx 46$ nm is calculated for our films from $\lambda(0) = 0.64\,\lambda_L(0)\sqrt{\xi_o/\ell}$, with $\lambda_L(0)=37$ nm [1]. The GL parameter $\kappa = \lambda/\xi = 1.2$, which implies that these Pb films are Type-II superconductors.

The critical fields $H^*_{c3}(T)$ are shown in Fig. 4.22 (solid lines). The $H^*_{c3}(T)$ values were determined by measuring the resistive transition as a function of temperature at a fixed magnetic field. The used criterion was set at 10 % of the normal state resistance at 8 K. The $H^*_{c3}(T)$ curves show the following structure: in *low fields*, a square-root $H^*_{c3}(T) \propto (1 - T/T_{c0})^{1/2}$ background, corresponding to a parabolic suppression of $T_c(H)$ (Eq. (4.48)), is seen that is superimposed with *collective oscillations* with a periodicity given by one flux quantum per AL unit cell, Φ_0/d^2 (=2.07 mT for $d=1\,\mu$m and 0.517 mT for $d=2\,\mu$m). At *higher fields*, there is a crossover from the square-root to a linear $H^*_{c3}(T)$ background. Superimposed with this linear background, *single-object-like oscillations* [40], reminiscent of the individual antidot oscillatory $T_c(H)$, are observed (most clearly seen in Fig. 4.22 (b) and in Fig. 4.27 (a)). Their period is in agreement with approximately one flux quantum per antidot area, Φ_0/a^2. The straight dashed and dotted lines in Fig. 4.22 correspond to $H_{c2}(T)$ and $H_{c3}(T)$, respectively. The dash-dotted line shows the square-root background of $H^*_{c3}(T)$, which appears due to the finite strip width w (Eq. (4.48)), and is only relevant in the low field regime.

4.7.1   Commensurability Effects at Low Magnetic Fields

Fig. 4.23 (a) shows the low magnetic field parts of the $H^*_{c3}(T)$ (or $T_c(H)$) phase boundaries for samples A-D with an AL. We observe a square-root

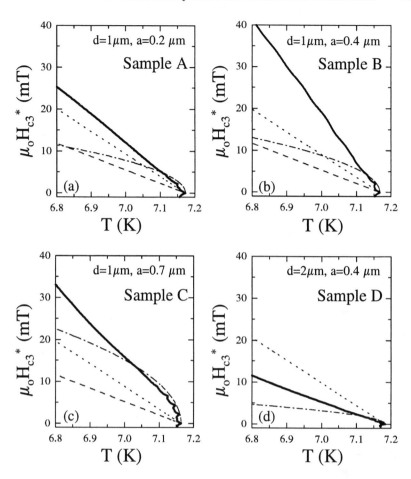

Fig. 4.22. *Solid lines* : measured critical fields $H_{c3}^*(T)$ in a perpendicular magnetic field for Pb films with a square antidot lattice with period d and antidot size a. *Dotted lines* : expected third critical field $H_{c3}(T)$ for a plane superconductor-vacuum boundary with the magnetic field along the boundary ($\eta = H_{c3}^*(T)/H_{c2}(T)=1.69$). *Dashed lines* : Upper critical field of the reference film E without antidots (Eq. (4.23)). *Dash-dotted curves* : square root-like behavior (Eq. (4.48)), expected for a wire network with line thickness $w = d - a$. (a)-(c) Results for samples A-C with dense antidot arrays, where the enhancement of H_{c3}^* (solid lines) is much higher ($\eta =H_{c3}^*/H_{c2} > 1.69$) than for sparse antidot arrays. (d) Results for sample D with a sparse antidot array, where the enhancement of $\eta=H_{c3}^*(T)/H_{c2}$ is smaller than 1.69

background of $H_{c3}^*(T)$ (shown with a dashed line for sample C) that is superposed with a cusplike structure, periodic in field. The background suppression of T_c is caused by the finite width of the superconducting strips between the antidots and follows Eq. (4.48) for a thin strip in a perpendicular magnetic field. The low field cusps in the $H_{c3}^*(T)$ curve (Fig. 4.23 (a)) are reminiscent of the $T_c(H)$ phase boundary for superconducting networks, as studied previously [80–83].

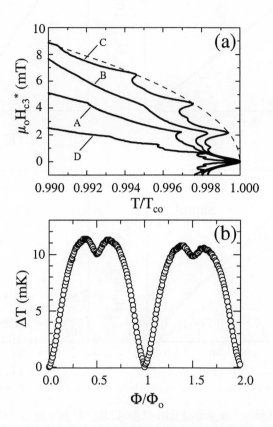

Fig. 4.23. (a) The low field part of the $H_{c3}^*(T)$ phase boundaries for samples A-D illustrating the appearance of 'collective' oscillations and the presence of the square-root $H_{c3}^*(T)$ background (Eq. (4.48)), shown by the dashed line for sample C. (b) After subtraction of the square-root background, the $T_c(H)$ phase boundary (shown for sample C) resembles the lowest level of the Hofstadter butterfly

For these superconducting networks, the lowest Landau level of the problem is the lowest level of the Hofstadter butterfly [84] which was nicely demonstrated by the experiments of Pannetier *et al.* [80]. The lowest level of the Hofstadter butterfly, and therefore also the $T_c(H)$ phase boundary of a square superconducting network, has a remarkable structure (Fig. 4.24). Cusps in $T_c(H)$ show up at specific values of the magnetic field, as a result of commensurability between the vortex lattice and the underlying periodic

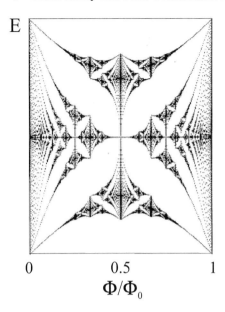

E

0   0.5   1

Φ/Φ_0

Fig. 4.24. Hofstadter butterfly for a square lattice (after Ref [84]): the lowest energy level corresponds to the $T_c(H)$ phase boundary for a square superconducting network

network. At integer applied flux $\Phi/\Phi_0 = L$ (where Φ is defined per area of a unit cell of the network) the most pronounced cusps appear. At rational flux $\Phi/\Phi_0 = p/q$ smaller cusplike minima are observed, corresponding to stable vortex configurations, like for example the checker-board pattern at $p/q=1/2$. When we subtract the montonic background, related to the finite width of the strips between the antidots, from the low field part of our measured $T_c(\Phi)$ curve (Fig. 4.23 (a)), we see the shape of the lowest level of the Hofstadter butterfly appearing (see Fig. 4.23 (b)).

The existence of cusplike minima at integer fluxes can also be interpreted as follows. When placing a superconducting network in a magnetic field that generates a non-integer number of flux quanta per plaquette, two different states $L\Phi_0$ and $(L+1)\Phi_0$ will inevitably be realized. Then, in the simplest London limit, for $L < \Phi/\Phi_0 < L+1$, the kinetic energy E, defining the shift of T_c in a magnetic field, becomes

$$E \propto \left(\frac{\Phi}{\Phi_0} - L\right)\left(L+1 - \frac{\Phi}{\Phi_0}\right)^2 + \left(L+1 - \frac{\Phi}{\Phi_0}\right)\left(L - \frac{\Phi}{\Phi_0}\right)^2 \quad (4.52)$$

$$\propto \left(\frac{\Phi}{\Phi_0} - L\right)\left(L+1 - \frac{\Phi}{\Phi_0}\right)$$

Here the factors $\Phi/\Phi_0 - L$ and $L+1 - \Phi/\Phi_0$ are the corresponding numbers of $(L+1)\Phi_0$ and $L\Phi_0$ states, respectively. While for a single loop parabolic minima are present in $T_c(H)$ at integer flux $\Phi/\Phi_0 = L$ (see dashed line in Fig. 4.25), Eq. (4.52) describes the set of intersecting parabolae forming cusps at integer flux for a network. These cusps compose the backbone of the Hofstadter butterfly which adds to the dependence, given by Eq. (4.52), new smaller cusps at rational fields.

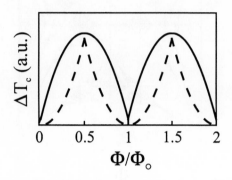

Fig. 4.25. Schematic representation of the $T_c(\Phi)$ phase boundary for a single loop, the Little-Parks oscillatory $T_c(\Phi)$ curve (*dashed line*), and for a network composed of a large amount of such loops (*solid line*)

Following the 'red thread' throughout this chapter, it is interesting to take a look at the evolution of the $T_c(\Phi)$ phase boundary when we increase the size of a 2D network starting from a single nanoplaquette (a loop) and ending with a huge array composed of such plaquettes (an infinite network) (see Fig. 4.26). Also here, it is seen that the parabolic minima at integer flux $\Phi/\Phi_0 = L$ change into cusplike minima [85]. At the same time, the maximum at $\Phi/\Phi_0 = 1/2$ transforms into a smaller cusplike minimum when the size of the array is increased.

The parabolic background superimposed with the $T_c(H)$ oscillations (see Fig. 4.23 (a)) gives us the possibility to check that, in nanostructured samples with an AL, the coherence length remains the same as in the reference non-perforated sample. Indeed, from the square-root envelope of the $H_{c3}^*(T)$ curve close to T_c (see the dash-dotted lines in Fig. 4.22) we determine $\xi(0)$ from Eq. (4.48), which is applicable when $w < \xi(T)$ ($w \equiv d - a$ is the width of the strips between the antidots). From this analysis we obtain the same coherence length $\xi(0)=(38\pm5)$ nm for all samples A-D. Moreover, the latter coincides with $\xi(0)=38$ nm determined earlier from the linear slope of the $H_{c2}(T)$ dependence of the reference film. This proves that the important parameter $\xi(0)$ was not influenced by nanopatterning. Therefore, the very large values of the enhancement factor (up to $\eta=3.6$) that we will show further on, cannot be attributed just to the reduction of $\xi(0)$ due to the possible presence of defects created due to nanostructuring.

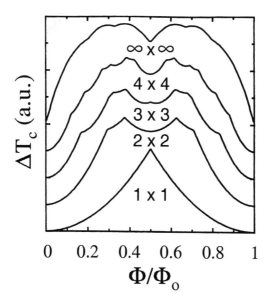

Fig. 4.26. Evolution of the $T_c(H)$ phase boundary (London limit) when increasing the number of plaquettes in a 2D network from one single loop (1×1) to an infinite array ($\infty \times \infty$)

4.7.2   The η-enhancement in High Magnetic Fields

At higher fields, a crossover to a different regime takes place. In this regime, the $H_{c3}^*(T)$ phase boundary resembles the calculated phase boundary for a single antidot. It shows weak oscillations, quasi-periodic in field, superimposed on a linear background. However, the slope of this linear background appears to be much steeper than expected. In all films A-C with *dense* antidot arrays, the linear background contribution to $H_{c3}^*(T)$ curve lies *substantially higher* than the $H_{c2}(T)$ line (Eq. (4.23)) of the reference film E (see the dashed lines in Fig. 4.22). For *sparse* antidot arrays (sample D) the $H_{c3}^*(T)$ curve is very close to $H_{c2}(T)$ of the reference sample E. This can be seen more clearly in Fig. 4.27, where the enhancement $\eta = H_{c3}^*(T)/H_{c2}(T)$ is plotted for samples B ($d=1\,\mu$m and $a=0.4\,\mu$m, dense array) and D ($d=2\,\mu$m and $a=0.4\,\mu$m, sparse array). The dashed line indicates the value $\eta = 1.69$ which, according to the theory, should be the maximum possible nucleation field enhancement for a single antidot in a superconducting plain film.

Surprisingly, the enhancement factor η found for dense antidot arrays (samples A-C) reaches values from 2.8 to 3.6, i.e. up to more than 200 % of the expected maximum enhancement. In sparse antidot arrays, the usual enhancement $\eta < 1.69$ was retrieved.

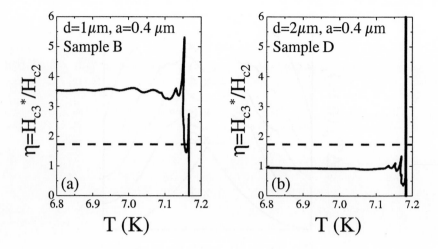

Fig. 4.27. Enhancement factor $\eta=H_{c3}^*(T)/H_{c2}(T)$ as a function of temperature obtained from the data in Fig. 4.22 (**b**) and Fig. 4.22 (**d**) after division by $H_{c2}(T)$ of the reference non-perforated Pb film (dashed line in Fig. 4.22). The dashed line shows the value $\eta=1.69$. (**a**) for the dense AL (sample B) the enhancement η is much larger than the expected maximum value 1.69. (**b**) For a sparse AL (sample D) η stays well below the $\eta = 1.69$ line

In the high field regime, the estimated width of the superconducting ring around two neighboring antidots, i.e. $\xi(T)$, is definitely smaller than the width w of the strips between the antidots. This implies that a simple overlap of these rings cannot be the cause of the transition to the superconducting state, seen as a single resistance drop. Moreover, the $H_{c3}^*(T)$ line was measured using a sufficiently low criterion (10 % of the normal-state resistance R_n). Therefore, the observed enhancement of $\eta = H_{c3}^*(T)/H_{c2}(T)$ in the high field regime can only be explained by invoking a nucleation of superconductivity in the bulk area between the antidots at much higher fields than expected. For a tentative explanation of the large enhancement of η in that part of the superconducting films with dense antidot arrays, the following scenario might be valid. For a bulk superconductor the lowest Landau level $E_{LLL}(H)$ is the well-known lowest level $\hbar\omega/2$ (i.e. $\eta=1$) in a single parabolic potential well (see Fig. 4.28 (**a**)). Adding one plane superconductor/vacuum interface creates a mirror image potential (Fig. 4.28 (**b**)) thus resulting in a lower level $0.59\,\hbar\omega/2$ (i.e. $\eta^{-1}=1/1.69=0.59$) in a double potential (Fig. 4.28 (**b**)). It is reasonable to expect then a further reduction of η^{-1} if the potential is formed by many overlapping parabolic potentials (see Fig. 4.28 (**c**)). The latter might explain the large reduction of η^{-1} and consequently a large increase of η in superconducting films with an AL, since the antidots introduce many superconductor/vacuum interfaces needed to form the potential shown schematically in Fig. 4.28 (**c**) . To confirm these tentative explanations, however, calculation of the lowest Landau level $E_{LLL}(H)$ is needed for dense AL, including the presence of broad superconducting strips between the antidots.

Since the $H_{c3}^*(T)$ line is measured using a criterion of 10 % of the normal state resistance R_n, it is clear that there is a global development of the superconducting state in the area between the antidots at substantially higher fields than expected. In this case, for calculating the flux, the integration area in $\Phi = \int b(\mathbf{r})d\mathbf{S}$ ($b(\mathbf{r})$ is the local magnetic field) is reduced to the effective area of the antidots. This might explain the crossover from the low field regime (where the whole nanoplaquette is transparent for the applied field) to the high field regime, where the flux lines are located only at the antidots. Such a change of the flux penetration pattern corresponds to the change in the $T_c(\Phi)$ period from Φ_0/d^2 to Φ_0/a^2.

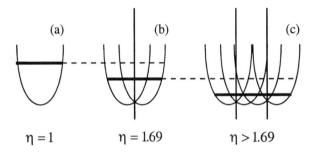

Fig. 4.28. A possible explanation of the $\eta = H_{c3}^*(T)/H_{c2}(T)$ enhancement. The solid horizontal lines represent the lowest Landau level $E_{LLL} = \eta^{-1}\hbar\omega/2$ for the given geometry. (**a**) bulk superconductor with no interfaces (η=1); (**b**) semi-infinite superconducting slab (η=1.69); (**c**) superconducting film with a dense AL. The antidots introduce many superconductor/vacuum interfaces ($\eta > 1.69$)

Since the strips forming a network are too narrow to allow the formation of vortices in the lines (*interstitial* vortices), the connectivity C of a superconducting network is not changed in the presence of magnetic field. When the strips get broader, and the opening in the periodically repeated nanoplaquette becomes smaller, much more superconducting material is left between the openings, thus realizing a regular pinning array rather than a network. In a regular array of antidots as pinning centers, for each characteristic size of antidots a there is a saturation number $n_s \approx a/\xi(T)$ [86]. This number gives the maximum number of flux quanta which can be trapped by an antidot with size a. If the applied field generates a flux per plaquette (i.e. per antidot in antidot lattices) which is smaller than $n_s \Phi_0$, then all the vortices are pinned by the antidots themselves, leading to the formation of Φ_0, $2\Phi_0$, ..., $n_s \Phi_0$ pinned vortex lattices [87]. In this case the connectivity remains the same, and it is given by the number of antidots in the film. If the generated flux per plaquette exceeds $n_s \Phi_0$, then the antidots are saturated and the connectivity is spontaneously increased due to the formation of interstitial vortices [8,9].

To summarize the experimental observations, a dramatic difference is found for the enhancement factor $\eta = H_{c3}^*(T)/H_{c2}(T)$ in films with a sparse and a dense AL. In the latter η exceeds by far (more than 200 %) the value 1.69, which is the expected maximum value for a single antidot. We think that this enhancement of η may be related to the reduction of the lowest Landau level in a potential formed by many overlapping parabolae (see Fig. 4.28 (**c**)).

Conclusions

Topological aspects of the nucleation of the superconducting order parameter Ψ play an essential role in defining the flux confinement in Type-II superconductors. We have analyzed the interplay between the $|\Psi|$ connectivity and the nucleation critical field $H_{c3}^*(T)$ in different superconducting structures. We started from a homogeneous bulk superconductor, where the connectivity is determined by positions of the zeros of $|\Psi|$ and their degeneracy. The Meissner state can then be considered as the singly connected state, while the Abrikosov mixed state exemplifies the case of ultimate connectivity, with a maximum possible number of zeros of $|\Psi|$ coinciding with the normal cores of Φ_0-vortices.

Further on we argued that, anticipating the formation of the normal cores, we can prefabricate microholes at the known locations, hereby changing the connectivity accordingly. Nanostructuring can be made by assembling nanoplaquettes in different ways: from single (loops, dots, etc.) plaquettes, over 1D or 2D clusters, to huge arrays of nanoplaquettes, like nanoengineered antidot lattices which cover macroscopically large areas.

We have demonstrated the substantial change of the $T_c(H)$ boundary in all these structures, in comparison with reference homogeneous bulk samples. In loops and in 1D and 2D clusters of loops, the vortices in general obey the conditions imposed by the precise sample topology, resulting in a strongly topology dependent critical field. It means that the phrase 'the critical fields of a superconductor are determined by the material', often mentioned in textbooks on superconductivity, is not correct, since *for the same material the $T_c(H)$ line is substantially dependent on the topology of the flux and condensate confinement.*

We have shown a spontaneous change of connectivity C in the 1D cluster of loops, namely in the double loop system. In a certain field interval, this type of sample spontaneously evolves from connectivity C = 2 to connectivity C = 1, by creating a normal spot in the middle strip.

In 2D clusters of loops and dots, we have analyzed the interplay at the phase boundary between the degenerate GVS and the Φ_0-vortices, where vortices are confined in the prefabricated pattern of microholes (antidots). Finally, we analyzed the effect of connectivity on $H_{c3}^*(T)$ by considering huge arrays of antidots. In dense antidot lattices, a substantial increase of $H_{c3}^*(T)$ above the expected critical field for a single antidot can be obtained, while for sparse antidot arrays the ratio $H_{c3}^*(T)/H_{c2}(T)$ always stays below the expected value 1.69. All examples, treated in this chapter, clearly demonstrate the importance of connectivity and sample topology in defining the critical parameter $T_c(H)$.

Acknowledgements

The authors are thankful to the FWO-Vlaanderen, the Flemish Concerted Action (GOA), the Bilateral TOURNESOL project, the Belgian Inter- University Attraction Poles (IUAP), and the ESF programme VORTEX for the financial support. Discussions with H. J. Fink, J. O. Indekeu, M. J. Van Bael, T. Puig, E. Rosseel, J. G. Rodrigo, J. T. Devreese, V. M. Fomin, F. M. Peeters, and Y. Bruynseraede are gratefully acknowledged.

References

1. P.-G. de Gennes: *Superconductivity of Metals and Alloys* (Benjamin, New York, 1966)
2. M. Tinkham: *Introduction to Superconductivity* (McGraw Hill, New York, 1975)
3. A.A. Abrikosov: *Fundamentals of the Theory of Metals* (North-Holland, Amsterdam, 1988)
4. D. Saint-James, P.-G. de Gennes: Phys. Lett. **7**, 306 (1963)
5. V.V. Moshchalkov: Solid State Comm. **77**, 389 (1991)
6. J.J. Palacios: Phys. Rev. B **58**, R5948 (1998)
7. V.A. Schweigert, F.M. Peeters, P.S. Deo: Phys. Rev. Lett. **81**, 2783 (1998)
8. M. Baert, V.V. Metlushko, R. Jonckheere, V.V. Moshchalkov, Y. Bruynseraede: Phys. Rev. Lett. **74**, 3269 (1995)
9. E. Rosseel, M.J. Van Bael, M. Baert, R. Jonckheere, V.V. Moshchalkov, Y. Bruynseraede: Phys. Rev. B **53**, R2983 (1996)
10. P. Michael, D.S. McLachlan: J. Low Temp. Phys. **14**, 607 (1974)
11. R.B. Dingle: Proc. R. Soc. London, Ser. A **211**, 500 (1952)
12. M. Abramowitz, I.A. Stegun: *Handbook of Mathematical Functions* (Dover, New York, 1970)
13. L. Landau: Z. Phys. **64**, 629 (1930)
14. V.V. Moshchalkov, M. Dhallé, Y. Bruynseraede: Physica C **207**, 307 (1993)
15. A.A. Abrikosov: Sov. Phys. JETP **5**, 1174 (1957)
16. W.M. Kleiner, L.M. Roth, S.H. Autler: Phys. Rev. **133**, A1226 (1964)
17. S. Takács: Czech. J. Phys. B **19**, 1365 (1969)
18. H.J. Fink, W.C.H. Joiner: Phys. Rev. Lett. **23**, 120 (1969)
19. S.V. Yampolskii, F.M. Peeters: Preprint
20. E. Montevecchi, J.O. Indekeu: Preprint
21. J.O. Indekeu, J.M.J. van Leeuwen: Phys. Rev. Lett. **75**, 1618 (1995)
22. J.O. Indekeu, J.M.J. van Leeuwen: Physica C **251**, 290 (1995)

23. R.B. Dingle: Proc. R. Soc. London, Ser. A **212**, 47 (1952)
24. W.A. Little, R.D. Parks: Phys. Rev. Lett. **9**, 9 (1962)
25. R.D. Parks, W.A. Little: Phys. Rev. **133**, A97 (1964)
26. F. London: *Superfluids* (John Wiley & Sons and Inc., New York, 1950)
27. M. Daumens, C. Meyers, A.I. Buzdin: Phys. Lett. A **248**, 445 (1998)
28. D. Saint-James: Phys. Lett. **15**, 13 (1965)
29. O. Buisson, P. Gandit, R. Rammal, Y.Y. Wang, B. Pannetier: Phys. Lett. A **150**, 36 (1990)
30. V.V. Moshchalkov, X.G. Qiu, V. Bruyndoncx: Phys. Rev. B **55**, 11 793 (1997)
31. V.V. Moshchalkov, L. Gielen, C. Strunk, R. Jonckheere, X. Qiu, C. Van Haesendonck, Y. Bruynseraede: Nature (London) **373**, 319 (1995)
32. A. Bezryadin, A.I. Buzdin, B. Pannetier: Phys. Lett. A **195**, 373 (1994)
33. V.M. Fomin, V.R. Misko, J.T. Devreese, V.V. Moshchalkov: Phys. Rev. B **58**, 11 703 (1998)
34. A. Houghton, F.B. McLean: Phys. Rev. B **19**, 172 (1965)
35. A.P. van Gelder: Phys. Rev. Lett. **20**, 1435 (1968)
36. V.M. Fomin, J.T. Devreese, V.V. Moshchalkov: Europhys. Lett. **42**, 553 (1998)
37. V.M. Fomin, J.T. Devreese, V.V. Moshchalkov: Europhys. Lett. **46**, 118 (1999)
38. V.A. Schweigert, F.M. Peeters: Phys. Rev. B **60**, 3084 (1999)
39. S.N. Klimin, V.M. Fomin, J.T. Devreese, V.V. Moshchalkov: Solid State Comm. **111**, 589 (1999)
40. A. Bezryadin, B. Pannetier: J. Low Temp. Phys. **98**, 251 (1995)
41. R. Benoist, W. Zwerger: Z. Phys. B **103**, 377 (1997)
42. V. Bruyndoncx, L. Van Look, M. Verschuere, V.V. Moshchalkov: Phys. Rev. B **60**, 10 468 (1999)
43. B.J. Baelus, F.M. Peeters, V.A. Schweigert: cond-mat/9910030
44. R.P. Groff, R.D. Parks: Phys. Rev. **176**, 567 (1968)
45. H.J. Fink: Phys. Rev. **177**, 732 (1969)
46. H.A. Schultens: Z. Phys. **232**, 430 (1970)
47. V. Bruyndoncx, C. Strunk, V.V. Moshchalkov, C. Van Haesendonck, Y. Bruynseraede: Europhys. Lett. **36**, 449 (1996)
48. S. Alexander, E. Halevi: J. Phys. (Paris) **44**, 805 (1983)
49. C.C. Chi, P. Santhanam, P.E. Blöchl: J. Low Temp. Phys. **88**, 163 (1992)
50. H.J. Fink, A. López, R. Maynard: Phys. Rev. B **26**, 5237 (1982)
51. P.-G. de Gennes: C. R. Acad. Sci. Ser. II **292**, 279 (1981)
52. S. Alexander: Phys. Rev. B **27**, 1541 (1983)
53. J.P. Straley, P.B. Visscher: Phys. Rev. B **26**, 4922 (1982)
54. S.B. Haley, H.J. Fink: Phys. Lett. **102A**, 431 (1984)
55. C. Strunk, V. Bruyndoncx, V.V. Moshchalkov, C. Van Haesendonck, Y. Bruynseraede, R. Jonckheere: Phys. Rev. B **54**, R12 701 (1996)
56. C. Ammann, P. Erdös, S.B. Haley: Phys. Rev. B **51**, 11 739 (1995)
57. J.I. Castro, A. López: Phys. Rev. B **52**, 7495 (1995)
58. R.D. Parks: Science **146**, 1429 (1964)
59. E.M. Horane, J.I. Castro, G.C. Buscaglia, A. López: Phys. Rev. B **53**, 9296 (1996)
60. J. Berger, J. Rubinstein: Phys. Rev. Lett. **75**, 320 (1995)
61. J. Berger, J. Rubinstein: Phil. Trans. R. Soc. Lond. A **355**, 1969 (1997)
62. J. Berger, J. Rubinstein: Physica C **288**, 105 (1997)
63. J. Berger, J. Rubinstein: Phys. Rev. B **56**, 5124 (1997)

64. J. Simonin, D. Rodrigues, A. López: Phys. Rev. Lett. **49**, 944 (1982)
65. H.J. Fink, S.B. Haley: Phys. Rev. B **43**, 10 151 (1991)
66. H.J. Fink, V. Grünfeld: Phys. Rev. B **31**, 600 (1985)
67. H.J. Fink, O. Buisson, B. Pannetier: Phys. Rev. B **43**, 10 144 (1991)
68. H.J. Fink: Phys. Rev. B **45**, 4799 (1992)
69. H.J. Fink, S.B. Haley: Phys. Rev. Lett. **66**, 216 (1991)
70. J. Simonin, C. Wiecko, A. López: Phys. Rev. B **28**, 2497 (1983)
71. T.M. Larson, S.B. Haley: Physica B **194-196**, 1425 (1994)
72. T.M. Larson, S.B. Haley: Appl. Superconductivity **3**, 573 (1995)
73. T.M. Larson, S.B. Haley: J. Low Temp. Phys. **107**, 3 (1997)
74. T. Puig, E. Rosseel, M. Baert, M.J. Van Bael, V.V. Moshchalkov, Y. Bruynser-aede: Appl. Phys. Lett. **70**, 3155 (1997)
75. T. Puig, E. Rosseel, L. Van Look, M.J. Van Bael, V.V. Moshchalkov, Y. Bruynseraede, R. Jonckheere: Phys. Rev. B **58**, 1998 (1998)
76. H.J. Fink, A.G. Presson: Phys. Rev. **151**, 219 (1966)
77. V. Bruyndoncx, J.G. Rodrigo, T. Puig, L. Van Look, V.V. Moshchalkov, R. Jonckheere: Phys. Rev. B **60**, 4285 (1999)
78. A.K. Geim, I.V. Grigorieva, S.V. Dubonos, J.G.S. Lok, J.C. Maan, A.E. Filippov, F.M. Peeters: Nature (London) **390**, 259 (1997)
79. H.T. Jadallah, J. Rubinstein, P. Sternberg: Phys. Rev. Lett. **82**, 2935 (1999)
80. B. Pannetier, J. Chaussy, R. Rammal, J.C. Villegier: Phys. Rev. Lett. **53**, 1845 (1984)
81. C.W. Wilks, R. Bojko, P.M. Chaikin: Phys. Rev. B **43**, 2721 (1991)
82. R. Théron, J.B. Simond, C. Leemann, H. Beck, P. Martinoli, P. Minnhagen: Phys. Rev. Lett. **71**, 1246 (1993)
83. J.E. Mooij, G. Schön: in *Single Charge Tunneling*, ed. by H. Grabert, M.H. Devoret (Plenum Press, New York, 1992), and references therein
84. D.R. Hofstadter: Phys. Rev. B **14**, 2239 (1976)
85. E. Rosseel: (1998), *Critical parameters of superconductors with an antidot lattice*, PhD thesis, Katholieke Universiteit Leuven, Leuven
86. G.S. Mkrtchyan, V.V. Shmidt: Sov. Phys. JETP **34**, 195 (1972)
87. V.V. Moshchalkov, M. Baert, V.V. Metlushko, E. Rosseel, M.J. Van Bael, K. Temst, R. Jonckheere, Y. Bruynseraede: Phys. Rev. B **54**, 7385 (1996)

5 Zero Set of the Order Parameter, Especially in Rings

Jorge Berger

Ort Braude College, 21982 Karmiel, Israel

Abstract. We discuss the conditions and the positions where the order parameter vanishes in a multiply connected sample. The first sections are extensions of the de Gennes–Alexander formalism, but most of the chapter deals with samples of finite width. There are several surprising predictions, e.g.: a vortex can be much thinner than the coherence length, and its position may radically change with the fluxoid number; there are critical points inside the superconducting area in the phase diagram. Experimental verification is proposed.

5.1 Introduction: The Price of Connectivity

The behavior of superconductors can be described by means of the Ginzburg–Landau (GL) formalism. This formalism was reviewed in Sects. 1.2 and 2.1.1, but in order to introduce our notation we rewrite the starting equations. According to this formalism, the equilibrium magnetic potential and the superconducting order parameter are such that they minimize the GL potential

$$G = \int \left[\mu(-|\psi|^2 + |\psi|^4/2) + |(i\nabla - \widetilde{\mathbf{A}})\psi|^2 \right] + (\kappa^2/\mu) \int |\nabla \times \widetilde{\mathbf{A}} - 2\pi \mathbf{H}_e/\Phi_0|^2 . \tag{5.1}$$

Here $\psi = |\psi|e^{i\varphi}$ is the order parameter, with the normalization introduced after (2.4), $\mathbf{H}_e$ is the applied field, Φ_0 the quantum of flux and κ the GL parameter; μ is an abbreviation for ξ^{-2}, where ξ is the coherence length, and $\widetilde{\mathbf{A}} = 2\pi \mathbf{A}/\Phi_0$, with $\mathbf{A}$ the magnetic potential. The first integral in (5.1) is over the sample volume and the second integral is over the entire space. Note that in many sources $\widetilde{\mathbf{A}}$ appears with the opposite sign; these sources disregard the awkward fact that the electron charge is negative. The normalizations in (5.1) take for granted that the temperature is below T_c, the normal/superconducting transition temperature when no magnetic field is applied.

The first GL equation is obtained by minimizing G with respect to ψ. Variation gives

$$(i\nabla - \widetilde{\mathbf{A}})^2 \psi = \mu(1 - |\psi|^2)\psi \tag{5.2}$$

inside the sample, with the boundary requirement that the normal component of $(i\nabla - \widetilde{\mathbf{A}})\psi$ vanishes. The second GL equation (obtained by minimizing G w.r.t. $\mathbf{A}$) is not required when the sample is very thin compared to $\kappa\xi$, and will be postponed until Sect. 5.4.1.

Inspection of the Ginzburg–Landau potential G shows that it contains only two terms that depend on the phase φ, and both are non-negative: $|(i\nabla - \widetilde{\mathbf{A}})\psi|^2$, which can be further decomposed into $(\nabla|\psi|)^2 + |(\nabla\varphi + \widetilde{\mathbf{A}})\psi|^2$, and H_i^2, with $\mathbf{H}_i = \nabla \times \mathbf{A} - \mathbf{H}_e$. It is therefore clear that, if there is some φ such that the φ-dependent terms vanish, this φ will be the most appropriate candidate for minimization.

One situation in which this happens is when the applied field vanishes within the sample: then we can pick φ such that $\nabla\varphi = -\widetilde{\mathbf{A}}$. (In this case, the current, H_i and the total field also vanish.) This case is considered in detail in Chap. 3. Another opportunity like this appears when the sample is a narrow wire (or a narrow shell parallel to the magnetic field), as in Fig. 5.1. Since the normal component of $(i\nabla - \widetilde{\mathbf{A}})\psi$ is required to vanish at the boundary, in the case of a thin wire we may take this component as negligible everywhere, so that we only have to require that the tangential component vanishes too. This is achieved if

$$\varphi(s') - \varphi(s) = -\int_s^{s'} \widetilde{\mathbf{A}} \cdot d\mathbf{r} , \qquad (5.3)$$

where s is the arc length and the integration is performed along the wire. Again, we get rid of the two phase dependent terms and are left with a minimization over $|\psi|$ only.

However, let us imagine that we fill the gap between a and a' with superconducting material. Instead of enhancing superconductivity, we add a constraint: $\psi(a)$ and $\psi(a')$, and in particular their phases (up to a multiple of 2π), must be the same. The minimizer of the unconstrained problem may therefore be ruled out, and superconductivity will have to compromise to a minimizer in the set $\{\psi(a') = \psi(a)\}$. This means that connectivity is a thermodynamic burden. Our question is whether there are cases in which

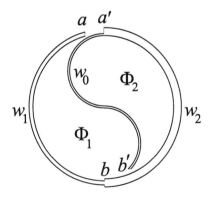

Fig. 5.1. A narrow superconducting wire, with several possible connectivities. The three branches have the same length L and their cross sections are w_0, w_1 and w_2

superconductivity – and therefore connectivity – are broken at some place in order to get rid of this burden.

Sections 5.2–5.4 consider situations in increasing order of generality in the following sense: in Sect. 5.2, wires are thin and have uniform (or piecewise uniform) cross sections; in Sect. 5.3, cross sections will be nonuniform and, in Sect. 5.4, wires will turn into wide surfaces. However, the formalisms of each section are quite independent, so that their consecutive reading is not essential.

5.2   The de Gennes–Alexander (dGA) Approach

The easiest situation for evaluation of the price of connectivity is provided by the dGA case, since the differential equations have already been integrated. The dGA formalism is applicable to networks of thin branches with constant cross section, at a bifurcation from the normal state. For further simplicity, we shall consider cases in which all the branches have the same length, L. Since the free energy decreases with μ, minimization of the free energy is translated here into minimization of the value of μ for which there is a non trivial solution.

Our analysis will be based on Chap. 2, but a word of warning is required. The solution of (2.16) is

$$\psi(s) = \exp(-i\gamma(s))[k_1 \exp(i\sqrt{\mu}s) + k_2 \exp(-i\sqrt{\mu}s)] \tag{5.4}$$

where γ is the integral in (5.3) (without the $-$ sign) and k_1 and k_2 are constants, which are chosen to comply with the values of ψ at the extremes of the branch, $s = 0$ and $s = L$. In general, these are 2 independent conditions and we obtain (2.19); but, if $\sqrt{\mu}L = \pi$ or $\sqrt{\mu}L = 0$, then we obtain a condition for $\psi(L)/\psi(0)$:

$$\psi(L) \exp(i\gamma(L))/\psi(0) = 1 \quad \sqrt{\mu}L = 0 \tag{5.5a}$$
$$\psi(L) \exp(i\gamma(L))/\psi(0) = -1 \quad \sqrt{\mu}L = \pi . \tag{5.5b}$$

If (5.5a) or (5.5b) is fulfilled, there is a family of degenerate solutions. Note that $\sqrt{\mu}L$ is non negative by definition and will never be greater than π, since we can always find a smaller μ with the same value of $\cos(\sqrt{\mu}L)$. If (2.19) is valid, the condition for having a point where $\psi(s) = 0$ along the branch $0 < s < L$ is

$$\psi(L) \exp(i\gamma(L))/\psi(0) < 0 . \tag{5.6}$$

It is interesting to compare this with (5.5b).

5.2.1   One Loop

Figure 5.1 with $a' = a$ has three nodes: b', a and b. The Alexander nodal equations (2.21), after elimination of $\psi(b')$, become

$$[(w_0 + w_1 + w_2)X - w_0/X]\psi(a) - (w_1 e^{i\gamma_1} + w_2 e^{-i\gamma_2})\psi(b) = 0 , \tag{5.7a}$$

$$-(w_1 e^{-i\gamma_1} + w_2 e^{i\gamma_2})\psi(a) + (w_1 + w_2)X\psi(b) = 0 \ , \ (5.7b)$$

where X is shorthand for $\cos(\sqrt{\mu_0}L)$, w_i is the cross section of branch i and γ_i is $\gamma(L)$ evaluated along this branch in the positive mathematical sense; μ_0 is the lowest eigenvalue of this system of equations and gives the value of μ at which the bifurcation from the normal state occurs. Here,

$$\cos(\sqrt{\mu_0}L) = ([w_0 + |w_1 e^{i(\gamma_1+\gamma_2)} + w_2|^2/(w_1+w_2)]/(w_0+w_1+w_2))^{1/2} \ . \ (5.8)$$

Let us now investigate under what conditions ψ vanishes at some point in the branches 1 or 2. From (5.7b), $\psi(b)e^{i\gamma_1}/\psi(a) < 0$ would require $w_1 < w_2$ together with $e^{i(\gamma_1+\gamma_2)} = -1$, which occurs when the flux $\Phi_1 + \Phi_2$ is an integer plus half number of quanta Φ_0. The same analysis can be repeated for branch 2. $\psi(b)$ would vanish if $w_1 = w_2$ and $e^{i(\gamma_1+\gamma_2)} = -1$. For $w_0 > 0$, $\psi(a)$ can never vanish for the lowest eigenvalue. In summary, there is a zero in the order parameter only when the flux enclosed by the circle is an integer plus half, and then the zero appears at the thinner branch. In the symmetric case, the zero appears at the point opposite to the branch ab'. At the branch ab' itself, it is easily seen that $\psi(b')e^{i\gamma_0}/\psi(a) > 0$ and the phase obeys (5.3) along the entire branch, as is physically obvious from the fact that this branch carries no current.

We know already that, for appropriate flux, ψ vanishes somewhere in the thinner outer branch. Let us now check the following: could there be a situation where it is thermodynamically favorable for superconductivity to spontaneously break the connectivity in order to allow (5.3) to be fulfilled everywhere, even though the dGA formalism does not require ψ to vanish? In other words, would it pay to replace the constraint that the sample encloses a given flux by the constraint that ψ has to vanish at some point? The answer seems to be negative: if $e^{i(\gamma_1+\gamma_2)} = -1$, then both constraints are compatible and the solutions coincide; changing $e^{i(\gamma_1+\gamma_2)}$ to a different value will cause μ_0 in (5.8) to decrease (dGA case), whereas the case with broken connectivity is insensitive to the flux and remains at its high value.

Actually, ψ does not have to vanish in order to cope with the price of connectivity. It is enough that ψ be small in some region to enable a fast change of φ in that region at a low energy price. In [6] it is found that this price is proportional to the minimum of $|\psi|$ (provided that this minimum is small).

The case of a loop without the extra branch is obtained if we set $w_0 = 0$. The expression for the eigenvalue now takes the simpler form

$$\cos(\sqrt{\mu_0}L) = |w_1 e^{i(\gamma_1+\gamma_2)} + w_2|/(w_1 + w_2) \ .$$

The conclusions remain qualitatively the same, except that now, for $w_1 = w_2$ and $\gamma_1 + \gamma_2 = \pi$ (mod 2π), $\cos(\sqrt{\mu_0}L) = 0$ and $\psi(a)$ and $\psi(b)$ are undetermined.

5.2.2   Two Loops

Let us now increase the connectivity in Fig. 5.1 by connecting also b and b'. There are now two nodes, and Alexander's equation becomes

$$(w_0 + w_1 + w_2)X\psi(a) - (w_0e^{i\gamma_0} + w_1e^{i\gamma_1} + w_2e^{-i\gamma_2})\psi(b) = 0 \text{ ,(5.9a)}$$
$$-(w_0e^{-i\gamma_0} + w_1e^{-i\gamma_1} + w_2e^{i\gamma_2})\psi(a) + (w_0 + w_1 + w_2)X\psi(b) = 0 \text{ ;(5.9b)}$$

μ_0 is given by

$$\cos(\sqrt{\mu_0}L) = |w_0e^{i\gamma_0} + w_1e^{i\gamma_1} + w_2e^{-i\gamma_2}|/(w_0 + w_1 + w_2) \ .$$

The value of μ_0 here is generically greater than in (5.8), as expected from the larger connectivity. From (5.9b) we see that, except for a positive factor,

$$\psi(b)e^{i\gamma_0}/\psi(a) \propto (w_0 + w_1e^{i(\gamma_0-\gamma_1)} + w_2e^{i(\gamma_0+\gamma_2)}) \ .$$

For simplicity, let us consider the case $\Phi_1 = \Phi_2 = \phi\Phi_0$, so that

$$\psi(b)e^{i\gamma_0}/\psi(a) \propto (w_0 + w_1e^{2\pi i\phi} + w_2e^{-2\pi i\phi})$$

and, similarly,

$$\psi(b)e^{i\gamma_1}/\psi(a) \propto (w_0e^{-2\pi i\phi} + w_1 + w_2e^{-4\pi i\phi}) \ ,$$

$$\psi(b)e^{-i\gamma_2}/\psi(a) \propto (w_0e^{2\pi i\phi} + w_1e^{4\pi i\phi} + w_2) \ .$$

If $w_1 \neq w_2$, we see that (5.6) can be fulfilled only if ϕ is an integer plus half. In this case, ψ will have a zero in the inner branch if $w_0 < w_1 + w_2$ and will have zeros at both outer branches if $w_0 > w_1 + w_2$. In the symmetric case $w_1 = w_2$, there is a zero in the middle branch for the entire region $\cos(2\pi\phi) < -w_0/(w_1 + w_2)$.

We see that there is a qualitative difference between the symmetric case here (or, say, the case of the double yin-yang in Sect. 2.2.2), where there is a zero at a fixed position for a finite region of ϕ, and the nonsymmetric case, or the case of the single loop, where zeros are present for isolated values of ϕ. In the first cases, the zero may be regarded as a consequence of a geometric symmetry; in the latter cases, it may be regarded as a consequence of the symmetry of Alexander's equation under the transformation $\psi(a) \rightarrow \overline{\psi(a)}, \psi(b) \rightarrow e^{-2i\gamma_0}\overline{\psi(b)}$, where the overline denotes complex conjugation.

5.3   Perturbational Approach

In this section we still deal with a wire (or shell) which is sufficiently narrow, so that ψ will depend only on the arc length and the induced field will be neglected. We first develop a formalism that will enable us to extend the dGA method to samples with nonuniform width and beyond the onset of superconductivity.

For a narrow wire, the boundary condition that the normal component of $\mathrm{Re}[\overline{\psi}(i\nabla - \widetilde{\mathbf{A}})\psi]$ vanishes may be replaced by the physical condition that the current

$$I = -w|\psi|^2(\varphi' + \widetilde{A}) \tag{5.10}$$

will be constant along the wire. In Gaussian units, the current is $c\Phi_0\mu I$ $/[2(2\pi\kappa)^2]$. Here $|\psi|$, φ and the cross section w are functions of the arc length s and $'$ denotes derivative with respect to it; $\widetilde{A}$ is the tangential component of $\widetilde{\mathbf{A}}$. We can use (5.10) to eliminate φ from the free energy. This gives

$$G = \int \left[w\left(-\mu|\psi|^2 + \frac{\mu}{2}|\psi|^4 + (|\psi|')^2\right) + \frac{I^2}{w|\psi|^2} \right] ds , \tag{5.11}$$

from which we obtain the Euler-Lagrange equation

$$(w|\psi|')' + w\mu(|\psi| - |\psi|^3) - I^2/(w|\psi|^3) = 0 . \tag{5.12}$$

In this equation the absolute value of ψ has been isolated from its phase, which enters the equation only through the constant I. This form was used in [6]. However, for the purpose of extension to the case of finite width, we reintroduce φ. Introducing (5.10) into (5.12) gives

$$(w|\psi|')' + w\mu(|\psi| - |\psi|^3) - w|\psi|(\varphi' + \widetilde{A})^2 = 0 . \tag{5.13}$$

In order to determine the two real functions $|\psi|$ and φ, we require an additional equation. This is obtained by taking the derivative of (5.10), which gives

$$w|\psi|(\varphi'' + \widetilde{A}') + (\varphi' + \widetilde{A})(2w|\psi|' + w'|\psi|) = 0 . \tag{5.14}$$

It is interesting to note that in obtaining (5.13) and (5.14) we have not made any assumption as to the value of $\mathrm{Im}[\overline{\psi}(i\nabla - \widetilde{\mathbf{A}})\psi]$ at the boundary. This value controls the normal derivative of $|\psi|$, but since ψ has no room to vary anyway, this condition has no influence in the one dimensional limit.

It can be verified by direct evaluation that the system of equations (5.13) and (5.14) is equivalent to the complex equation

$$(i\partial_s - \widetilde{A})^2\psi + (iw'/w)(i\partial_s - \widetilde{A})\psi = \mu(\psi - |\psi|^2\psi) , \tag{5.15}$$

which is obtained in [15] as the limiting case of a two dimensional problem.

In the following we shall be interested in a loop which encloses flux $\phi\Phi_0$. In this section we choose a gauge for which A is constant, define the perimeter as $2\pi R$ and $\theta = s/R$; we also redefine $'$ as derivative with respect to θ, and μ as $(R/\xi)^2$. With these adimensionalized definitions, (5.15) becomes

$$\mathcal{H}_0\psi + (iw'/w)(i\partial_\theta - \phi)\psi = \mu(\psi - |\psi|^2\psi) , \tag{5.16}$$

with $\mathcal{H}_0 = (i\partial_\theta - \phi)^2$.

5.3.1   Uniform Cross Section

If we set $w' \equiv 0$ in (5.16), we find various solutions. One family is

$$\psi = \sqrt{1 - (m - \phi)^2/\mu}\, e^{-mi\theta} , \tag{5.17}$$

with m integer. A solution like this exists for $\mu \geq (m - \phi)^2$. Another family of solutions is

$$|\psi|^2 = q - (2n\nu^2/\mu)\,\mathrm{cn}^2(\nu\theta, n) , \tag{5.18}$$

with $\nu = K(n)/\pi$ (or a multiple of it), $q = (2/3)[1 + (2n - 1)\nu^2/\mu]$ and

$$\varphi = \sqrt{\frac{q}{8\mu}[\mu^2(q - 2)^2 - 4\nu^4]} \int_0^\theta \frac{d\theta'}{|\psi(\theta')|^2} - \phi\theta . \tag{5.19}$$

Here K is the elliptic integral and cn a Jacobian elliptic function. The value of n has to be such that either ψ vanishes at some place or $\varphi(2\pi) = \varphi(0)$ (mod 2π). This family coalesces with a member of (5.17) for $n = 0$ and the case of broken connectivity is obtained for $(1 + n)\nu^2 = \mu$. By continuously varying n and ϕ, and changing the value of $\varphi(2\pi) - \varphi(0)$ when connectivity is broken, it is possible to bridge continuously between solutions in (5.17) with different winding numbers m. The coalescence of families (5.17) and (5.18) occurs at the lines $6(\phi - m)^2 - 2\mu = 1$. The region $n < 0$ does not give new solutions. For additional families of solutions and cases in which $\nu = iK(1 - n)/\pi$, see [4]. Clearly, it is always possible to add constants to θ and to φ.

Solutions in family (5.17) are local minima of G for $6(\phi - m)^2 - 2\mu < 1$ and saddle points beyond this range [13]; a numerical search indicates that solutions in family (5.18), and in particular the solution that breaks connectivity, are never local minima.

5.3.2   Weakly Nonuniform Cross Section

We saw that for the case of uniform cross section connectivity is never broken. However, we shall see that small deviations from uniformity generically stabilize broken connectivity.

Using (5.17), (5.11) and (5.10), we find that the free energy increases with $|\phi - m|$, so that for $\phi = m + 1/2$ there will be a degenerate ground state, since the states with winding numbers m and $m + 1$ share the lowest free energy. We want to analyze what happens near this situation and therefore write $\phi = m + 1/2 + \epsilon\phi_1$, where ϵ will be our parameter of smallness. For the shape of the cross section we write

$$w(\theta) = w_0 \left(1 + \frac{\epsilon}{2} \sum_{j \neq 0} \beta_j e^{ji\theta}\right) , \tag{5.20}$$

where w_0 is the average width and $\beta_{-j} = \overline{\beta}_j$. We will limit this study to the region close to the onset of superconductivity; more precisely, we require $|\psi|$ to be at most of order $\epsilon^{1/2}$. For $\phi = m + 1/2$ and uniform cross section, this onset occurs at $\mu = 1/4$ and we therefore write $\mu = 1/4 + \epsilon\mu_1$.

We rewrite (5.16) in the form

$$(\mathcal{H}_0 + V)\psi = \mu\psi \tag{5.21}$$

where $\mathcal{H}_0$ is now $(i\partial_\theta - m - 1/2)^2$ and

$$V = -2i[\phi - (m + \tfrac{1}{2})]\partial_\theta + \phi^2 - (m + \tfrac{1}{2})^2 + \frac{iw'}{w}(i\partial_\theta - \phi) + \mu|\psi|^2 .$$

Equation (5.21) is not linear, since V involves $|\psi|$, which will have to be determined self consistently. Since in the limit $\epsilon \to 0$ (5.21) is linear and its general groundstate is any linear combination of $e^{-mi\theta}$ and $e^{-(m+1)i\theta}$, we write

$$\psi = \epsilon^{1/2}\eta(e^{-mi\theta} + ge^{-(m+1)i\theta}) + \epsilon^{3/2}\psi_1 , \tag{5.22}$$

where, without loss of physical generality, we may take η real and non negative. Keeping only terms to $O(\epsilon)$, V becomes

$$V = \epsilon\left\{(2\phi_1 + \frac{1}{2}\sum j\beta_j e^{ji\theta})(-i\partial_\theta + m + \frac{1}{2}) + \frac{\eta^2}{4}|1 + ge^{-i\theta}|^2\right\} . \tag{5.23}$$

Substituting (5.22) and (5.23) into (5.21), collecting the coefficients of $e^{-mi\theta}$ and $e^{-(m+1)i\theta}$ and using $\eta \neq 0$ we obtain

$$\phi_1 - \mu_1 + \frac{\eta^2}{4}(1 + 2|g|^2) = \frac{1}{4}\beta_1 g ,$$

$$-\phi_1 - \mu_1 + \frac{\eta^2}{4}(2 + |g|^2) = \frac{1}{4}\frac{\overline{\beta}_1}{g} . \tag{5.24}$$

From here we learn that only the first harmonic of the shape of $w(\theta)$ is important and that $\beta_1 g$ is real. The temperature for the onset of superconductivity is obtained by setting $\eta \to 0$ and taking the lowest solution in (5.24). This gives $\mu = 1/4 - \sqrt{\phi_1^2 + |\beta_1|^2/16}$.

From (5.24), it is easy to express ϕ_1 and η as functions of $|g|$:

$$\frac{\phi_1}{|\beta_1|} = \frac{1 - |g|^2}{12}\left(\frac{4\mu_1}{|\beta_1|(1 + |g|^2)} - \frac{1}{|g|}\right)$$

$$\frac{\eta^2}{|\beta_1|} = \frac{1}{3}\left(\frac{8\mu_1}{|\beta_1|(1 + |g|^2)} + \frac{1}{|g|}\right) . \tag{5.25}$$

Figure 5.2 shows $|g|$ as a function of the flux for different values of the parameter $\mu_1/|\beta_1|$.

We see from this figure that there are two different regimes: for $\mu_1 < |\beta_1|/2$ (high temperature regime), $|g|$ is a single valued function of the flux, whereas

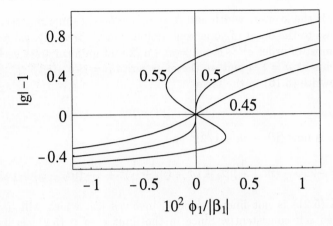

Fig. 5.2. g determines the shape of the order parameter in Eq. (5.22); ϕ_1 measures the deviation of the flux from $(m + 1/2)\Phi_0$. Each curve is marked by its value of $\mu_1/|\beta_1|$. The curve $\mu_1/|\beta_1| = 1/2$ has infinite slope at $\phi_1 = 0$, indicating the presence of a critical point

for $\mu_1 > |\beta_1|/2$ there is a reentrance near $\phi \approx m + 1/2$. The fact that the transition between both regimes occurs at $\mu_1 = |\beta_1|/2$ can be verified by evaluating $d\phi_1/d|g|$ and $d^2\phi_1/d|g|^2$ and noting that both vanish at the same point $|g| = 1$. Therefore, there is a *critical point* at $(\phi = m + 1/2, \mu = 1/4 + \epsilon|\beta_1|/2)$ and we dub it P_2. (The index "2" suggests that this is a second transition as the temperature is lowered; the first one is from normal to superconducting.)

Let us first discuss the case $\mu_1 < |\beta_1|/2$. The factor $(1 - |g|^2)$ ensures that $|g| = 1$ for $\phi_1 = 0$. This means that $e^{-mi\theta} + ge^{-(m+1)i\theta}$ in (5.22) vanishes for $\theta = \pi + \arg g$. We know already that $\beta_1 g$ is real. For $|g| = 1$ and $\phi_1 = 0$, (5.24) gives $\beta_1 g = -4\mu_1 + 3\eta^2$, which is positive at the onset of superconductivity and, by continuity, has to remain positive along the line $\phi_1 = 0$ in the entire high temperature regime. It follows that the leading term of ψ vanishes at $\theta = \pi - \arg \beta_1$. If the shape of $w(\theta)$ is dominated by the first harmonic, this means that the leading term of ψ vanishes at the thinnest part of the loop.

We are still left with the question of whether the entire order parameter vanishes at some point or whether the small contribution $\epsilon^{3/2}\psi_1$ remains. In [6] it is shown that the entire ψ vanishes at $\phi_1 = 0$. This can also be seen from continuity arguments: for $|g|$ appreciably smaller than 1, the winding number of ψ is m, whereas for $|g|$ appreciably greater than 1, its winding number is $m + 1$. Since ψ changes continuously with $|g|$, whereas its winding number does not, there must be some $|g|$ for which the winding number is undefined.

Note the qualitative difference between the cases of uniform and nonuniform cross section: in the first case, ψ is either proportional to $e^{-mi\theta}$ or to

$e^{-(m+1)i\theta}$, and it is impossible to pass continuously from some m to the next; in the case of nonuniform cross section, ψ is a sum over all harmonics, with a higher weight at some m. As the flux increases and crosses $m + 1/2$, this weight is gradually transferred to $m + 1$ by increasing $|g|$ from $|g| \ll 1$ to $|g| \gg 1$.

Let us now consider the case $\mu_1 > |\beta_1|/2$. In this case there is a region (which for $\mu_1 \approx |\beta_1|/2$ is $|\phi_1| \lesssim 4|\beta_1|[\mu_1/(3|\beta_1|) - 1/6]^{3/2}$) where (5.25) has three solutions. At each extreme of this region, two solutions of the Euler-Lagrange equations coalesce, implying that the second variation of the free energy vanishes. This suggests that at these points the solution found becomes unstable. It is natural to conclude that the portion of Fig. 5.2 with negative slope, which is the continuation of the "bridging solution" of Sect. 5.3.1, is the unstable one. It follows that for $\mu_1 > |\beta_1|/2$ the order parameter will "jump" from winding number m to $m + 1$ as the flux is suitably increased, without going through a stable solution that breaks connectivity.

One of the most easily measurable quantities is the current I. Using (5.25) we obtain that, to leading order in ϵ, it is proportional to $(|g|^2 - 1)(1 + |g|^2 + 8|g|\mu_1/|\beta_1|)/(|g| + |g|^3)$. A usual experimental technique measures the a.c. susceptibility χ, which is proportional to $dI/d\phi$. It follows that χ is proportional to $d|g|/d\phi_1$ and, therefore, diverges linearly at the stability limits and quadratically at the critical point P_2. For experimental implications see [5].

Equations (5.24) are obtained by considering the mth and $(m + 1)$th harmonics in (5.21). The other harmonics give the leading terms of ψ_1 in (5.22), except for the harmonics m and $m + 1$; in order to obtain these, perturbation to order $O(\epsilon^2)$ is required.

The reason that only β_1 appears in (5.24) is that only $e^{i\theta}$ can link between $e^{-mi\theta}$ and $e^{-(m+1)i\theta}$ in first order perturbation; in second order, the linkage could be made by any term of the form $\beta_{j+1}\beta_{-j}e^{i\theta}$. However, if the shape of $w(\theta)$ has an n-fold axis, so that all the harmonics of $w(\theta)$ are multiples of n, $e^{-mi\theta}$ and $e^{-(m+1)i\theta}$ cannot be linked in any order of perturbation. We may expect this to be related to a failure of assumption (3.10) when $w(\theta)$ has n-fold symmetry and the flux is half-integer.

5.4 Rings with Finite Width

In the previous sections we assumed that the "radial" dimension of the sample was very small compared to the coherence length ξ. In that case ψ depended only on the arc length and the problem became one-dimensional. In this section we release this assumption. One of the questions we want to answer is whether the zero which we found in previous sections when the flux is an integer plus half still breaks connectivity. Explicitly, does ψ now vanish on an entire surface that connects the inner and outer boundary, or only at a line parallel to the magnetic field, as in the case of vortices? We also want to find out whether the critical point P_2 of Sect. 5.3.2 and the singularities associated with it still exist, or whether these are smeared by the finite width of the shell.

In the zeroth order approximation we consider a cylindrical shell with inner radius R_i, outer radius $R_o = R_i + w$ and heigth h. In the typical situation we shall consider a uniform applied field $H_e \widehat{\mathbf{z}}$, where z is the axis of the shell.

For a thick shell the definition of the "enclosed" flux requires some convention, since the flux through the sample itself is not negligible. We shall adopt some conventions in the following subsections and in general define $\phi(r) = \pi H_e r^2 / \Phi_0$, the flux through a circle of radius r, in quantum units.

If the order parameter is small, the GL equation (5.2) becomes a linear separable equation, and was solved in Sect. ??. (See also [2].) The Neuman boundary conditions allow for lowest energy solutions that are independent of z, so that we are left with a two-dimensional problem which is independent of the height h of the sample. Writing the order parameter in the form

$$\psi_0^{(m)} = \mathcal{R}_m(r) e^{-mi\theta} \tag{5.26}$$

and choosing the gauge such that $\mathbf{A} = A(r)\widehat{\theta}$, where θ is the angular cylindrical coordinate, the linearized GL equation becomes

$$-\mathcal{R}_m'' - \frac{\mathcal{R}_m'}{r} + \left(\frac{m}{r} - \widetilde{A}\right)^2 \mathcal{R}_m = \mu \mathcal{R}_m , \tag{5.27}$$

where we have switched back to the dimensional notation $\mu = \xi^{-2}$. This has the solution

$$\mathcal{R}_m(r) = r^{|m|} e^{-\phi(r)/2} [k_M M(\alpha, |m|+1, \phi(r)) + k_U U(\alpha, |m|+1, \phi(r))] , \tag{5.28}$$

where M and U are Kummer's confluent hypergeometric functions, $\alpha = \frac{1}{2}(1 + |m| - m - \frac{1}{2}\mu_0/b)$, μ_0 (which depends on m and H_e) is the lowest eigenvalue for μ and $b = \pi H_e / \Phi_0$. The boundary conditions require that the derivative of $\mathcal{R}_m(r)$ vanish; these conditions can be written as $a_M k_M + a_U k_U = 0$ with

$$
\begin{aligned}
a_M &= 2(|m| + 1 - \alpha) M(\alpha - 1, |m| + 1, \phi(r)) \\
&\quad + (2\alpha + \phi(r) - 2 - |m|) M(\alpha, |m| + 1, \phi(r)) \\
a_U &= -2U(\alpha - 1, |m| + 1, \phi(r)) + (2\alpha + \phi(r) - 2 - |m|) U(\alpha, |m| + 1, \phi(r))
\end{aligned}
\tag{5.29}
$$

at $r = R_i$ and $r = R_o$. The constants k_M, k_U and the eigenvalue μ_0 are determined by three equations: $a_M k_M + a_U k_U = 0$ at $r = R_i$ or $r = R_o$, the determinant $a_M(R_i) a_U(R_o) - a_M(R_o) a_U(R_i)$ has to vanish, and some normalization (with suitable units) can be chosen. The winding number m has to be chosen so that the lowest value of μ_0 is obtained. It should be born in mind that these values are obtained from a system of nonlinear equations that has several solutions; we have made an effort to always pick the relevant solution, but the possibility of convergence to some other solution cannot be discarded.

A well known feature of the solution (5.28) is that there exist magnetic fields for which the linearized GL equation is degenerate. In the limit of a thin shell this occurs when the enclosed flux equals $(m + 1/2)\Phi_0$: for each of these fluxes, the eigenvalues μ_0 obtained for m and for $m + 1$ are the same and the lowest possible. If the shell is not thin, there still exist fluxes where the values of μ_0 for $\psi_0^{(m)} = \mathcal{R}_m(r)\exp(-mi\theta)$ and for $\psi_0^{(m+1)} = \mathcal{R}_{m+1}(r)\exp(-[m+1]i\theta)$ coalesce, and we denote them by $\Phi_*^{(m)}$. As the magnetic field is swept across these fluxes, the equilibrium order parameter, and measurable quantities such as the current around the shell, change discontinuously.

5.4.1 Order of the Normal–Superconducting Transition

The opening of Sect. 5.4 describes the minimizer of the linearized GL potential, with the implicit assumption that the magnetic field is the same as the applied field. Linearization of this potential is a good approximation only if the order parameter is small. A natural question is whether there exists a situation where $|\psi|$ is small. This situation will occur at a bifurcation from the normal state, i.e. at a second order phase transition. In the following, we shall check the stability of the solution described above, namely, whether it is a local minimum of the entire GL potential.

The GL potential is a functional of two fields, ψ and $\mathbf{A}$, but we can first minimize w.r.t. $\mathbf{A}$ and obtain

$$\nabla \times \nabla \times \widetilde{\mathbf{A}}_i = \mathbf{J} \,, \tag{5.30a}$$

$$\mathbf{J} = \frac{\mu}{\kappa^2}\mathrm{Re}[\overline{\psi}(i\nabla - \widetilde{\mathbf{A}})\psi] \,, \tag{5.30b}$$

where we have separated $\widetilde{\mathbf{A}}$ into an external part $\widetilde{\mathbf{A}}_e$, due to the applied field, and an induced part $\widetilde{\mathbf{A}}_i$, due to the supercurrents, which vanishes far from the sample; (5.30a) is Ampère's law and $\mathbf{J}$ is the supercurrent density. (In Gaussian units, the current density is $c\Phi_0\mathbf{J}/[2(2\pi)^2]$.) From here we learn that, for small $|\psi|$, $\widetilde{\mathbf{A}}_i$ is quadratic in ψ, decreases with κ^2 and, if the current has to circulate through some cross section, we expect $\widetilde{\mathbf{A}}_i$ to be an increasing function of it.

The form (5.1) of the GL potential guarantees that the order parameter of the minimizer is bounded. Moreover, it has been shown [14] that $|\psi|$ is always bounded by 1. Having in mind a small $|\psi|$, we denote by $\mathbf{J}_2$ the value of $\mathbf{J}$ which is obtained by replacing $\widetilde{\mathbf{A}}$ with $\widetilde{\mathbf{A}}_e$ in (5.30b), and then expand $\widetilde{\mathbf{A}}_i = \widetilde{\mathbf{A}}_{i2} + \widetilde{\mathbf{A}}_+$, where $\nabla \times \nabla \times \widetilde{\mathbf{A}}_{i2} = \mathbf{J}_2$ and $\widetilde{\mathbf{A}}_+$ fulfills

$$\nabla \times \nabla \times \widetilde{\mathbf{A}}_+ = -(\mu/\kappa^2)|\psi|^2(\widetilde{\mathbf{A}}_+ + \widetilde{\mathbf{A}}_{i2}) \,. \tag{5.31}$$

For the cylindric shells considered above, it is easy to evaluate $\widetilde{\mathbf{A}}_{i2}$ in the limits $h \gg R_o$ and $h \ll \kappa\xi$. In the first case we have a two dimensional problem and, using cylindrical symmetry and writing $\widetilde{\mathbf{A}}_{i2} = \widetilde{A}_{i2}(r)\widehat{\theta}$, we are left with an ordinary differential equation for $\widetilde{A}_{i2}$, which can be solved numerically. The boundary conditions are $\nabla \times \widetilde{\mathbf{A}}_{i2} = 0$ at R_o, which gives $\widetilde{A}'_{i2}(R_o) = -\widetilde{A}_{i2}(R_o)/R_o$, and continuity of $\widetilde{\mathbf{A}}_{i2}$ at $r = 0$ together with continuity of $\nabla \times \widetilde{\mathbf{A}}_{i2}$ at R_i, which require $\widetilde{A}'_{i2}(R_i) = \widetilde{A}_{i2}(R_i)/R_i$. In the three dimensional problem, $\widetilde{\mathbf{A}}_{i2}(\mathbf{r})$ can be evaluated by integrating the Green's function $\mathbf{J}_2(\mathbf{r}')/(4\pi|\mathbf{r} - \mathbf{r}'|)$. For $h \ll \kappa\xi$, $\mathbf{r}$ and $\mathbf{r}'$ are required only on the same plane; in this case we can perform the integration over θ' analytically and are left with a numeric integration over r'. The integrand has a singularity at $r' = r$, which has to be treated separately.

Let us now regard $\widetilde{\mathbf{A}}_i$ and $\mathbf{J}$ as known functions of ψ. Noting that $|\nabla \times \widetilde{\mathbf{A}}_i|^2 = \nabla \cdot (\widetilde{\mathbf{A}}_i \times (\nabla \times \widetilde{\mathbf{A}}_i)) + \widetilde{\mathbf{A}}_i \cdot (\nabla \times \nabla \times \widetilde{\mathbf{A}}_i)$, using (5.30a) and (5.30b) and the fact that $\widetilde{A}_i$ decays sufficiently fast far from the sample, the GL potential can be written in the form

$$G = G_2 + G_4 + G_+ \tag{5.32}$$

with

$$
\begin{aligned}
G_2 &= \int \left(-\mu|\psi|^2 + \left|\left(i\nabla - \widetilde{\mathbf{A}}_e\right)\psi\right|^2 \right) \\
G_4 &= \mu \int \left(\frac{1}{2}|\psi|^4 - \kappa^{-2}\frac{\kappa^2\widetilde{\mathbf{A}}_{i2}}{\mu} \cdot \frac{\kappa^2\mathbf{J}_2}{\mu} \right) \\
G_+ &= -\frac{\kappa^2}{\mu}\int \widetilde{\mathbf{A}}_+ \cdot \mathbf{J}_2 ,
\end{aligned}
\tag{5.33}
$$

where the integrals are over the sample. Following (5.30a) and (5.30b), μ and κ have been factored out in the expression for G_4. G_n depends on the shape of ψ and is proportional to the nth power of the "size of ψ".

In the linearized GL formalism, G is replaced by G_2. Clearly, for small μ and $\widetilde{\mathbf{A}}_e \neq 0$, $G_2 \geq 0$ and the minimizer is the normal state $\psi = 0$. As μ increases, there are two possibilities: if $G_4 > 0$, then, for μ such that $G_2 < 0$, G_2 will dominate for small sizes of ψ, forcing a bifurcation from $\psi = 0$, but for large sizes, G_4 will dominate, keeping ψ at a limited size; this size increases continuously with μ from $\psi = 0$, so that the transition is of second order. On the other hand, if $G_4 < 0$ for some shape of ψ, it may pay to pass over to this ψ even before G_2 becomes negative. However, the positive G_2 will act as a potential barrier and prevent ψ from leaving the normal state, unless it is sufficiently large. Therefore, there will be a discontinuous (first order) transition, and the size of ψ will be limited by G_+.

It follows that the order of the transition depends on the sign of G_4. For a wide range of parameters, we always found $\int \widetilde{\mathbf{A}}_{i2} \cdot \mathbf{J}_2 > 0$. ($\widetilde{\mathbf{A}}_i$ tends to be in the same direction as the current that generates it.) Therefore, the transition from normal to superconducting state will be of first order if κ is sufficiently

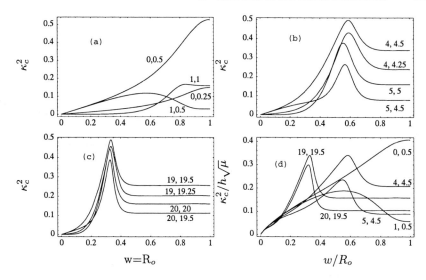

Fig. 5.3. Lower bounds for κ_c as functions of the width of the shell. Below these lines, the normal/superconducting transition is of first order. Each line is marked with two numbers: the first is the winding number and the second, the "adjusted flux" ϕ_{ad}. **(a)–(c)**: long shells; **(d)**: flat rings

small and of second order if κ is sufficiently large. We shall denote by κ_c the limiting value between both regimes. The expression for G_4 suggests that κ_c will be larger under conditions that yield larger currents: larger cross section, more connectivity, magnetic fluxes which are not integer.

Figures 5.3(a)–(c) show lower bounds for κ_c as functions of the ratio $(R_o - R_i)/R_o$ between the width of the shell and its outer radius, for the case $h \gg R_o$. These were evaluated by using the shape of ψ that minimizes G_2, i.e. (5.28), to check the sign of G_4. G_4 must be positive for a second order transition, since otherwise the order parameter predicted by assuming a second order transition would not be stable. κ_c has been calculated for different values of the winding number and the flux. In order to have a "fair comparison" of the "enclosed" fluxes among shells that differ in width, we define the "adjusted flux" ϕ_{ad} as follows: in the absence of magnetic field, $\phi_{ad} = 0$; at $\Phi_*^{(m)}$ ($m \geq 0$), $\phi_{ad} = m + 1/2$; between these values, ϕ_{ad} is a linear function of the flux. In the limit of a thin shell ($R_i \approx R_o$), ϕ_{ad} is just the flux in units of Φ_0.

For thin shells, the current density is proportional to $\phi_{ad} - m$, the induced potential to $(\phi_{ad} - m)w$, and κ_c^2, to $(\phi_{ad} - m)^2 w$. For an almost full cylinder, a situation with $m < \phi_{ad}$ has higher κ_c than a similar situation with $m > \phi_{ad}$; this may reflect the fact that a larger winding number shrinks the width of the region where currents circulate and also favors competition between paramagnetic and diamagnetic currents. κ_c does not increase monotonically

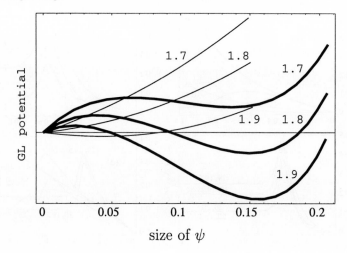

Fig. 5.4. Superconducting energy of a cylinder as a function of the size of the order parameter, for fixed flux and various values of the temperature and the winding number. Here $\phi(R_o) = 2.1$ and $\kappa^2 = 0.1$; the lines are marked by the value of μR_o^2. Thick lines: $m = 0$; thin lines: $m = 1$.

with the width, but has a peak; for a given ϕ_{ad}, this peak is located roughly at the width for which $\phi(R_o)$ is maximum. The height of this peak remains nearly constant at about $\sqrt{1/2}$, the known limit between superconductivities of types I and II; we have no intuitive explanation for this result. For large fluxes, κ_c becomes independent of the width shortly beyond the peak. The reason for this is that superconductivity is appreciable only close to the outer boundary, so that the inner part of the shell becomes irrelevant.

Figure 5.3(d) shows κ_c for the case $h \ll \kappa\xi$. The results are qualitatively the same as for the long shells, but κ_c^2 has to be scaled by $h\sqrt{\mu}$.

We already know when the normal-superconducting transition must be discontinuous. It would be interesting to have at least an estimate for the value of κ for which it *could* be discontinuous. It is tempting to argue that such an estimate could be obtained by evaluating G_4 for the shape of ψ that minimizes G_4 itself, rather than G_2. However, this criterion gives very poor estimates, especially for high fluxes and large ratio $(R_o - R_i)/R_o$.

Figure 5.4 shows the GL potential G as a function of the size of the order parameter, for several values of the temperature, for a long full sample ($h \gg R_o$, $R_i = 0$). In the figure, the "size" has been defined as the average of $|\psi|^2$, $\int |\psi|^2 r \mathrm{d}r / \int r \mathrm{d}r$, and the shape of ψ has been taken so as to minimize G for the chosen size. Unlike the case of Figs. 5.3, the shape of ψ minimizes the entire potential, and not only G_2. In the calculations we have made two assumptions: first, we assume that ψ has the symmetry of the problem, i.e. is a radial function times $\mathrm{e}^{-mi\theta}$ and, second, we have kept in G only the contributions up to $O(|\psi|^6)$, i.e. we dropped $\mathbf{A}_+$ at the r.h.s. of (5.31). For the values of the flux considered $[\phi(R_o) = 2.1]$, the best candidates for a minimum of G are $m = 0$ and $m = 1$.

For high temperature ($\mu R_o^2 = 1.7$), both curves have their minima at $\psi \equiv 0$, so that the sample has to be in the normal state. As the temperature is lowered to $\mu R_o^2 = 1.8$, G could be lowered if the sample had an order parameter with winding number 0 and size of about 0.15. In order to pass from the normal to this new state, ψ would have to "jump", so that the normal state is still a local minimum. When the temperature is lowered to $\mu R_o^2 = 1.9$, the normal state becomes unstable, not with respect to $m = 0$, but rather with respect to $m = 1$. In summary, the shape of the order parameter obtained when the sample is cooled from $\mu R_o^2 = 1.7$ to $\mu R_o^2 = 1.9$ depends on how it is cooled: if it is cooled carefully, so that the jump is avoided, we must obtain $m = 1$; if it is left for a long time at $\mu R_o^2 = 1.8$ and forced to jump, we must obtain $m = 0$.

Figure 5.4 is good as an illustration, but the assumptions that led to it are not necessarily justified. The sizes of ψ are small in the range considered, but μ/κ^2 is large, so that $G_{2(n+1)}$ is smaller than G_{2n} just by a moderate factor. Actually, if $G_4 < 0$ and G_6 were negligible, then G would not have a finite minimum. As for the second assumption, since the GL equations are nonlinear, the minimizer may not have the entire symmetry of the problem. It is known [3] that, for appropriate parameters in cylindric samples, ψ would vanish along a single line that does not coalesce with the cylinder axis.

If h is of the same order of magnitude as R_o, one might expect to have an intermediate situation between the two extremes considered here. However, a complication appears: at the corners ($r = R_o, z = \pm h/2$) a very large demagnetizing field develops that forces these corners to remain normal [8].

In the following subsections we shall consider transitions of second order, so that we may deal with situations where expansions on powers of ψ can provide a good description. In view of Fig. 5.3(d), these are frequently encountered situations, since in many experiments $h\sqrt{\mu}$ is small.

5.4.2   Nonuniform Cross Section – Nondegenerate Perturbation

Our purpose now is to extend the treatment of Sect. 5.3.2 to rings of finite width. The main objective of this subsection will be to evaluate the temperature for the onset of superconductivity. This temperature is a monotonic function of the eigenvalue μ of the linearized GL equation. We write

$$\mu = \mu_0 + \epsilon\mu_1 + \epsilon^2\mu_2 + \dots,$$
$$\psi = \psi_0 + \epsilon\psi_1 + \dots, \tag{5.34}$$

where ψ_0 is given by (5.26) and (5.28) [we omit here the index (m)] and μ_0 is determined as described after (5.29). We now want to determine μ_1 and μ_2. Introducing (5.34) into the linearized GL equation gives to first order in ϵ

$$(\mathcal{H}_0 - \mu_0)\psi_1 = \mu_1\psi_0 , \tag{5.35}$$

where the operator $\mathcal{H}_0$ is now

$$\mathcal{H}_0 = \left(i\nabla - br\widehat{\theta}\right)^2 \tag{5.36}$$

and b was defined after (5.28).

Let us now define a metric by integrating in the *unperturbed* region:

$$(\phi_1, \phi_2) = \int_{-h/2}^{h/2} dz \int_0^{2\pi} d\theta \int_{R_i}^{R_o} r dr \overline{\phi}_1 \phi_2 . \tag{5.37}$$

In the subspace of functions that have Neuman boundary conditions $\mathcal{H}_0$ is Hermitic, but in general, it has the property

$$(\phi_1, \mathcal{H}_0\phi_2) - (\mathcal{H}_0\phi_1, \phi_2) = \int_{\partial\Omega} \left(\phi_2\nabla\overline{\phi}_1 - \overline{\phi}_1\nabla\phi_2\right) \cdot \widehat{\nu}_0 , \tag{5.38}$$

where $\int_{\partial\Omega}$ denotes integration over the surface of the unperturbed sample and $\widehat{\nu}_0$ is the unit vector normal to this surface. For example, if the only perturbation in the shape of the sample is in its outer boundary and, in addition, this perturbation is independent of z, (5.38) becomes

$$(\phi_1, \mathcal{H}_0\phi_2) - (\mathcal{H}_0\phi_1, \phi_2) = hR_o \int_0^{2\pi} \left(\phi_2\frac{\partial\overline{\phi}_1}{\partial r} - \overline{\phi}_1\frac{\partial\phi_2}{\partial r}\right) d\theta . \tag{5.39}$$

In the following, we shall consider only this situation: the inner boundary of the shell will be located at $r = R_i$ and the outer one at $r = R_i + w(\theta)$, where $w(\theta)$ is given by (5.20).

Projecting Eq. (5.35) onto ψ_0 and using (5.39) we obtain

$$\mu_1 = -\frac{hR_o}{(\psi_0, \psi_0)} \int_0^{2\pi} \overline{\psi}_0 \frac{\partial\psi_1}{\partial r} d\theta .$$

$\partial\psi_1/\partial r$ is obtained by requiring that the normal component of $(i\nabla - \widetilde{\mathbf{A}})(\psi_0 + \epsilon\psi_1)$ vanish to order ϵ at the boundary $r = R_i + w(\theta)$. Expanding around $r = R_o$ gives

$$\frac{\partial\psi_1}{\partial r} = -\frac{w}{2}\sum_{j\neq 0}\beta_j e^{(j-m)i\theta}\left(\frac{d^2}{dr^2} + jb - \frac{jm}{r^2}\right)\mathcal{R}_m(r), \tag{5.40}$$

evaluated at $r = R_o$. We see that $\partial\psi_1/\partial r$ does not contain the $(-m)$-th harmonic and therefore $\mu_1 = 0$. It follows that the differential equation (5.35) for ψ_1 is the same as the equation we had for ψ_0, but the boundary condition has been modified. We write $\psi_1 = \sum_{l\neq m}\mathcal{Y}_l(r)e^{-li\theta}$. The solution of (5.35) for every harmonic is of the form (5.28), but at $r = R_o$ the Neuman condition has to be replaced by (5.40). We find

$$\mathcal{Y}_l(r) = \frac{1}{2}\frac{w\beta_{m-l}c_{l,m}\mathcal{R}_m(R_o)}{a_M(l; R_o)a_U(l; R_i) - a_M(l; R_i)a_U(l; R_o)}\left(\frac{r}{R_o}\right)^{|l|}e^{b(R_o^2-r^2)/2} \times$$
$$[a_U(l; R_i)M(\alpha, |l| + 1, br^2) - a_M(l; R_i)U(\alpha, |l| + 1, br^2)], \tag{5.41}$$

where $a_{M,U}(l;r)$ are the functions defined in (5.29), with l substituted into m, and $c_{l,m} = R_o[\mu_0 - (bR_o - l/R_o)(bR_o - m/R_o)]$. The second derivative which appears in (5.40) was eliminated using the differential equation for $\mathcal{R}_m$.

μ_2 is obtained by the same arguments that lead to μ_1. We obtain

$$\mu_2 = -\frac{w\mathcal{R}_m(R_o)}{2B_m}\left\{\sum_{l\neq m}\beta_{l-m}c_{l,m}\mathcal{Y}_l(R_o)+\right.$$

$$\left.\frac{w\mathcal{R}_m(R_o)}{4}\left[\mu_0 - (bR_o - \frac{m}{R_o})(3bR_o + \frac{m}{R_o})\right]\sum_{j\neq 0}|\beta_j|^2\right\}, \quad (5.42)$$

with

$$B_m = \int_{R_i}^{R_o} r\mathcal{R}_m^2 dr . \quad (5.43)$$

Note that $\mathcal{R}_m$ depends on the applied field and so does B_m.

5.4.3   Degenerate Perturbation

At the degeneracy flux $\Phi_*^{(m)}$, the determinant $a_M(R_i)a_U(R_o) - a_M(R_o)a_U(R_i)$ vanishes simultaneously for the winding numbers m and $m+1$. Thus, $\mathcal{Y}_{m+1}$ in (5.41) diverges. To handle this problem we have to use degenerate perturbation theory.

An additional difference between this and the previous subsection is that now we will be interested in knowing the order parameter for some region beyond the onset of superconductivity; therefore, the nonlinear term in (5.2) is not entirely neglected. We shall nevertheless neglect the induced magnetic potential. We expect this to be justified when $\mu hw\kappa^{-2} \ll 1$, so that the current is small. Still, we shall consider the region where $|\mu - \mu_P^{(m)}|$ is at most of order ϵ, where $\mu_P^{(m)}$ is the value of μ_0 for $\Phi = \Phi_*^{(m)}$.

Let us set the convention that "the" flux Φ is that through the outer boundary and let us write $\Phi = \Phi_*^{(m)} + \epsilon\phi_1\Phi_0$. Then the magnetic potential will be $\widetilde{\mathbf{A}} = (b_m + \epsilon\phi_1/R_o^2)r\widehat{\theta}$, with $b_m = \Phi_*^{(m)}/(R_o^2\Phi_0)$. We denote by $\mathcal{H}_P$ the operator $\mathcal{H}_0$ in (5.36) with b substituted by b_m. For the order parameter we write

$$\psi = \sqrt{\eta\epsilon}(\psi_0^{(m)} + g\psi_0^{(m+1)} + \epsilon\psi_1 + ...) \quad (5.44)$$

where η and g still have to be determined by the perturbation procedure. Introducing these expressions into (5.2) we obtain, to lowest order in ϵ,

$$(\mathcal{H}_P - \mu_P^{(m)})\psi_1 = f_m(r)e^{-mi\theta} + gg_m(r)e^{-(m+1)i\theta} + \eta \times \text{(other harmonics)} \quad (5.45)$$

with

$$
\begin{aligned}
f_m &= [2(\phi_1/R_o^2)(m - b_m r^2) \\
&\quad +(\mu - \mu_P^{(m)})/\epsilon - \eta\mu_P^{(m)}(\mathcal{R}_m^2 + 2g^2\mathcal{R}_{m+1}^2)]\mathcal{R}_m \ , \\
g_m &= [2(\phi_1/R_o^2)(m + 1 - b_m r^2) \\
&\quad +(\mu - \mu_P^{(m)})/\epsilon - \eta\mu_P^{(m)}(2\mathcal{R}_m^2 + g^2\mathcal{R}_{m+1}^2)]\mathcal{R}_{m+1} \ .
\end{aligned}
$$

$$(5.46)$$

We project this expression onto $\psi_0^{(m)}$ and use (5.39), and then repeat this procedure for $\psi_0^{(m+1)}$. This leads to a system of equations for g and η:

$$
\begin{aligned}
A_m\phi_1 + B_m(\mu - \mu_P^{(m)})/\epsilon &= (Q_m + g^2 S_m)\eta - C_m\beta_1 g \ , \\
A'_{m+1}\phi_1 + B'_{m+1}(\mu - \mu_P^{(m)})/\epsilon &= (g^2 Q'_{m+1} + S_m)\eta - C_m\overline{\beta}_1/g \ ,
\end{aligned} \quad (5.47)
$$

where B_m was defined in (5.43) and

$$A_m = 2R_o^{-2}\int_{R_i}^{R_o}(mr - b_m r^3)\mathcal{R}_m^2\, dr \ , \tag{5.48a}$$

$$C_m = \frac{w}{2}c_{m+1,m}\mathcal{R}_m(R_o)\mathcal{R}_{m+1}(R_o) \ , \tag{5.48b}$$

$$Q_m = \mu_P^{(m)}\int_{R_i}^{R_o} r\mathcal{R}_m^4\, dr \ , \tag{5.48c}$$

$$S_m = 2\mu_P^{(m)}\int_{R_i}^{R_o} r\mathcal{R}_m^2\mathcal{R}_{m+1}^2 \ . \tag{5.48d}$$

The primes in the second equation in (5.47) mean that b_m and $\mu_P^{(m)}$ are kept unchanged when the index m is substituted by $m+1$. Likewise, $c_{m+1,m}$ [defined after (5.41)] is evaluated at $b = b_m$. Note that the coefficients A_m, A'_m, B_m, B'_m, Q_m, Q'_m and S_m do not depend on the deviation of the shape of the sample from perfect axial symmetry. Eq. (5.48b) was obtained for a deviation at the outer boundary which depends only on θ; in the general case, $C_m\beta_1$ has to be replaced by

$$\frac{1}{h}\int_{\partial\Omega}\left[\psi_0^{(m+1)}\mu_P^{(m)}\overline{\psi_0^{(m)}} - (i\nabla - \tilde{\mathbf{A}})\psi_0^{(m+1)}\cdot\overline{(i\nabla - \tilde{\mathbf{A}})\psi_0^{(m)}}\right]\nu \ , \quad (5.49)$$

where $\epsilon\nu$ denotes how far out one has to move from the unpertubed sample (in the direction normal to it) until the corresponding point in the boundary of the perturbed sample is reached.

The values of these coefficients depend on the normalizations of $\psi_0^{(m)}$ and $\psi_0^{(m+1)}$; accordingly, this affects the values of η and g so that the order parameter itself and the transition temperature are independent of the normalizations. As a final observation, we note that $g\beta_1$ has to be real. The phase of β_1 depends on the origin of the axial angle θ and we choose it such that $\beta_1 < 0$. If $w(\theta)$ is symmetric about $\theta = 0$ and $\theta = \pi$ and monotonic for $0 \le \theta \le \pi$, this choice means that the shell is thinnest at $\theta = 0$. It follows that g will be real.

In the remainder of this subsection, we push the perturbation procedure further for the onset of superconductivity, where $\eta = 0$. The line for the onset of superconductivity in the temperature - magnetic flux plane will be denoted by Γ_1. We write $\mu = \mu_P^{(m)} + \epsilon\mu_1 + \epsilon^2\mu_2 + \dots$ (Unlike μ_0, $\mu_P^{(m)}$ does not vary with the flux.) The first order correction μ_1 is obtained by solving (5.47). There are two solutions and we choose the lowest:

$$\mu_1 = -\frac{1}{2}\left[\left(\frac{A_m}{B_m} + \frac{A'_{m+1}}{B'_{m+1}}\right)\phi_1 + \sqrt{\frac{(2C_m\beta_1)^2}{B_mB'_{m+1}} + \left(\frac{A_m}{B_m} - \frac{A'_{m+1}}{B'_{m+1}}\right)^2\phi_1^2}\right].$$
(5.50)

From here and (5.47) it follows that $C_m\beta_1g/B_m = -\mu_1 - (A_m/B_m)\phi_1$ is always positive and, since $B_m > 0$ and $\beta_1 < 0$,

$$C_mg < 0 \qquad\qquad (5.51)$$

along the transition line Γ_1 and close to $\Phi_*^{(m)}$. By continuity, and since we expect g to depend mainly on the flux and only marginally on the temperature, this inequality is expected to hold in the entire region of order ϵ investigated here.

For every l different from either m or $m+1$, $\mathcal{Y}_l$ obeys the same differential equation as for the unperturbed case. However, now the leading term in (5.44) contains two contributions. Both contributions will be present in $\partial\psi_1/\partial r$, which appears in the boundary conditions for $\mathcal{Y}_l$, and therefore Eq. (5.41) is modified and contains a second term. The case of l equal to either m or $m+1$ requires special treatment. We describe here the case $l = m$; $l = m+1$ is entirely analogous.

$\mathcal{Y}_m$ satisfies the inhomogeneous equation

$$\mathcal{Y}_m'' + \frac{1}{r}\mathcal{Y}_m' + \left[\mu_P^{(m)} - \left(b_mr - \frac{m}{r}\right)^2\right]\mathcal{Y}_m = -f_m , \qquad (5.52)$$

where $(\mu - \mu_P^{(m)})/\epsilon$ is replaced by μ_1 and η by 0. This is subject to the conditions that $\mathcal{Y}_m'$ vanishes at $r = R_i$ and leads to the appropriate value of $\partial\psi_1/\partial r$ at $r = R_o$. Since μ_1 has been picked as an eigenvalue, if one of these conditions is fulfilled, the second follows automatically. One solution of (5.52) with the appropriate boundary conditions, which we denote by $y_m(r)$, can be obtained by solving (5.52) numerically, with $y_m(0) = y_m'(0) = 0$. This is not the only solution: any function $\mathcal{Y}_m(r) = y_m(r) + K_m\mathcal{R}_m(r)$ will be a solution, too.

We proceed now to second order perturbation (at the onset of superconductivity). We obtain $(\mathcal{H}_P - \mu_P^{(m)})\psi_2 = (\mu_2 - \phi_1^2r^2/R_o^4)\psi_0 + \mu_1\psi_1$, with

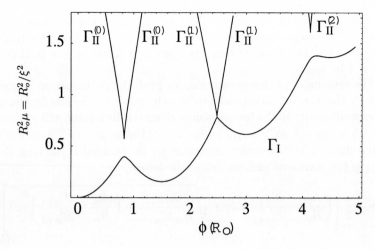

Fig. 5.5. Phase diagram in the temperature - magnetic flux plane ($\mu \propto T_c - T$). The sample is normal below the line Γ_I and superconducting above it. At the left branch of $\Gamma_{II}^{(m)}$, the state with winding number $m + 1$ becomes unstable and must decay into m; at its right branch, m decays into $m + 1$. The points $P_1^{(m)}$ are the maxima of Γ_I and $P_2^{(m)}$ are the minima of Γ_{II}. For the evaluation of this phase diagram we have used $w/R_o = 0.426$ (as in Refs. [17,18]), $\epsilon|\beta_1| = 0.15$ and $\beta_i = 0$ for $i > 1$. Γ_I was evaluated to $O(\epsilon^2)$ and Γ_{II}, to $O(\epsilon)$. For this ratio w/R_o, C_1 is small, so that Γ_I and $\Gamma_{II}^{(1)}$ almost touch

$\psi_0 = \psi_0^{(m)} + g\psi_0^{(m+1)}$. Projecting onto $\psi_0^{(m)}$ and $\psi_0^{(m+1)}$, using (5.39) and keeping terms up to $O(\epsilon^2)$ leads to two equations for μ_2, K_m and K_{m+1}. The third equation is provided by some normalization criterion, which is usually taken as $(\psi_1, \psi_0) = 0$. From here, μ_2 and the missing information for ψ_1 are obtained.

The onset for superconductivity occurs at the lowest line in Fig. 5.5 (Γ_I). This line was obtained by seven different algorithms: near $\Phi = \Phi_*^{(m)}$ ($m = 0, 1, 2$) we used degenerate perturbation; far from $\Phi = \Phi_*^{(m)}$ we used (5.42).

5.4.4   Critical Points

In Section 5.3.2 we found that thin loops have two critical points in the temperature - magnetic flux diagram, which we denote by P_1 and P_2. Here we extend this result to shells which are not necessarily thin.

$P_1^{(m)}$ is defined as the point on the critical line Γ_I such that the sample always remains supeconducting as the field is swept close to $\Phi = \Phi_*^{(m)}$ for temperatures below that of $P_1^{(m)}$, whereas it passes through the normal state if the temperature is higher than that of $P_1^{(m)}$. (See Fig. 5.5.) $P_2^{(m)}$ is defined as the point on a critical line $\Gamma_{II}^{(m)}$ such that, for temperatures below that of

$P_2^{(m)}$, there is a first order transition when the field is swept close to $\Phi = \Phi_*^{(m)}$, whereas the order parameter changes smoothly if the temperature is higher than that of $P_2^{(m)}$. The analysis of this subsection will be kept to first order in ϵ.

$P_1^{(m)}$ is obtained by maximizing (5.50). This gives

$$\phi_1 = \frac{A_m B'_{m+1} + A'_{m+1} B_m}{A_m B'_{m+1} - A'_{m+1} B_m} \frac{|C_m \beta_1|}{\sqrt{-A_m A'_{m+1}}}$$

$$\mu_1 = \frac{2|C_m \beta_1|\sqrt{-A_m A'_{m+1}}}{A_m B'_{m+1} - A'_{m+1} B_m} . \tag{5.53}$$

If w/R_o is sufficiently small, $A_m < 0$ and $A'_{m+1} > 0$. However, as w/R_o increases, A'_{m+1} eventually becomes negative and the temperature for the onset of superconductivity becomes a monotonic function of the flux close to $\Phi = \Phi_*^{(m)}$. The middle column of Table 5.1 shows the values of w/R_o for which A'_{m+1} vanishes. Generically, Γ_{I} has cusplike maxima for low fluxes and rises monotonically for high fluxes; for a full (or nearly full) disk, Γ_{I} is monotonic in the entire range, and for a very thin shell, Γ_{I} has maxima up to very high fluxes.

Table 5.1. Values of the width/radius ratio, w/R_o, for which A'_{m+1} or C_m vanishes. If w/R_o is smaller than the value for which $A'_{m+1} = 0$, then the transition value of μ has a local maximum near $\Phi = \Phi_*^{(m)}$; if w/R_o is smaller than the value for which $C_m = 0$, then vortices pass through the thin part of the sample.

m	$A'_{m+1} = 0$	$C_m = 0$
0	0.8673	1.0000
1	0.6137	0.4436
2	0.5138	0.2940
3	0.4537	0.2197
4	0.4119	0.1753
5	0.3806	0.1458
6	0.3558	0.1248
7	0.3356	0.1091
8	0.3187	0.0969
9	0.3042	0.0872
10	0.2916	0.0792

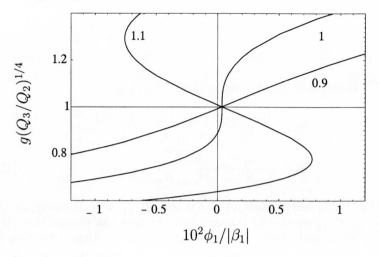

Fig. 5.6. $|g|$ as a function of the flux, for several values of the coherence length. (Q_m and $Q'_{(m+1)}$ provide a normalization constant.) For each curve, $(\mu - \mu_P^{(m)})/\epsilon$ equals the value of μ_1 given in (5.54), times the number shown next to the curve. The behavior is the same as in the case of an infinitesimally thin shell. For the evaluation of these curves we have used $w/R_o = 0.426$ and $m = 2$. (For these values, $g > 0$.)

For given values of $(\mu - \mu_P^{(m)})$ and ϕ_1, η and g can be obtained from equations (5.47). Figure 5.6 shows $|g|$ as a function of the flux for different values of the parameter $(\mu - \mu_P^{(m)})$. Qualitatively, we recover the behavior of Fig. 5.2. We see that, for $(\mu - \mu_P^{(m)})$ sufficiently large, there is an unstable branch; when it is reached as the flux is swept, g jumps to the other stable branch. The stability limit is characterized by $d\phi_1/dg = 0$, and its locus in the temperature - flux diagram is the curve $\Gamma_{II}^{(m)}$ in Fig. 5.5. The lowest point in this curve, $P_2^{(m)}$, is obtained by requiring in addition $d^2\phi_1/dg^2 = 0$. At this point we find $|g| = (Q_m/Q'_{m+1})^{1/4}$,

$$\mu_1 = \frac{2|C_m|\beta_1(Q_mQ'_{m+1})^{1/4}(A_m\sqrt{Q'_{m+1}} - A'_{m+1}\sqrt{Q_m})}{(A_mB_{m+1} - A'_{m+1}B_m)(\sqrt{Q_mQ'_{m+1}} - S_m)} , \tag{5.54}$$

and $\phi_1 = -\mu_1(B_m\sqrt{Q'_{m+1}} - B_{m+1}\sqrt{Q_m})/(A_m\sqrt{Q'_{m+1}} - A'_{m+1}\sqrt{Q_m})$. Figure 5.7 shows the temperatures of $P_1^{(m)}$ and $P_2^{(m)}$, as functions of the width of the shell, for small values of m.

Most experiments are performed at constant temperature, while the magnetic field is varied, i.e., along a horizontal line in Fig. 5.5. The experimental response will be a continuous function of the flux if the temperature is between those of $P_1^{(m)}$ and $P_2^{(m)}$, but will exhibit hysteresis if the temperature is below that of $P_2^{(m)}$.

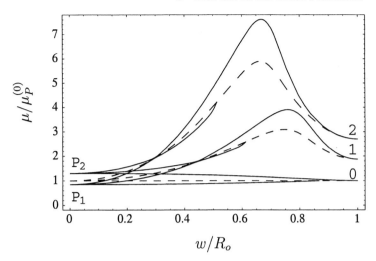

Fig. 5.7. Comparison among the values of $\mu = \xi^{-2}$ at the critical points, as functions of the width w of the sample. The dashed lines describe $\mu_P^{(m)}$; the values of μ for P_1 and P_2 are below and above each of them. The winding numbers m are shown at the right. The distances between the values of μ at $P_{1,2}^{(m)}$ and $\mu_P^{(m)}$ are proportional to $\epsilon|\beta_1|$, which is taken here as 0.15. They are also proportional to C_m, and therefore the three lines touch each other when $C_m = 0$; this happens at the value of w/R_o given in Table 5.1, and again at $w = R_o$. When $A'_{m+1} = 0$, the line for $P_1^{(m)}$ collapses into that of $\mu_P^{(m)}$. For $w \ll R_o$, $P_{1,2}$ and μ_P become independent of m

Figure 5.5 suggests that discontinuous transitions will ususally be encountered for low fluxes, whereas close to $\Gamma_{\rm I}$ continuous behavior prevails. However, anomalous behavior is also possible: the temperature of $P_2^{(m)}$ usually decreases with m, because $\mu_P^{(m)}$ increases with m; on the other hand, μ_1 in (5.54) is not a monotonic function of m, and usually has a minimum near the value of m where C_m changes sign. It follows that, for sufficiently large values of $\epsilon|\beta_1|$, the temperature for some $P_2^{(m)}$ may be higher than that of $P_2^{(m-1)}$.

One of the experimental tools in the study of mesoscopic samples is a.c. magnetic susceptibility (see e.g. [9,18]). In this technique the response is proportional to the derivative of the average magnetization with respect to the applied flux. The average magnetization is a complicated expression, but among other things, it depends on g. Therefore, the susceptibility diverges as $\Gamma_{\rm II}^{(m)}$ is approached, and diverges quadratically at $P_2^{(m)}$. This divergence, together with some calculated curves, was obtained in [5] for thin rings. However, it is surprising that the divergence is not smeared in a thick shell, where the "enclosed" flux appears to be poorly defined.

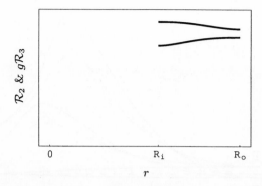

Fig. 5.8. Radial dependence of the leading terms in (5.44), including their relative weights, at a flux slightly lower than the degeneracy flux $\Phi_*^{(m)}$. (Here, $|g|\mathcal{R}_{m+1}(r)$ is the lower curve.) In this evaluation we used $m = 2$, $(R_o - R_i)/R_o = 0.426$ and $\epsilon\beta_1 = -0.15$. The "y-axis" starts from the lower edge of the frame

5.4.5   Vortices

Figure 5.8 shows $\mathcal{R}_m(r)$ and $|g|\mathcal{R}_{m+1}(r)$ for a situation in which $\Phi \approx 0.997\Phi_*^{(m)}$. If R_i/R_o is not very small and the flux is not very large, these shapes are typical. ($\mathcal{R}_m(r)$ typically changes from increasing to decreasing as Φ is changed from $\Phi_*^{(m-1)}$ to $\Phi_*^{(m)}$.) As we see from Fig. 5.6, for temperatures higher than that of $\mathrm{P}_2^{(m)}$, $|g|$ increases with the flux. This means that, as the flux is increased close to $\Phi_*^{(m)}$, a situation will be reached such that $\mathcal{R}_m(R_o) = |g|\mathcal{R}_{m+1}(R_o)$; for this flux, the leading term in (5.44) vanishes at a point on the outer boundary, at the angle for which $\mathrm{e}^{-mi\theta}$ and $g\mathrm{e}^{-(m+1)i\theta}$ have opposite phases. As the flux is increased further, the point where $\mathcal{R}_m(r) = |g|\mathcal{R}_{m+1}(r)$ and ψ vanishes will occur in the interval $R_i < r < R_o$. Finally this point reaches the inner boundary and, for still larger fluxes, $|\psi|$ will be positive everywhere [until a flux is reached such that $\mathcal{R}_{m+1}(R_o) = |g|\mathcal{R}_{m+2}(R_o)$]. The point (or actually the line parallel to the magnetic field) where $\psi = 0$ is called a *vortex*.

Figure 5.9 shows the contour plots of $|\psi|$ and the corresponding current densities for $\Phi = \Phi_*^{(0)}$ and for $\Phi = \Phi_*^{(2)}$. In both cases, the darkest region contains a vortex. It is remarkable that for $\Phi = \Phi_*^{(0)}$ the vortex is at the thinnest part of the ring and, for $\Phi = \Phi_*^{(2)}$, at its thickest part. The reason is that $C_0 > 0$, so that it follows from (5.51) that $g < 0$ and the lowest order terms in (5.44) cancel for $\theta = 0$; on the other hand, $C_2 < 0$, so that $g > 0$ and $\psi_0^{(2)} + g\psi_0^{(3)} = \mathrm{e}^{-2i\theta}(\mathcal{R}_2(r) + g\mathcal{R}_3(r)\mathrm{e}^{i\theta})$ can vanish for $\theta = \pi$. This is a general feature: for low fluxes, $C_m > 0$ and the vortex is located at the thin side, whereas the opposite situation occurs for large m. The values of m for which C_m becomes negative decrease with the relative width of the ring. The last column in Table 5.1 shows the values of w/R_o at which C_m changes sign. C_m

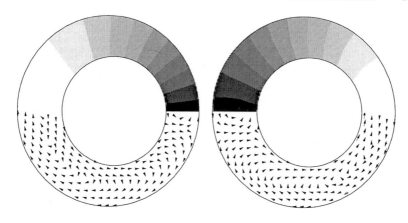

Fig. 5.9. The shape of this sample corresponds to $(R_o - R_i)/R_o = 0.426$, $\epsilon\beta_1 = -0.15$; the other harmonics are taken as 0. In the upper half of the sample we show a contour plot of $|\psi|$; in the lower half, the current density field. Left: $\Phi = \Phi_*^{(0)}$; right: $\Phi = \Phi_*^{(2)}$

does not decrease monotonically with w/R_o; for a full cylinder $(w/R_o = 1)$, $C_m = 0$ for every m. C_0 is always positive (except for $w/R_o = 1$). The effect of the higher order terms in (5.44) is a shift in the position of the vortex by a distance of order ϵ.

We have not found an intuitive argument that explains why the vortex appears at the opposite side of the ring when m crosses the limiting value in Table 5.1, although we have found contradictory handwaving arguments in the literature, unrelated to m. One of them claims that superconductivity should be broken at the thin side, which plays the role of a "defect". The other argument says that, the thinner a layer, the more it favors surface superconductivity and, therefore, the thin side is the one where superconductivity survives.

According to "common wisdom", vortices have a core of size of the order of $\xi = \mu^{-1/2}$, and cannot exist in a region which is narrower than this size. This argument can be used to qualitatively explain the upper critical field in type II superconductivity: the density of vortices is H/Φ_0; this density cannot be much larger than μ, and therefore there is an upper critical field of the order of $\mu\Phi_0$. For a thin film parallel to the field, its width must be at least $1.8\mu^{-1/2}$ [16] in order to be able to accomodate vortices. For a disk, it is similarly found that its radius has to be at least $1.3\mu^{-1/2}$ in order to contain a vortex. Moreover, old experiments have shown that vortices cannot pass through superconducting bridges which are too narrow.

Contrary to the previous paragraph, our formalism predicts that connectivity can force a vortex to be present in the ring, regardless of its width.

Even if the order parameter does not have axial symmetry, its winding number can still be defined as the integral of its phase, i.e. $(i/2\pi) \oint (\nabla\psi \cdot d\mathbf{r}/\psi)$. However, if there is a vortex in the sample, the value of this integral will depend on the path, being larger by one when the vortex is enclosed than when it is not. At first sight, it might seem impossible to pass continuously from an order parameter with winding number m to another with winding number $m+1$, but, from the previous argument, this is what happens as a vortex moves from the outer to the inner boundary: the set of paths with winding number $m+1$ grows at the expense of the set with winding number m, until it fills the entire sample.

Figure 5.10 is a schematic drawing of the streamlines in the sample, close to a degeneracy flux $\Phi_*^{(m)}$. The figure at the right corresponds to a flux slightly higher than that at the left. There are three distinct classes of streamlines. The circuit at the left in each figure represents the screening (Meissner) currents; the field they induce tends to expel the applied field. The circuit at the right represents the currents around the vortex; they circulate in the paramagnetic sense. The circuit that encloses the hole of the sample represents the current that circulates around the ring. If the vortex is located near the outer (respectively inner) boundary, then the current that circulates around the ring has diamagnetic (respectively paramagnetic) sense. If the flux is not sufficiently close to $\Phi_*^{(m)}$, then the vortex and the currents around it are not present. If the flux is quite apart from degeneracy, then both extremes of the screening circuit merge, giving rise to currents that circulate around the ring, in the diamagnetic (respectively paramagnetic) sense close to the outer (respectively inner) boundary. These paramagnetic and diamagnetic regions are separated by a line where the current vanishes; this line moves inwards as the flux increases, until it reaches the inner boundary and a new screening circuit is formed.

Let us finally consider the case of temperatures lower than that of $P_2^{(m)}$. From Fig. 5.6 we see that, as the flux is increased, an unstable branch will be reached and $|g|$ will increase discontinuously. This means that the vortex will jump inwards. If the discontinuity in g is too large (as is the case for temperatures much lower than that of $P_2^{(m)}$) the winding number will jump from m to $m+1$, without the presence of any vortex in the sample.

5.4.6   Inhomogeneous Fields

If the applied field is not uniform, but does have cylindric symmetry, we can still use the formalism exposed at the opening of Sect. 5.4, and (5.27) is still valid. In the present case

$$A(r) = \frac{1}{r} \int_0^r H_e(r')r'dr' . \tag{5.55}$$

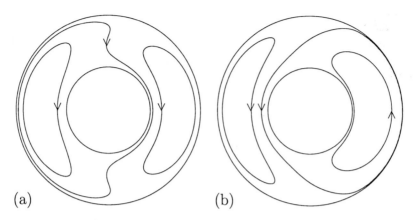

(a)   (b)

Fig. 5.10. Near a degeneracy flux, and provided that the temperature is not much lower than that of P$_2$, there are three types of streamlines. At the left side in each figure, we show the screening (clockwise) current. At the right side we show the vortex currents: they surround a point where the order parameter vanishes and the integral of the gradient of the phase is nonzero along these streamlines. For the circulating current (encloses the hole of the sample) there are two possibilities: **(a)** the vortex is close to the outer boundary, then the circulating current does not enclose the vortex and circulates in the same sense as the screening current; **(b)** opposite case

In general, (5.27) is an ordinary eigenvalue differential equation and can be solved numerically. However, we consider below a few cases in which analytic treatment may be useful.

Zero Field through the Sample In this case, the magnetic potential in the sample is due only to the flux $\phi\Phi_0$ enclosed within the hole $r < R_i$ and $\widetilde{A}(r) = \phi/r$ for $R_i \leq r \leq R_o$. This leads to

$$\mathcal{R}_m(r) = k_J J_{|\phi-m|}(\sqrt{\mu}r) + k_Y Y_{|\phi-m|}(\sqrt{\mu}r) , \qquad (5.56)$$

where $J.(.)$ (respectively, $Y.(.)$) is the Bessel function of the first (respectively, second) kind and the eigenvalue μ and the ratio k_J/k_Y are such that the Neuman condition is fulfilled. Since m enters (5.56) only through $|\phi - m|$, it follows that $\mathcal{R}_m$ and $\mathcal{R}_{m+1}$ have the same eigenvalue for $\phi = m + 1/2$. This means that $\Phi_*^{(m)} = (m+1/2)\Phi_0$, as in the case of a thin shell. Moreover, $\mathcal{R}_m$ and $\mathcal{R}_{m+1}$ become identical for $\Phi = \Phi_*^{(m)}$. Their common expression can be written in the form

$$\mathcal{R}_* = (k_J \sin(\sqrt{\mu}r) + k_Y \cos(\sqrt{\mu}r))/\sqrt{r} . \qquad (5.57)$$

Instead of the case of Fig. 5.8 we have a new scenario: as $|g|$ increases from the situation with winding number m to that with winding number

$m + 1$, $\mathcal{R}_m$ and $|g|\mathcal{R}_{m+1}$ will not become equal at a single point, but rather for all r in the sample. This means that instead of a vortex there will be a *cut* with vanishing ψ across the sample, from the inner to the outer boundary, i.e. the superconducting region becomes singly connected, as we found in the one-dimensional problem. This is a particular case of Theorem 4 in Chap. 3.

Piecewise Constant Field Let us now consider a magnetic field of the form

$$H_e(r) = \begin{cases} h_1 & r \leq R_M \\ h_2 & r > R_M \end{cases}$$

with $R_i < R_M < R_o$.

Following (5.55), for $r \leq R_M$ we recover the case of a homogeneous field $H_e = h_1$; for $r > R_M$ we obtain the same form as for $H_e = h_2$, but m has to be replaced with $m - \phi_M$, where $\phi_M \Phi_0$ is the flux of a field $h_1 - h_2$ through a disk of radius R_M. At $r = R_M$, we require continuity of $\mathcal{R}_m$ and $\mathcal{R}'_m$.

Figure 5.11 shows $\mathcal{R}_m$ and $|g|\mathcal{R}_{m+1}$ for a case $h_2 = -h_1$. The sign reversal of the magnetic field causes a reversal in the growing trends of $\mathcal{R}_m$ and $|g|\mathcal{R}_{m+1}$ and as a result, for $|g|$ in the appropriate range, $|g|\mathcal{R}_{m+1}(r) > \mathcal{R}_m(r)$ far from the boundaries and the opposite holds near the boundaries. This means that we have a different scenario than in Fig. 5.8, since now ψ vanishes at two points. For paths which enclose the hole of the sample and cross the line $\theta = 0$ between these two points, the winding number is $m + 1$; for paths that cross $\theta = 0$ skipping the line between these two points, the winding number is m. The currents around the inner point where ψ vanishes induce a positive field at it, as usual in vortices; the currents around the outer point induce a negative field and we call this point an *antivortex*.

In summary, as $|g|$ increases, a vortex - antivortex pair appears in the middle of the sample $R_i \leq r \leq R_o$ and then the vortex moves towards the inner boundary while the antivortex towards the outer one, until $|\psi| > 0$ everywhere and the winding number is $m + 1$ for every path.

Thin Shells If $(R_o - R_i) \ll R_o$, we can write the flux as the sum of the contribution from the regions inside and outside R_i, and treat the latter as a perturbation. For example, it is interesting to find a condition for the field at which the degeneracy of m and $m+1$ occurs. Writing $\tilde{A} = (m+1/2+\phi_1(r))/r$, the condition for degeneracy, to first order pertubation and for linearized GL, is

$$\int_{R_i}^{R_o} \frac{\phi_1(r)\mathcal{R}_*^2 dr}{r} = 0 , \tag{5.58}$$

where $\mathcal{R}_*$ was defined in (5.57). As a test to (5.58) we have applied it to the degeneracy field of Fig. 5.11 (in spite of the fact that the considered shell is

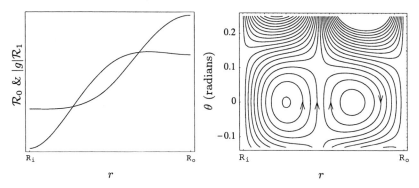

Fig. 5.11. Left: Radial dependence of the leading terms in (5.44), for a case in which the magnetic field switches sign at $r = R_M$. The "y-axis" starts far below the lower edge of the frame. Right: Streamlines at a radial strip of the sample, close to $\theta = 0$. (If the first harmonic is dominant, this is the region where the sample is narrowest.) Along the line $\theta = 0$ there is a vortex (respectively, antivortex) at the side close to the inner (respectively, outer) boundary. For $\theta \gtrsim 0.15$, the screening currents are seen; again, their sense depends on the direction of the local field. For these figures we have used $(R_o - R_i)/R_o = 0.426$ and $R_M = (R_i + R_o)/2$; the degeneracy between winding numbers 0 and 1 occurs when the flux through $r \leq R_M$ is $0.684\Phi_0$

not thin). We obtain

$$\int_{R_i}^{R_o} \frac{\phi_1(r)\mathcal{R}_*^2 \mathrm{d}r}{r} \bigg/ \int_{R_i}^{R_o} \frac{|\phi_1(r)|\mathcal{R}_*^2 \mathrm{d}r}{r} = -3.6 \times 10^{-6} \ .$$

5.4.7   Experimental and Numerical Evidence

Experimental Phase Diagram Kanda *et al.* [10] obtained the phase diagram for a mesoscopic Al ring (Fig. 5.12), by means of voltage-induced tunneling measurements. The ring was intended to have uniform cross section, but, clearly this is never perfectly true in a real experiment. As far as one could expect, their result is in agreement with Fig. 5.5: close to the onset of superconductivity (Γ_I), there is a continuous passage between consecutive winding numbers, whereas far from Γ_I the passage is discontinuous and hysteretic.

ac Susceptibility For narrow rings, the ac susceptibility is proportional to the derivative of the current with respect to the flux. Zhang and Price [17] measured the ac susceptibility for a mesoscopic Al ring. As in the previous case, the ring was intended to have uniform cross section. Again, they found continuous passage between m and $m+1$ close to Γ_I and hysteretic behavior

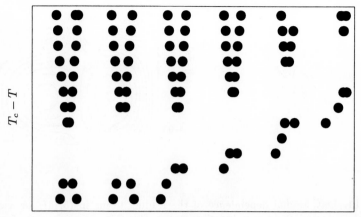

$T_c - T$

magnetic field

Fig. 5.12. Experimental phase diagram [10]. The lower points indicate the normal/superconducting transition; the points at the left (respectively, right) of the "V"-shaped formations show the stability limit of the state with winding number $m + 1$, where it decays into m (respectively, decay from m to $m + 1$). In this experiment, $w/R_o = 0.38$ and β_1 is unknown

far from this line. Moreover, they found magnetic signatures of the critical points $P_2^{(m)}$. Close to Γ_I, the passage between winding numbers shows paramagnetic ac susceptibilities, as predicted. This feature was discussed in [5]; note, however, that for a wide ring as used in this experiment, evaluation of the induced flux as being due to the current around the ring may not be a good approximation, and the contributions of the entire current distribution ought to be taken into account.

The divergence of the ac susceptibility at $P_2^{(m)}$, predicted at the end of Sect. 5.4.4, is consistently found. In Fig. 5.13 we show the ac susceptibility for the temperatures at which this divergence appears. We emphasize that these are precisely the limiting temperatures between the continuous and hysteretic regimes for each m. Setting $w/R_o = 0.426$ (as reported in this experiment) in Fig. 5.7, we obtain that the values of μ at $P_2^{(0)}$, $P_2^{(1)}$ and $P_2^{(2)}$ are in the ratio 1:1.4:2.8 for a deviation from axial symmetry $|\beta_1| = 0.15$. (In the experiment, $|\beta_1|$ was not determined.) The ratios among the values of μ are to be compared with those of the temperatures of the experimental candidates for $P_2^{(m)}$, measured from T_c. Taking the reported value $T_c = 1.266$ K, these temperature differences are in the ratio 1:1.5:2.0, in semiquantitative agreement with theory.

Far from Axial Symmetry The case of a sample with cylindric eccentric boundaries was treated in [7], using linearized GL and a variational approach.

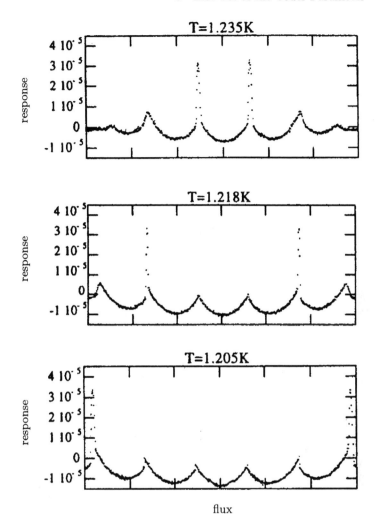

Fig. 5.13. ac susceptibility (which is proportional to the experimental "response") as a function of the flux. (Reproduced from Ref. [18]; $w/R_o = 0.426$). Each temperature corresponds to the limit between continuous and discontinuos transitions between consecutive winding numbers, $0 \leftrightarrow 1$, $1 \leftrightarrow 2$ and $2 \leftrightarrow 3$. The flux vanishes in the middle of the graphs and every tick corresponds to Φ_0 through the average radius of the sample. Note that, in each graph, for winding numbers lower (respectively, higher) than those involved in the divergence, the transition is discontinuous (respectively, continuous)

The same scenario obtained here near axial symmetry, shows up for large eccentricity. For small winding numbers, vortices pass through the sample across its thinnest part, whereas for large m they pass across the opposite side. For a sample with relative width $(R_o - R_i)/R_o = 0.31$, the first vortex that crosses through the thickest side occurs for the passage from $m = 3$ to $m = 4$. According to Table 5.1, this is expected for relative width in the range $0.22 \leq (R_o - R_i)/R_o \leq 0.29$, indicating that eccentricity tends to lower the effective relative width.

The results obtained in this section are based on the assumption that the induced potential can be neglected. This is expected for $\mu h(R_o - R_i)\kappa^{-2} \ll 1$ in the case of thin samples, or $\sqrt{\mu}(R_o - R_i)/\kappa \ll 1$, in the case of long samples. In the opposite extreme, we expect bulk behavior, i.e. Meissner or Abrikosov state. We would like to have an estimate for the value of $\mu h(R_o - R_i)\kappa^{-2}$ at which our scenario breaks down. This is provided by the results of Baelus *et al.* [1]. They also considered disks with a circular hole at an asymmetric position, but used the entire GL equations, which they solved numerically.

In one of their examples, $\mu h(R_o - R_i)\kappa^{-2} \sim 0.1$. They did obtain hysteresis at the transition between $m = 0$ and $m = 1$, which at the temperature they considered is far from the superconductivity onset, and did not find hysteresis for the transitions at higher m (which are closer to Γ_I; see Fig. 5.5). For the continuous transitions ($m \geq 1$), they always found the vortex at the thick side of the ring, as expected from Table 5.1 and their ratio $(R_o - R_i)/R_o = 0.75$. For large eccentricity, the transition between $m = 0$ and $m = 1$ becomes continuous; in this regime the vortex is found at the thin side of the ring [12], again as expected. Surprisingly, for large eccentricity the transition between $m = 1$ and $m = 2$ becomes discontinuous. Apparently, the effective relative width is reduced to values close to 0.44, so that $C_1 \sim 0$ and the anomalous situation discussed in p. 161 is encountered. This situation could be checked by varying the temperature: if $C_1 \sim 0$, $\Gamma_{II}^{(1)}$ and Γ_I should be very close to each other.

For their example with $\mu h(R_o - R_i)\kappa^{-2} \sim 0.2$, *two* vortices appear simultaneously, indicating that our formalism is no longer valid.

Nonsmooth Boundaries We have studied the case of a long (i.e. independent of z) sample with eccentric squared boundaries. Half of its cross section is shown in Fig. 5.14. The GL equations were solved numerically, by means of the program described in Chap. 8. This time the induced potential was not neglected. The results are qualitatively the same as for the case of circular boundaries. Far from the superconducting/normal transition, the winding number changes hysteretically as the magnetic field is swept, whereas close to this transition the change is mediated by the continuous passage of a vortex. When the winding number changes from 0 to 1, or from 1 to 2, the vortex passes through the thin part of the sample; when the winding number changes from 2 to 3, the vortex passes through the thick part. Making the natural identification of $R_{i,o}$ as a length proportional to the respective perimeter, the studied sample obeyed $(R_o - R_i)/R_o = 1/2$; according to Table 5.1 the effec-

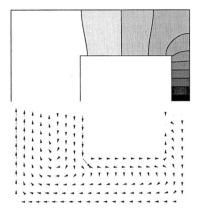

Fig. 5.14. Numerically evaluated contour plot of $|\psi|$ and field plot of the current density. There is a vortex in the middle of the narrow branch. Most of the current is screening current, but a few arrows are seen around the vortex and the circulating current is seen in the counterclockwise sense along the inner boundary. Here each outer edge equals 1.27 coherence lengths, the flux enclosed by the outer boundary is $1.07\Phi_0$ and $\kappa = 0.4$

tive value of $(R_o - R_i)/R_o$ was somewhat lower, as in the case of eccentric circular boundaries.

The shape of the magnetization we found is similar to that in Fig. 20b in [1]. The maximum of the magnetization is not necessarily found at the first ($m = 0$) peak.

The presentation of Fig. 5.14 suggests that $|\psi|$ and the current have the same mirror symmetry as the sample. For the magnetic fields at which there was a vortex in the sample, this appeared to be indeed the case, but in general, this does not seem to be true.

5.4.8 Josephson Junctions

In this section we have considered the situation where $T \sim T_c$ and, in addition, the shape of the sample is not extremely far from a uniform cross section. We shall now review the opposite situation: $T \ll T_c$ and the sample contains a "weak link", which is the only segment where the current density is not zero in the interior of the sample. This weak link is called a Josephson junction; the structure which consists of a loop with a weak link is called "rf-SQUID", and was studied long ago [11].

The behavior of the rf-SQUID is basically the same as that obtained in Sect. 5.4.4: there are two possible regimes, one with continuous and the other with discontinous passage between consecutive winding numbers. However, whereas in Sect. 5.4.4 the requirement for continuous passage is a low value of $\mu_1/(|\beta_1|\mu_P^{(m)})$ [this can be obtained from (5.54), estimating A_m, B_m, C_m, etc. from their definitions], in the rf-SQUID case the condition is $2\pi L J_c \leq \Phi_0$, where L is here the self-inductance of the loop and J_c the maximum current which can pass through the junction.

Increasing $|\beta_1|$ produces a narrower constriction at the thin side of the ring and, therefore, a lower possible current through this constriction. In this sense, J_c may be regarded as a continuation of the effect provided by $1/|\beta_1|$. A large value of $\mu_1/\mu_P^{(m)}$ means strong superconductivity and hence, large J_c. In this sense, J_c may also be regarded as the continuation of the influence of the ξ-dependence of μ_1. However, the rf-SQUID is not influenced by the length of the loop, whereas $\mu_P^{(m)} \propto R_o^{-2}$. This comparison suggests that, once the perimeter is much larger than the coherence length, its precise value is not important. At the other extreme, near the onset of superconductivity, the induced field is negligible and thus the self-inductance is unimportant.

It would be interesting to have a study that bridges between these two extreme situations.

5.5   Main Conclusions

A mesoscopic superconducting ring exhibits qualitatively different behaviors, depending on whether it has or does not have an axis of rotation symmetry. This difference shows up for magnetic fields in ranges close to the degeneracy fields, at which the energies of states with consecutive winding numbers coincide.

We have mostly studied the case of samples with an eccentric hole. In this case and near the onset of superconductivity, the state of the sample can pass continuously from a given winding number to the next. This occurs by means of a vortex which crosses the sample. This vortex may cross either through the thinnest or through the thickest part the sample, depending on the winding number and on the ratio between the width and the sample radius. Table 5.1 tells us which scenario occurs.

An additional feature, is the existence of the critical point P_2, discussed in Sect. 5.4.4. At P_2, the magnetic susceptibility diverges quadratically.

Several predictions stem from the present study, and they cry out for experimental verification. The experiments reviewed in Sect. 5.4.7 were not designed to check the present predictions. Rather, they were intended to check unrelated aspects of the ideal case where there is axial symmetry, and the qualitative agreement with our predictions was revealed by accident, thanks to experimental imperfections.

The feature which seems to be the most easily measurable is the magnetic susceptibility near P_2. This has been done by means of SQUID microsusceptometry [17,18] and experimentally equivalent situations have been coped with by means of ballistic Hall junctions [9]. A possibility which seems especially attractive would be to measure the magnetization of half of the sample, since this could reveal in which half the vortex is found.

In order to measure the order parameter directly, the most appropriate technique seems to be scanning tunneling microscopy/spectroscopy. It is hard to use this tecnique in Al (which is popular in the fabrication of mesoscopic samples, due to its large coherence length). However, successful measurements have been performed in In disks. It is also possible to measure the superconducting gap at fixed tunneling points, as in [10]. In this experiment, there were only two contact points, so that no information was obtained on the shape of the order parameter. If the same kind of experiment were repeated with several leads, the shape of the order parameter could be mapped.

Acknowledgements

I wish to thank Akinobu Kanda for sending me the data for Fig. 5.12, Gustavo Buscaglia for sending me the computer program used for the case of squared boundaries, John Price for sending me Xianxian Zhang's thesis and unpublished data, and François Peeters for sending me unpublished results.

References

1. B. J. Baelus, F. M. Peeters, V. A. Schweigert: Phys. Rev. B **61**, 9734 (2000)
2. R. Benoist, W. Zwerger: Z. Phys. B **103**, 377 (1997)
3. J. Berger: Physica C **332**, 281 (2000)
4. J. Berger, J. Rubinstein: Physica C **288**, 105 (1997)
5. J. Berger, J. Rubinstein: Phys. Rev. B **56**, 5124 (1997)
6. J. Berger, J. Rubinstein: SIAM J. Appl. Math. **58**, 103 (1998)
7. J. Berger, J. Rubinstein: Phys. Rev. B **59**, 8896 (1999)
8. E. H. Brandt: Phys. Rev. B **58**, 6506 (1998)
9. A. K. Geim *et al.*: Appl. Phys. Lett. **71**, 2379 (1997); A. K. Geim, S. V. Dubonos, J. G. S. Lok, M. Henini, J. C. Maan: Nature **396**, 144 (1998)
10. A. Kanda, M. C. Geisler, K. Ishibashi, Y. Aoyagi, T. Sugano: Physica B **284**, 1870 (2000)
11. O. V. Lounasmaa: *Experimental Principles and Methods Below* 1K. (Academic, London 1974)
12. F. Peeters: unpublished
13. G. Richardson: Quart. Appl. Math., in press
14. J. Rubinstein: 'Six lectures on superconductivity'. In: *Proc. of the CRM Summer School on Boundaries, Interfaces and Transitions*, ed. by M. Delfour (American Mathematical Society, 1998) pp. 163–184
15. J. Rubinstein, M. Schatzman: J. Math. Pure Appl. **77**, 801 (1998)
16. D. Saint-James, G. Sarma, E. J. Thomas: *Type II superconductivity* (Pergamon, Oxford 1969)
17. X. Zhang, J. C. Price: Phys. Rev. B **55**, 3128 (1997)
18. X. Zhang: SQUID Microsusceptometry of Mesoscopic Superconducting Rings. PhD Thesis, University of Colorado, Boulder (1996)

6   Persistent Currents in Ginzburg–Landau Models

Luís Almeida[1] and Fabrice Bethuel[2]

[1] Laboratoire J.A. Dieudonné - CNRS (UMR 6621), Université de Nice - Sophia-Antipolis, Parc Valrose, F-06108 NICE Cédex 02, France
[2] Laboratoire d'Analyse Numérique - CNRS (UMR 7598), Université Pierre et Marie Curie, Boite Courrier 187, 4, Place Jussieu, F-75252 PARIS Cédex 5, France

6.1   Introduction

Persistent currents in superconducting rings were predicted in the early 1950's (see [17]), and then experimentally observed in the early 1960's (see [12] and the notes that follow it). Moreover, it was shown that they are extremely stable, lasting for years!

The original experiment was performed with type I superconductors. Roughly speaking, it went as follows: they took a ring-shaped sample at ambient temperature. They applied an external magnetic field, decreased the temperature past the superconducting transition temperature and finally removed the field. A persistent current was then observed: moreover, the flux of the magnetic field through the inner hole was quantized. Note that our analysis leads us to believe that, at least in the case of high κ and low applied field, if one would proceed differently, for instance by decreasing the temperature first and only then turning on the external field, nothing should be observed .

A natural problem is to analyze this phenomenon is the framework of Ginzburg-Landau models and, if possible, to derive (rigorous) mathematical proofs. There are at least two important questions which should be considered:

1) Explain why some flux remains trapped in the ring.

2) Explain the very strong stability (persistent character) of these states. As we will see below, we have a rather satisfactory mathematical explanation for this property in the case of high κ (extreme type II) superconductors.

6.2   Onset of Currents - Trapped Fluxes

In this section we briefly discuss the first question. In the Ginzburg-Landau theory the carriers of superconductivity (i.e. the cooper pairs) are modeled by a complex-valued function whose squared norm, $|\psi|^2$, represents the cooper pair density. In the presence of an external constant applied magnetic field

h_{ex}, the Gibbs energy is given by

$$F(u, A) = \int_{\mathbb{R}^2} \frac{|h - h_{\mathrm{ex}}|^2}{8\pi} + \int_{\Omega} \frac{1}{2m^*} |(\frac{\hbar}{i}\nabla - \frac{e^*}{c}A)u|^2 + \int_{\Omega} V(u) , \qquad (6.1)$$

where Ω is a 2-dimensional domain and

$$V(u) = (\frac{\alpha}{2}|u|^2 + \frac{\beta}{4}|u|^4)$$

$$= \frac{\beta}{4}(|u|^2 + \frac{\alpha}{\beta})^2 - \frac{\alpha^2}{4\beta},$$

Here, m^*, e^* stand for the mass and the charge of the cooper pair. It is conventional to take $e^* = 2e$ and $m^* = 2m$, where e and m denote the mass and charge of a free electron. The first of these identities is firmly established on an experimental level. However, the second one is deduced from a free-electron approximation, and is only a crude estimate in the case of a real material. In fact, m^* depends on the material and the experimental determination of its exact value is highly involved (see [21], section 4.1).

Note that critical points of $F(u, A)$ satisfy the corresponding Euler–Lagrange equations, called Ginzburg-Landau equations,

$$\begin{cases} \dfrac{1}{m^*}(\dfrac{\hbar}{i}\nabla - \dfrac{e^*}{c}A)^2 u = \alpha u + \beta|u|^2 u, \\ -\dfrac{1}{4\pi}\nabla^T h = (iu, (\hbar\nabla - i\dfrac{e^*}{c}A)u). \end{cases} \qquad (6.2)$$

where $\nabla^T = i\nabla$.

The parameters α and β depend on the material and on the temperature. To emphasize this temperature dependence we will often write $\alpha(T)$ and $\beta(T)$. Stable configurations will be (local) minimizers for $F(u, A)$. The parameter β is positive, whatever T, whereas α changes sign for a critical temperature $\alpha = T_c$. More precisely,

$$\begin{aligned} \alpha(T) \leq 0, \quad \text{if } T \leq T_c , \\ \alpha(T) \geq 0, \quad \text{if } T \geq T_c , \end{aligned} \qquad (6.3)$$

Near T_c, α behaves like $C(T - T_c)$, where C is some strictly positive constant, and therefore is monotone. We will assume throughout that β is a constant (which is a good approximation in the vicinity of T_c).

Clearly, when $T > T_c$, α being positive, the potential $V(u)$ is convex and hence so is the functional $F(u, A)$. Therefore it has a unique (local) minimizing configuration, namely

$$u = 0, \quad h = h_{\mathrm{ex}} , \qquad (6.4)$$

for which $F = 0$. The vector potential A_{ex} corresponding to this configuration is taken as $A_{\mathrm{ex}} = h_{\mathrm{ex}}A_0$, where A_0 is any solution of the equation

$$h_0 = \operatorname{curl} A_0 = 1.$$

For instance, we may choose $A_0 = \dfrac{1}{2}(-y, x) = \dfrac{i}{2} r \exp(i\theta)$.

For $T < T_c$, the potential V is no longer convex and thus the trivial solution (6.4), although still critical, is not necessarily minimizing. Hence, the first interesting question is to determine when the trivial solution ceases to be minimizing, i.e., when the (first) transition occurs. This question has already been addressed, from different points of view, in various papers (e.g. [10], [5] and [19]).

Remark: For minimizers a simple comparison principle (maximum principle) yields

$$|u|^2 \le \frac{|\alpha|}{\beta}.$$

6.2.1 The Transition Temperature

Near the trivial solution (6.4), the quadratic part of the functional is

$$Q(u, \tilde{A}) = \int_{\mathbb{R}^2} \frac{|\operatorname{curl} \tilde{A}|^2}{8\pi} + \frac{1}{2m^*} \int_\Omega |(\frac{\hbar}{i}\nabla - \frac{e^*}{c} h_{\mathrm{ex}} A_0) u|^2 + \frac{\alpha(T)}{2} \int_\Omega |u|^2 ,$$

Consider next the eigenvalue problem for the operator $\frac{1}{m^*}(\frac{\hbar}{i}\nabla - \frac{1}{2m^*}\frac{e^*}{c} h_{\mathrm{ex}} A_0)^2$ with Neumann boundary conditions (where we have the strictly positive first eigenvalue λ_1 and associated eigenfunction ψ_1), i.e.

$$\begin{cases} \dfrac{1}{m^*}(\dfrac{\hbar}{i}\nabla - \dfrac{e^*}{c} h_{\mathrm{ex}} A_0)^2 \psi_1 = \lambda_1 \psi_1 , & \text{in } \Omega \\[2mm] n \cdot (\dfrac{\hbar}{i}\nabla - \dfrac{e^*}{c} h_{\mathrm{ex}} A_0) = 0, & \text{on } \partial\Omega. \end{cases} \tag{6.5}$$

Note that the eigenvalue λ_1 depends on m^*, h_{ex} and the domain, whereas ψ_1 depends only on h_{ex} and the domain Ω. We renormalize ψ_1 so that

$$\int_\Omega |\psi_1|^2 = 1, \tag{6.6}$$

and recall that ψ_1 is a solution of the minimization problem

$$\lambda_1 = \inf\{\frac{1}{m^*} \int_\Omega |(\frac{\hbar}{i}\nabla - \frac{e^*}{c} h_{\mathrm{ex}} A_0)\psi|^2, \int_\Omega |\psi|^2 = 1\}.$$

Our claim is the following.

Proposition 1 *Let T_1 be such that $\alpha(T_1) = -\lambda_1$. Then, for*

i) $T \ge T_1$, *the trivial solution (6.4) is (locally) minimizing.*

ii) $T < T_1$, *the trivial solution (6.4) is no longer minimizing.*

Therefore, the transition occurs at $T = T_1$.

Remark: Notice that if $\Omega = \mathbb{R}^2$, then, $\lambda_1 = 0$, and thus $T_1 = T_c$.

Proof of the proposition: Since λ_1 is the first eigenvalue, we have that for any admissible configuration u

$$\frac{1}{m^*} \int_\Omega |(\frac{\hbar}{i}\nabla - \frac{e^*}{c}h_{\mathrm{ex}}A_0)u|^2 \geq \lambda_1 \int_\Omega |u|^2 ,$$

and therefore,

$$F(u,\widetilde{A}) \geq \frac{\alpha + \lambda_1}{2} \int_\Omega |u|^2 .$$

If $\alpha(T) + \lambda_1 \geq 0$, this last expression is positive and thus the quadratic form is positive definite and assertion *i)* is proved.

If $\alpha(T) + \lambda_1 < 0$, take $u = \mu\psi_1$ and $A = h_{\mathrm{ex}}A_0$. Then, we have

$$F(u,A) = \frac{1}{2m^*} \int_\Omega \mu^2 |(\frac{\hbar}{i}\nabla - \frac{e^*}{c}h_{\mathrm{ex}}A_0)\psi_1|^2 + \frac{\alpha\mu^2}{2} \int_\Omega |\psi_1|^2 + \frac{\beta\mu^4}{4} \int_\Omega |\psi_1|^4$$

$$= \mu^2 \frac{\alpha + \lambda_1}{2} \int_\Omega |\psi_1|^2 + \frac{\beta\mu^4}{4} \int_\Omega |\psi_1|^4 .$$

$$(6.7)$$

Since the first factor $(\lambda_1 + \alpha)$ is negative, choosing $\mu > 0$ sufficiently small, we can make $F(u,A) < 0$, and thus the conclusion follows.   ∎

Proposition 2 *In the vicinity of T_1, a minimizing configuration (u,A) behaves like*

$$\begin{cases} u = \mu_0\psi_1 + o(\varepsilon^3) \\ A = h_{\mathrm{ex}}A_0 + O(\varepsilon^2) \end{cases}$$

and the lower order terms can be made precise. Here, as before, ψ_1 denotes an eigenfunction for λ_1 and the constant μ_0 is given by

$$\mu_0 = C(h_{\mathrm{ex}},\psi_1)\varepsilon, \tag{6.8}$$

where the constant $C(h_{\mathrm{ex}},\psi_1)$ can be computed explicitly in terms of h_{ex} and ψ_1 (see the remark below), and

$$\varepsilon^2 := |\lambda_1 + \alpha(T)|.$$

Sketch of the Proof: We will sketch the proof for the simpler functional

$$E(u) = \frac{1}{2} \int_\Omega (|\nabla u|^2 + \alpha|u|^2 + \frac{\beta}{2}|u|^4),$$

and its minimizer $u_\star$ in H_0^1. Let λ_1, u_1, be the first eigenvalue and eigenfunction of $-\Delta$ in Ω. They satisfy

$$\begin{cases} -\Delta u_1 = \lambda_1 u, & \text{in } \Omega, \\ u_1 = 0, & \text{on } \partial\Omega, \\ |u_1|_{L^2} = 1. \end{cases}$$

The first step is to prove that $u_\star \to 0$ as $\alpha \to -\lambda_1^-$. This can be deduced from the fact that $E(u_\star) \to 0$ when $\alpha \to -\lambda_1^-$.

Next, let v_1 be the L^2 orthogonal projection of $u_\star$ on $\mathbb{R}u_1$ (the one dimensional eigenspace generated by u_1). Then, we may decompose,

$$u_\star = v_1 + w,$$

where $w \in (\mathbb{R}u_1)^\perp$. Note that, by orthogonality,

$$\|v_1\| \le \|u\|.$$

The Euler-Lagrange equation for $u_\star$ is,

$$-\Delta u_\star + \alpha u_\star = -\beta|u_\star|^2 u_\star,$$

which yields for w,

$$-\Delta w + \alpha w = -\beta|u_\star|^2 u_\star + \varepsilon^2 v_1.$$

Note that the restriction of the operator $(-\Delta + \alpha\, Id)$ to $(\mathbb{R}u_1)^\perp$ is elliptic and its image is $(\mathbb{R}u_1)^\perp$. It follows that

$$\|w\| \le C(\epsilon^2\|u_\star\| + \|u_\star\|^3), \tag{6.9}$$

and that

$$\int \beta|u_\star|^2 u_\star u_1 = \varepsilon^2 \int v_1 u_1 . \tag{6.10}$$

From (6.9) we obtain

$$\|w\| = o(1)\|u_\star\|$$

and writing $v_1 = \mu u_1$, we have

$$\|w\| = o(\mu)$$

and

$$u_\star = \mu v_1 + o(\mu) \tag{6.11}$$

Inserting this into (6.10), we obtain

$$\mu^3 \int \beta|v_1|^4 + o(\mu^3) = \varepsilon^2 \mu.$$

i.e.

$$\mu = \frac{\varepsilon}{[\beta \int |\psi_1|^4 + o(1)]^{1/2}} = \frac{\varepsilon}{[\beta \int |\psi_1|^4]^{1/2}} + o(\varepsilon). \tag{6.12}$$

Going back to (6.9) we deduce

$$\|w\| = o(\varepsilon^3).$$

Inserting this into (6.9) we obtain the desired conclusion.

Remarks:

1) Estimate (6.12) concerning the value of μ can also be derived by arguing that $u_\star$ is a minimizer. Following the same type of ideas, we will now show how the value of $C(h_{\mathrm{ex}}, \psi_1)$ in equation (6.8) can be made precise. We write, as before, the expansions

$$\begin{cases} u = \mu_0 \psi_1 + o(\mu_0^3) \\ A = h_{\mathrm{ex}} A_0 + \mu_0^2 \tilde{A} + l.o.t. \end{cases} \tag{6.13}$$

Set $\tilde{h} = \partial_{x^1} \tilde{A}_2 - \partial_{x^2} \tilde{A}_1$. In view of the Ginzburg-Landau equations, we deduce

$$\begin{cases} -\dfrac{\nabla^T \tilde{h}}{4\pi} = (i\psi_1, (\hbar\nabla - i\dfrac{e^*}{c}A)\psi_1), \\ \tilde{h} = 0, \quad \text{on } \partial\Omega. \end{cases}$$

so that

$$\begin{cases} -\dfrac{\Delta \tilde{h}}{4\pi} = \hbar(\partial_{x^1}\psi_1 \times \partial_{x^2}\psi_1) - \dfrac{e^*}{c}h_{\mathrm{ex}}|\psi_1|^2, \\ \tilde{h} = 0, \quad \text{on } \partial\Omega. \end{cases}$$

This equation completely determines $\tilde{h}$, and hence also $\tilde{A}$, by

$$\tilde{A} = -\nabla^T \tilde{\zeta}, \quad \text{where} \quad \Delta\tilde{\zeta} = \tilde{h} \quad \text{on } \mathbb{R}^2.$$

Inserting Ansatz (6.13) into the computation of $F(u, A)$, and neglecting terms of order higher than four in μ_0, we obtain

$$F(u, A) = \mu_0^2 \frac{\lambda_1 + \alpha}{2} \int_\Omega |\psi_1|^2$$
$$+ \mu_0^4 \left(\frac{\beta}{4} \int_\Omega |\psi_1|^4 + \frac{1}{8\pi} \int_\Omega |\tilde{h}|^2 + \frac{1}{2m^*} \int_\Omega ((\frac{\hbar}{i}\nabla - \frac{e^*}{c}h_{\mathrm{ex}}A_0)\psi_1, \tilde{A}\psi_1) \right) + \dots$$
$$= \frac{\mu_0^2 \varepsilon^2}{2} + \mu_0^4 \left(\frac{\beta}{4} \int_\Omega |\tilde{h}|^2 + \frac{1}{2m^*} \int_\Omega ((\frac{\hbar}{i}\nabla - \frac{e^*}{c}h_{\mathrm{ex}}A_0)\psi_1, \tilde{A}\psi_1) \right) + \dots$$

The value of μ_0 which minimizes this expression is then given by

$$\mu_0 = \left[\beta \int_\Omega |\psi_1|^4 + \frac{1}{2\pi} \int_\Omega |\tilde{h}|^2 + \frac{2}{m^*} \int_\Omega ((\frac{\hbar}{i}\nabla - \frac{e^*}{c}h_{\mathrm{ex}}A_0)\psi_1, \tilde{A}\psi_1) \right]^{-1/2} \varepsilon,$$

so that

$$C(h_{\text{ex}}, \psi_1) =$$

$$\left[\beta \int_\Omega |\psi_1|^4 + \frac{1}{2\pi} \int_\Omega |\tilde{h}|^2 + \frac{2}{m^*} \int_\Omega ((\frac{\hbar}{i}\nabla - \frac{e^*}{c}h_{\text{ex}}A_0)\psi_1 \ , \ \tilde{A}\psi_1) \right]^{-1/2} \quad (6.14)$$

A more rigorous proof of expansion (6.13) as well as the value of $C(h_{\text{ex}}, \psi_1)$ is given in [4].

2) In the case of multiple eigenfunctions (for λ_1), ψ_1 has to maximize

$$\left[\beta \int_\Omega |\psi_1|^4 + \frac{1}{2\pi} \int_\Omega |\tilde{h}|^2 + \frac{2}{m^*} \int_\Omega ((\frac{\hbar}{i}\nabla - \frac{e^*}{c}h_{\text{ex}}A_0)\psi_1 \ , \ \tilde{A}\psi_1) \right] \quad (6.15)$$

among all eigenfunctions. A more precise description of branches of solutions emanating (in that case) from the trivial on is given in [4].

Since $u \simeq \mu_0\psi_1$, an important issue is to determine the properties of ψ_1. This problem has been considered extensively. In particular, the precise form of ψ_1 has been given for special domains: St. James and de Gennes [20], Bolley and Helffer [7], Dauge and Helffer [11], and Del Pino, Felmer and Sternberg [13] for the half-plane, and Bauman, Phillips and Qi [5] for a disk.

The analysis is considerably simpler in case ψ_1 has no zeros. Unfortunately, this can not be excluded a priori. In the case of zero magnetic field inside the domain, it has been shown that for ring-shaped domains, ψ_1 must have zeros if the flux through the inner hole is half an odd integer multiple of the quantum of magnetic flux (see Berger and Rubinstein [6] and Helffer, Hoffman-Ostenhof, Hoffman-Ostenhof and Owen [14]). More precisely, the zero set consists of a line connecting the two components of the boundary (more general multiply-connected domains have also been considered in [14]). However, for generic values of the flux, we do not know if ψ_1 has zeros or not.

In the case we consider here, the field inside the domain is not zero: it is actually constant equal to h_{ex}. Therefore, the previous results do not apply (straightforwardly at least). Nevertheless, one might expect that they somehow remain valid for very thin ring-shaped domains (see works by Berger and Rubinstein in this direction). For general domains, we however are not aware of any result in that direction and this is indeed a very challenging problem. We believe that for certain specific values of h_{ex}, ψ_1 will have zeros: yet, for generic values of h_{ex} the zero set should be empty. We will assume this throughout – then, denote by d_1 the winding number of ψ_1. A natural question is then clearly:

(Q1) How does d_1 relate to the applied magnetic field?

In order to get some insight, we address first the question in the case Ω is an annulus

$$\Omega := B(R_1) \setminus B(R_0), \quad \text{for } 0 < R_0 < R_1,$$

where $B(R)$ is the disc of radius R.

6.2.2   The Case of an Annulus

Assuming ψ_1 has no zeros, we may write

$$\psi_1 = \rho \exp(i(\varphi + d_1\theta)), \quad \text{in } \Omega := B(R_1) \setminus B(R_0), \tag{6.16}$$

where the function φ is single-valued. We have

$$\int |\psi_1|^2 = \int |\rho|^2 = 1,$$

which is independent of d_1. In order to compute

$$\frac{1}{m^*} \int_\Omega |(\frac{\hbar}{i}\nabla - \frac{e^*}{c}A)\psi_1|^2$$

it is convenient to use polar coordinates and write, using $A = A_0 h_{ex} = h_{ex}(-y, x)/2$,

$$\int_\Omega |(\frac{\hbar}{i}\nabla - \frac{e^*}{c}A)\psi_1|^2$$
$$= \int_{[R_0,R_1]\times[0,2\pi]} r \left[|\hbar\frac{\partial\psi_1}{\partial r}|^2 + \rho^2[\frac{\hbar}{r}(\frac{\partial\varphi}{\partial\theta} + d_1) - \frac{e^*}{2c}rh_{ex}]^2\right] dr d\theta$$
$$= \int_\Omega \hbar^2[\rho^2|\nabla\varphi|^2 + |\frac{\partial\rho}{\partial r}|^2] + 2\pi \int_{R_0}^{R_1} r\rho^2[\frac{\hbar}{r}d_1 - \frac{e^*}{2c}rh_{ex}]^2]dr . \tag{6.17}$$

Since we have to minimize the previous quantity with respect to the constraint $\int |\psi|^2 = 1$, clearly we have to take $\varphi = $ constant, for instance, $\varphi = 0$. It follows that we have (as in Bauman, Phillips, Qi) the 1–dimensional minimization problem for the function φ defined on the interval $[R_0, R_1]$, i.e. we have to minimize

$$G(\rho) = \frac{\hbar^2}{m^*} \int_{R_0}^{R_1} r \left[|\frac{\partial\rho}{\partial r}|^2 + \frac{\rho^2}{r^2}[d_1 - \frac{\pi r^2 h_{ex}}{\varphi^*}]^2\right] dr ,$$

under the constraint

$$2\pi \int_{R_0}^{R_1} r\rho^2 dr = 1.$$

Here, both the function ρ and the integer d_1 are unknowns.

Setting $\phi_i = \pi R_i^2 h_{ex}$, for $i = 0, 1$ (i.e. ϕ_i is the magnetic flux through $B(R_i)$), it clearly appears from the previous minimization problem that we should have

$$\chi(\frac{\phi_0}{\phi_*}) \le d_1 \le \chi(\frac{\phi_0}{\phi_*}) + 1. \tag{6.18}$$

where $\chi(x)$ denotes the integer part of x. If R_0, R_1, $R_1 - R_0$ are all large, as well as h_{ex}, then the ratio $\pi r^2 h_{ex}/\phi_*$ is also large. This fact, and simple heuristic arguments related to the previous minimization problem, lead us to believe that ρ (and therefore also ψ_1) concentrate near the boundary of the annulus, and that d_1 is close to $\chi(\frac{\phi_0}{\phi_*})$. The discussion for small (or thin) annuli seems more delicate.

Remark: From (6.18) we see that d_1 increases with h_{ex}. In particular, each time there is a jump in d_1, the eigenvalue will be degenerate, and there will be an associated eigenfunction with a non-empty zero set. In this situation, the Ansatz (6.16) clearly does not make sense. Let us emphasize again nevertheless that we believe such a situation is not generic (in values of h_{ex}).

In case the eigenvalue is degenerate, we believe that the eigenspace corresponding to the first eigenvalue λ_1 is spanned by two radially symmetric eigenfunctions corresponding to different degrees. Each of them should be a maximizer of (6.15). By a convexity argument, the branch of minimizers of the energy should correspond to one of them. However, let us emphasize that these are only guesses and that the real picture might be more elaborated.

6.2.3   General Domains

Here we do not have any mathematical result. Nevertheless, we believe that (6.18) remains true (where now ϕ_0 and ϕ_1 represent the flux through the areas limited by the inner and outer boundaries of our domain, respectively). The analysis for large domains and large fields should also remain valid.

6.3   Temperature Below T_c

Recall that in the experiment the external field is kept constant as the temperature is lowered. We may assume that for each temperature the system is in a (stable) critical point (i.e. that the transformation is quasi-stationary). It is likely that along its evolution the system follows a branch of solutions (which bifurcates at $T = T_c$ from the trivial solution). Moreover, the winding number should be constant along this branch.

Consider again the case of an annulus and assume that λ_1 is non degenerate and that ansatz (6.16) holds. In this case we are ensured, by standard bifurcation theory, that the branch of minimizers, in the vicinity of $T = T_c$, consists of symmetric functions (of degree d_1).

The branch of symmetric solutions exists for all $T < T_c$. Actually they are minimizing in the class of radially symmetric configurations. If the inner hole of the annulus is large (when compared to the characteristic size of a giant vortex of degree equal to the degree associated to the branch considered), we believe they are stable (in contrast with the case of the disk - see [5]). In that case, as the temperature decreases, one might therefore expect that the system remains on this branch.

6.4   Stability of Currents once the External Field Is Removed

6.4.1   Rewriting F

At the end of the first stage of the experiment the temperature is decreased to its final value below T_c and kept fixed. Consequently, the parameters $\alpha < 0$ and $\beta > 0$ are also fixed and we may take the ratio $\alpha/\beta = -1$ (up to a change of units). Therefore, (6.1) becomes (up to an additive constant)

$$F(u, A) = \int_{\mathbb{R}^2} \frac{|h - h_{\text{ex}}|^2}{8\pi} + \int_\Omega \frac{1}{2m^*} |(\frac{\hbar}{i}\nabla - \frac{e^*}{c}A)u|^2 + \frac{\beta}{4} \int_\Omega (|u|^2 - 1)^2 . \quad (6.19)$$

By a further change of units, this functional is commonly written, in the mathematical literature

$$F_\varepsilon(u, A) = \int_{\mathbb{R}^2} |dA - h_{\text{ex}}|^2 + \int_\Omega |\nabla_A u|^2 + \frac{1}{4\varepsilon^2} \int_\Omega (|u|^2 - 1)^2 , \quad (6.20)$$

where we introduce the notation $dA = \text{curl } A$ and $\nabla_A = \nabla - iA$. We also use ε^2 as a paremter proportional to $1/\beta$. Note that the ε used in this section is not related to the ε of section 6.2.1.

6.4.2   Removing h_{ex}

Up to now we have decreased T, keeping h_{ex} constant. This has led us to follow a first branch of solutions, described in section 6.3. In particular, this branch has selected a winding number d (recall that we saw this thanks to linearization in the neighborhood of T_c).

Next, we remove the external field, i.e. we decrease continuously h_{ex} to zero. The final configuration has therefore to be a (stable) critical configuration for the functional

$$F_\varepsilon(u, A) = \int_{\mathbb{R}^2} |dA|^2 + \int_\Omega |\nabla_A u|^2 + \frac{1}{4\varepsilon^2} \int_\Omega (|u|^2 - 1)^2 , \quad (6.21)$$

We expect that the winding number d, which was chosen by the first branch remains conserved in that process. Actually, this can be established in a rigorous mathematical level in the case ε is small.

6.4.3   The Case Where ε Is Small

A large mathematical literature has been devoted in recent years to the asymptotic limit where ε goes to zero (the so called London limit in the

Physics literature). In this context the stability of permanent currents has been established in [1] and [2]. Next, we briefly describe the main results and sketch some ideas involved.

The configuration space is, as before, $H^1 \times H^1$ quotiented by the gauge invariance. We denote it by H_{inv}

For $\Lambda \in \mathbb{R}^+$ we define the energy level set F_ε^Λ as

$$F_\varepsilon^\Lambda := \{(u, A) \in H_{inv} \ : \ F_\varepsilon(u, A) < \Lambda\} \,. \tag{6.22}$$

In [1] (see also [16] and [18] for related results) it was shown that energy level sets of the Ginzburg-Landau functional contain several different connected components (called "topological sectors") which can be classified - up to a possible local fine-structure - by the topological degree of the condensate wave function. This degree, as we saw above, is intimately associated with the quantification of magnetic flux through the inner hole of our ring. The presence of the different path components gives us the existence of the associated energy wells, but gives no information about the height of the barriers between them.

In [2] a more thorough study was carried out in order to estimate the height of these barriers. It is shown that, as expected from the physical behavior, the barriers are at energy levels considerably higher than the bottom of the wells. The height of the barrier between two different topological sectors, corresponds to the infimum on the space of continuous paths in state-space linking the two sectors, of the maximum of the energy along each such path - we call it the threshold energy for such transition. Defined in this way, we immediately see that associated with the threshold energy is a mountain-pass solution of the Ginzburg-Landau equations (Euler-Lagrange equations for the energy functional). Thus, this procedure also produces solutions of Ginzburg-Landau which are not necessarily local minimizers of the corresponding energy.

The main results of [1] and [2] are summarized in the following

Theorem 1.
(1) *Given $\Lambda \in \mathbb{R}^+$, there exists $\varepsilon_0 > 0$, depending only on Λ, such that for $\varepsilon < \varepsilon_0$, we can define a continuous map*

$$\deg : F_\varepsilon^\Lambda \to \mathbb{Z}$$
$$\sigma = (u, A) \mapsto \deg(u, \Omega) \,, \tag{6.23}$$

such that this map coincides with the classical notion of degree when u has values in S^1 (i.e. when $u \in W^{1,2}(\Omega, S^1)$).

For each $n \in \mathbb{Z}$, $\mathrm{top}_n(F_\varepsilon^\Lambda) := \deg^{-1}(n) = \{\sigma = (u, A) \in F_\varepsilon^\Lambda \ : \ \deg(u, \Omega) = n\}$, is an open and closed subset of F_ε^Λ which we call the nth topological sector of F_ε^Λ.
(2) *Suppose $\sigma_0 \in \mathrm{top}_n(F_\varepsilon^\Lambda)$ and $\sigma_1 \in \mathrm{top}_{n+1}(F_\varepsilon^\Lambda)$, and let c_n be defined by*

$$c_n := \inf_{\gamma \in \mathcal{V}} \{ \max_{s \in [0,1]} F_\varepsilon(\gamma(s)) \} . \tag{6.24}$$

where $\mathcal{V}$ is the space of continuous paths in H_{inv} between σ_0 and σ_1. Then, there exists a state σ such that σ is a (mountain pass) critical point of F_ε and $F_\varepsilon(\sigma) = c_n$.

(3) For ε sufficiently small, there is a constant $\alpha_1(n)$ s.t.

$$\pi |\log \varepsilon| - \alpha_1 \leq c_n \leq \pi |\log \varepsilon| + \alpha_1 . \tag{6.25}$$

Idea of the proof: Part (1) can be proved by doing a parabolic regularization of our original function u to obtain a continuous function u^h having no charged singularities (a technique we developed in [3]). The degree of the original u can then be defined to be the degree of its regularization u^h.

Minimizing F_ε inside each component of F_ε^Λ, we may obtain solutions of Ginzburg-Landau which are locally minimizing, i.e. critical points of F_ε which are local minima. These are the stationary states that should be associated with permanent currents.

Part (2) is an immediate consequence of (1) and the fact that our functional satisfies the Palais-Smale condition. It gives the existence of mountain-pass points for F_ε (which correspond to mountain-pass type solutions of Ginzburg-Landau). Unlike the solutions obtained minimizing the energy inside each topological sector, the solutions obtained in part (2) will not necessarily be local minimizers of F_ε, and are probably unstable.

The number c_n defined by (6.24) is called the **threshold energy** for the transition from the state σ_0 to the state σ_1. It will be the infimum of the energies for which such a transition is possible - it is the height of the barrier separating these states. This concept should play a crucial role in the behavior of our system.

For part (3) we notice that, intuitively, a transition from $\text{top}_n(F_\varepsilon^\Lambda)$ to $\text{top}_{n+1}(F_\varepsilon^\Lambda)$ corresponds to passing one "quantum of vorticity" of u (in the F_ε model this would correspond to one quantum of magnetic flux $hc/2e$, see [12]) from the outside of our annulus into the inner hole. Since a degree one vortex of size η has an energy of at least the order of $\pi \log(\eta/\varepsilon)$, we can construct a path having this as leading term for the energy. Then, once we fix η to be some convenient constant, we will obtain that the maximum of the energy along the path (and, consequently, also our upper bound) is of the order of $\pi |\log \varepsilon|$, as claimed.

The lower bound is more difficult to establish since, we should prove these energy levels are attained by every path from σ_0 to σ_1. We would like to show that every such path will essentially contain a transition of the type of the one constructed above to obtain the upper bound for c_n, where an isolated vortex is brought from outside Ω into the inner hole.

What we actually prove is a slightly weaker statement: first, we succeed in defining the relative minimal connection (in the spirit of [9]) between any point $\gamma(s)$ in our path and the initial point, and in proving it is "approximately" continuous. This enables us to show that there exists $s \in [0,1]$ s.t $v = \gamma(s) \in H^1(\Omega, \mathbb{C})$ has, in some weak sense, a unique singularity of degree one situated in the interior of Ω, about half way between the two components of $\partial\Omega$. As a matter of fact, this suffices to prove our result since, as was shown in [2], this type of isolated singularity of degree one costs an energy of the order of $\pi|\log\varepsilon|$.

Remark: Theorem 1 is consistent with the physical behavior of the system described above, namely the existence and great stability of persistent currents. On the one hand, the existence of such non-trivial critical points indicates that our energy has several wells, apart from the one corresponding to the ground state - this is given by part 1 of the theorem. On the other hand, the great stability of these states means that the transition probabilities between the different wells are very low. The transition probability is expected to be exponentially decreasing with the height of the barriers. Therefore, for ε small, the low transition probabilities (high stability) are a direct consequence of the lower bound in (6.25).

6.4.4 The Case Where ε Is Not Small

In this case, the fact that there are "thresholds" between sectors corresponding to different values of d has not been established. Even if these barriers exist we do not know if the stability of the current might be explained by them alone.

In this context, recall that the value $\varepsilon = \sqrt{2}$ plays a very important role. Indeed, for $\Omega = \mathbb{R}^2$, it was observed by Bogomolny, that the second order equations might be replaced, for minimizers, by first order equations, due to self-duality (see also the book by Jaffe and Taubes [15], where this is used to construct mathematical solutions). On the physical level, this value of the parameter defines the transition from type I to type II superconductors.

In that setting, Boutet de Monvel, Georgescu and Purice (see [8]) have shown that there is no cost to move a vortex from one boundary of Ω to the other one, if one adds a Dirichlet type condition: $|u| = 1$ on $\partial\Omega$. This suggests the threshold energy might be very small in the self-dual case and one might have to seek for other explanations for stability in this setting.

References

1. L. Almeida *Topological sectors for Ginzburg-Landau energies*, Revista Matemática Iberoamericana **15** (1999) 487-546.
2. L. Almeida *Threshold transition energies for Ginzburg-Landau functionals*, Nonlinearity **12** (1999) 1389-1414.

3. L. Almeida and F. Bethuel *Topological methods for the Ginzburg-Landau equation*, J. Math. Pures Appl. **77** (1998) 1-49.
4. L. Almeida and F. Bethuel (work in progress).
5. P. Bauman, D. Phillips and Q. Tang *Stable nucleation for the Ginzburg-Landau System with an Applied Magnetic Field*, Arch. Rational Mech. Anal. **142** (1998) 1-43.
6. J. Berger and J. Rubinstein *On the zero set of the wave function in superconductivity*, Comm. Math. Phys. **202** (1999) 621-628.
7. C. Bolley and B. Helffer *An application of semi-classical analysis to the asymptotic study of the supercooling field of a superconducting material*, Ann. Inst. Henri Poincaré, Anal. Non Linéaire **58** (1993), 189-233.
8. A. Boutet de Monvel, V. Georgescu and R. Purice *a boundary value problem related to the Ginzburg-Landau model*, Comm. Math. Phys. **142** (1991) 1-23.
9. H. Brezis, J.M. Coron and E. Lieb *Harmonic maps with defects*, Comm. Math. Phys. **107** (1986) 649-705.
10. S. Chapman *Nucleation of superconductivity in decreasing fields I*, Euro. Jnl. of Applied Mathematics **5** (1994) 449-468.
11. M. Dauge and B. Helffer *Eigenvalues variation I, Neumann problem for Sturm-Liouville equations*, J.D.E. **104** (1993), 243-262.
12. B. Deaver B and W. Fairbank *Experimental evidence for quantized flux in superconducting cylinders*, Phys. Rev. Letters **7** (1961) 43-46.
13. M. del Pino, P. Felmer and P. Sternberg it Boundary concentration for eigenvalue problems related to the onset of superconductivity, to appear in Comm. Math. Physics.
14. B. Helffer, M Hoffmann-Ostenhoff, T. Hoffmann-Ostenhoff, M. Owen *Nodal sets for the ground state of the Schrödinger operator with zero magnetic field in a non simply connected domain*, Comm. Math. Phys. **202** (1999) 629-649.
15. A. Jaffe and C. Taubes *Vortices and Monopoles*, Birkhäuser, Boston, 1980.
16. S. Jimbo and Y. Morita *Ginzburg-Landau equation and stable solutions in a rotational domain*, SIAM J. Math. Anal. **27** (1996) 1360-1385
17. F. London, *Superfluids*, John Wiley and Sons, New York, 1950.
18. J. Rubinstein and P. Sternberg *Homotopy classification of minimizers of the Ginzburg-Landau energy and the existence of permanent currents*, Commun. Math. Phys. **179** ((1996) 257-263
19. J. Rubinstein and P. Sternberg *Second order phase transitions* Preprint.
20. D. Saint-James and P.G. de Gennes *Onset of superconductivity in decreasing fields*, Phys. Letters **7** (1963), 306-308.
21. M. Tinkham Introduction to Superconductivity, Mc Graw-Hill, 1996.

7 On the Normal/Superconducting Phase Transition in the Presence of Large Magnetic Fields[1]

Peter Sternberg

Department of Mathematics, Indiana University, Bloomington, IN 47405, USA

7.1 Formulation of the Problem

There has been much activity recently involving an examination of the phase transition between the normal and superconducting state when a sample is subjected to an applied magnetic field. The curve relating critical temperature to applied field marking this transition has in particular been the subject of numerous studies by experimental physicists, see e.g. [17] or [5]. In this article, we will explore some attempts to understand the transition analytically. The results described below involve a mixture of formal and rigorous mathematical analysis.

Our starting point will be the Ginzburg-Landau model (cf. [14,11]). We shall consider this problem in a setting examined by many experimentalists, namely the case of a thin, mesoscopic sample subjected to a constant applied field $\mathbf{H_e}$ directed orthogonal to the sample cross-section. In particular, we assume the sample thickness is small relative to the coherence length. This leads us to consider the Ginzburg-Landau theory for a two-dimensional sample. As we are focusing here on a bifurcation from the normal state, given by $\Psi = 0$ and magnetic potential $\mathbf{A}$ satisfying $\nabla \times \mathbf{A} = \mathbf{H_e}$, we shall concern ourselves with an investigation of the linearized Ginzburg-Landau equations about the normal state. Using a characteristic radius R of the sample to non-dimensionalize the Ginzburg-Landau equations (cf. [2]), we are lead to a study of the following eigenvalue problem with first eigenfunction $\Psi^{(1)} : \Omega \to \mathbb{C}$, where Ω has been scaled to have radius of $\mathcal{O}(1)$.

$$(i\nabla + \phi\mathbf{A_N})^2\Psi^{(1)} = \mu\Psi^{(1)} \text{ in } \Omega, \tag{7.1}$$
$$\nabla\Psi^{(1)} \cdot \widehat{\nu} \qquad = 0 \text{ on } \partial\Omega. \tag{7.2}$$

Here Ω is a smooth, bounded planar domain, ϕ is the reduced magnetic flux, $\widehat{\nu}$ is a unit normal to the boundary and $\mathbf{A_N}$ satisfies

$$\nabla \times \mathbf{A_N} = 1\widehat{z}, \quad \mathrm{div}\,\mathbf{A_N} = 0 \text{ in } \Omega, \quad \text{and} \quad \mathbf{A_N} \cdot \widehat{\nu} = 0 \text{ on } \partial\Omega.$$

We have taken non-reflecting boundary conditions which take on the simple form (7.2) in light of the gauge choice made above on $\mathbf{A_N}$. The eigenvalue

[1] Research supported by N.S.F. grant DMS-9322617

$\mu = \mu(\phi)$ is related to the transition temperature T by

$$\frac{T_c - T}{T_c} = \frac{\xi(0)^2}{R^2}\mu$$

where T_c is the critical temperature at zero field and $\xi(0)$ is the coherence length at zero temperature.

We note that this eigenvalue and corresponding eigenfunction can be characterized variationally as the minimum of the corresponding Rayleigh quotient

$$\mu(\phi) \equiv \inf_{\psi \in H^1(\Omega)} \frac{\int_\Omega |(i\nabla + \phi \mathbf{A_N})\psi|^2 \, dx \, dy}{\int_\Omega |\psi|^2 \, dx \, dy}, \tag{7.3}$$

where $H^1(\Omega)$ denotes the Sobolev space of square-integrable functions $\psi : \Omega \to \mathbb{C}$ having square-integrable first derivatives.

We shall be most interested here in pursuing an expansion for μ and the first eigenfunction $\Psi^{(1)}$ in the asymptotic regime of large magnetic flux, i.e. $\phi \gg 1$. Before pursuing this, let's make two basic observations regarding the large flux setting. First, simple scaling arguments reveal that in this context, the behavior of the bifurcating superconducting state on a bounded domain Ω in the presence of a large magnetic flux ϕ is very much related to the problems on the whole plane and on a half-plane with flux $\phi = 1$. Related to this first observation is the second one: one expects–based on experiment and general theory–that the eigenfunction will concentrate near the boundary of the sample. That is, one anticipates a phenomenon commonly referred to as 'surface superconductivity' in the physics literature ([8]). We shall return to this second point momentarily and indeed, we will offer a rigorous mathematical argument for its validity using Ginzburg-Landau theory, but first we focus on the setting of unbounded domains.

Consider first the case of (7.1) when $\Omega = \mathbb{R}^2$ and $\phi = 1$. In this case, it is well-known that the first eigenvalue $\mu = 1$ (see e.g. [9] for a more careful statement and appropriate references). For the case $\Omega = \mathbb{R}^2_+ = \{(x,y) : x > 0\}$, perhaps the first analysis was carried out by St. James and de Gennes [18]. (In this regard, see also [7].) In the case of the half-plane, they found that the first eigenfunction $\Psi^{(1)}$ takes the form

$$\Psi^{(1)}(x,y) = \psi_1(x)e^{i\beta^* y} \tag{7.4}$$

where $\psi_1 : [0,\infty) \to \mathbb{R}$ is the eigenfunction associated with the double minimization over scalar functions u and real numbers β given by

$$\lambda_1 \equiv \inf_\beta \inf_{u \in H^1([0,\infty))} \frac{\int_0^\infty (u')^2 + (x - \beta)^2 u^2 \, dx}{\int_0^\infty u^2 \, dx} \tag{7.5}$$

Then β^* in (7.4) is the (unique) value of the parameter β satisfying this infimum. As such, the eigenfunction ψ_1 is a solution to the problem

$$\mathcal{L}(\psi_1, \lambda_1) \equiv \psi_1'' + [\lambda_1 - (x - \beta^*)^2]\psi_1 = 0$$
$$\psi_1'(0) = 0 = \psi_1(\infty)$$

The formal analysis of the one-dimensional problem (7.5) carried out in [18] was made rigorous in [4] and [10] where it was shown that ψ_1 decays to zero exponentially as $x \to \infty$ and that the optimal β is indeed unique. In fact, one finds that β^* and λ_1 are related by $(\beta^*)^2 = \lambda_1$ and numerically one can find the approximate value $\lambda_1 \approx 0.59$. In particular, we note that onset in a half-plane occurs at a higher temperature (lower μ) than in the whole plane.

Two further results on the half-plane are useful–one for the formal analysis on a bounded sample to follow, and the other for a rigorous treatment to be described briefly later in the article. The first is given by the Lemma below, in which we have chosen the normalization $\psi_1(0) = 1$:

Lemma 1. *Define I_k as the kth moment of the first eigenfunction ψ_1:*

$$I_k \equiv \int_0^\infty x^k \psi_1^2 \, dx.$$

Then for every positive integer k, one can express I_k in terms of β^* and I_0. In particular,

$$I_1 = \beta^* I_0,$$
$$I_2 = \frac{3}{2}(\beta^*)^2 I_0 \ \left(= \frac{3}{2}\lambda_1 I_0\right) \quad and$$
$$I_3 = \frac{1}{6} + \frac{5}{2}(\beta^*)^3 I_0.$$

This is proved in the appendix to [3] with a slightly different scaling. Another useful fact involves the rigorous justification of the claim

$$\Psi^{(1)}(x, y) = \psi_1(x)e^{i\beta^* y}.$$

Formally, if one seeks the solution to (7.1)–(7.2) on $\mathbb{R}_+^2$ with the lowest value of μ for $\phi = 1$ via separation of variables, one can quickly arrive at this formula where ψ_1 and β^* are characterized through (7.5). However, it is not so trivial to rigorously justify that this solution is the *only* solution at $\mu = \lambda_1$, i.e. to prove that the first eigenvalue is simple for the half-plane. It turns out, though, that this is indeed the case as was shown recently in [9] (cf. Theorem 3.2).

7.2   Analysis of Onset in a Disc

As our first venture into a mathematical analysis of onset on a bounded domain, let us consider the case of a disc D with curvature κ (not to be confused

with the Ginzburg-Landau parameter which does not enter in this article due
to our choice of non-dimensionalization.) This problem has received the attention of numerous authors. It turns out that the linearized problem (7.1)
can be solved explicitly in terms of Kummer functions [6] and a complete
rigorous analysis of both the linear problem and the associated nonlinear bifurcation theory has been worked out in [1]. (See also e.g. [12].) What one
finds, both in experiment and via Ginzburg-Landau theory, is a tempera-
ture/field curve which oscillates as it increases, with sharp cusps emerging at
intervals whose length approaches a constant. In particular, one finds in [1]
the following rigorous expansion for μ in a disc:

$$\mu(\phi) \sim \lambda_1 \phi - \frac{1}{3I_0} \kappa \phi^{1/2} \text{ for } \phi \gg 1. \tag{7.6}$$

The presence of oscillations in the temperature/field curve for a disc, (which,
as we shall see, emerge in the next term of the expansion) though by now well-
acknowledged, were perhaps a surprise at first, since the common wisdom was
that such 'Little-Parks' type oscillations should be associated with a sample
having non-trivial topology such as a ring–not in a disc.

We choose to begin with the case of the disc despite the presence of
the above-mentioned work for two reasons. First, the formal analysis below
represents a different, and perhaps simpler way of capturing (7.6) as well
as the oscillations in the temperature/field curve. Secondly, it will serve as
a kind of 'warm-up' problem in which we introduce a method of attack to
be later applied to the case of a general bounded domain. We shall utilize
the method of formal matched asymptotic expansions under the assumption
$\phi \gg 1$. This will involve a boundary layer construction which is then to be
matched to the normal state $\Psi^{(1)} = 0$ in the interior of the disc.

To analyze (7.1), one first must determine $\mathbf{A_N}$. One easily solves

$$\nabla \times \mathbf{A_N} = \hat{\mathbf{z}}, \quad \text{div } \mathbf{A_N} = 0 \text{ in } D, \text{ and } \mathbf{A_N} \cdot \hat{\nu} = 0 \text{ on } \partial D,$$

to find

$$\mathbf{A_N} = \frac{1}{2}(-y, x).$$

However, this turns out to be inconvenient since $\mathbf{A_N} \neq 0$ on ∂D. (In this vein,
note that on a half-plane, the choice $\mathbf{A_N} = (0, x)$ happily *does* vanish on the
boundary, substantially simplifying the analysis of that case.) Therefore, in
a neighborhood of the boundary we instead decompose $\mathbf{A_N}$ as $\mathbf{A_N} = \mathbf{q} + \mathbf{p}$
where

$$\nabla \times \mathbf{q} = \hat{\mathbf{z}}, \quad \text{div } \mathbf{q} = 0 \text{ near } \partial D, \quad \mathbf{q} = 0 \text{ on } \partial D, \tag{7.7}$$

and $\mathbf{q}$ is purely tangential, while

$$\nabla \times \mathbf{p} = 0, \quad \text{div } \mathbf{p} = 0 \text{ near } \partial D, \quad \mathbf{p} \cdot \hat{\nu} = 0 \text{ on } \partial D. \tag{7.8}$$

Letting

$$\eta = \text{distance to } \partial D, \quad s = \text{arclength along } \partial D,$$

we find

$$\mathbf{q}(\eta, s) = -\eta\left(\frac{1 - \kappa\eta/2}{1 - \kappa\eta}\right)\mathbf{t}(s)$$

where $\mathbf{t}(s)$ denotes the unit tangent vector to ∂D. Then we make a change of variables (exploiting the gauge invariance):

$$\Psi^{(1)} = \psi e^{i\phi G} e^{i\phi^{1/2}\beta^* s}, \tag{7.9}$$

where G is a scalar function satisfying $\nabla G = \mathbf{p}$ and the presence of the factor $e^{i\phi^{1/2}\beta^* s}$ is motivated by the appearance of a similar term arising in the half-plane solution. We should emphasize that (7.9) is *not* assumed to be a polar decomposition for $\Psi^{(1)}$; that is, ψ is not assumed to be real.

The eigenvalue problem (7.1) then transforms (near ∂D) to:

$$\left(i\nabla - \phi\frac{(\eta(1 - \kappa\eta/2) + \phi^{-1/2}\beta^*)}{1 - \kappa\eta}\mathbf{t}\right)^2 \psi = \mu(\phi)\psi.$$

The solution to this problem must be matched to the normal state $\Psi^{(1)} = 0$ away from ∂D.

We stretch the normal variable η by introducing $\dot{\xi} = \phi^{1/2}\eta$ and seek a formal expansion for $\mu(\phi)$ and $\psi = \psi(\xi, s)$ in powers of $\phi^{1/2}$:

$$\mu = \mu_0\phi + \mu_1\phi^{1/2} + \mu_2 + \ldots,$$
$$\psi = \Psi_0(\xi, s) + \frac{1}{\phi^{1/2}}\Psi_1(\xi, s) + \frac{1}{\phi}\Psi_2(\xi, s) + \ldots$$

$$\tag{7.10}$$

At order $\mathcal{O}(\phi)$ we find:

$$\mathcal{L}(\Psi_0, \mu_0) \equiv \Psi_{0\xi\xi} + [\mu_0 - (\xi + \beta^*)^2]\Psi_0 = 0,$$
$$(\Psi_0)_\xi(0, s) = 0 = \Psi_0(\infty, s).$$

$$\tag{7.11}$$

Note that the condition $\Psi_0(\infty, s) = 0$ represents a matching condition to the normal state in the interior of the sample. We recognize this as the eigenvalue problem arising in the analysis of the half-plane. Hence, $\mu_0 = \lambda_1$ and

$$\Psi_0(\xi, s) = \psi_1(\xi)W_0(s)$$

with W_0 to be determined.

At order $\mathcal{O}(\phi^{1/2})$, we obtain an inhomogeneous equation with left-hand side again given by the self-adjoint operator $\mathcal{L}(\cdot, \lambda_1)$ now applied to Ψ_1. Through multiplication by the homogeneous solution ψ_1 and extensive use of the moment identities in Lemma 1, we obtain as a solvability condition the formula

$$\mu_1 = -\frac{\kappa}{3I_0}.$$

So far we have

$$\mu(\phi) \sim \lambda_1\phi - \frac{\kappa}{3I_0}\phi^{1/2}$$

(where $I_0 = \int \psi_1^2 \approx 1.312$) in agreement with [1], and

$$\Psi^{(1)}(\eta, s) \sim \psi_1(\sqrt{\phi}\eta)W_0(s)e^{i\phi G}e^{i\phi^{1/2}\beta^* s}$$

where W_0 remains to be determined.

At order $\mathcal{O}(1)$ the solvability condition takes the form of an eigenvalue problem for W_0 and μ_2:

$$W_{0ss} + ic_1W_{0s} + (c_2 - \frac{I_0}{\beta^*}\mu_2)W_0 = 0 \qquad (7.12)$$

for $-L/2 < s < L/2$ where $L = \frac{2\pi}{\kappa} =$ arclength of ∂D and $c_i = c_i(\kappa, \beta^*, I_0)$. One obtains a general solution to (7.12) through the obvious ansatz

$$W_0(s) = e^{as} \qquad (7.13)$$

for $a = a_{\pm}(I_0, \beta^*, \mu_2)$ arising as the roots of the quadratic coming from the substitution of (7.13) into (7.12).

Of course, (7.12) must be supplemented by boundary conditions and we now arrive at a key point: the boundary conditions for W_0 *are not periodic*. Instead, they come from the requirement that

$$\Psi^{(1)} = \psi e^{i\phi G}e^{i\phi^{1/2}\beta^* s} \sim \psi_1(\xi)W_0(s)e^{i\phi G}e^{i\phi^{1/2}\beta^* s} \qquad (7.14)$$

is single-valued. Thus, the boundary conditions are:

(1) The amplitude of $|W_0|$ is periodic, and
(2) The phase of W_0 jumps so that the change in phase of $\Psi^{(1)} = 2\pi k(\phi)$ for some integer $k(\phi)$.

This will ensure that $\Psi^{(1)}$ is single-valued. One immediate consequence of (1) is that the root a above must be purely imaginary. Carrying out the algebra, the conditions above lead to determination of the eigenvalue $\mu_2 = \mu_2(\phi)$ as a bounded, piecewise smooth, oscillatory function of ϕ which approaches a periodic function as $\phi \to \infty$.

In conclusion, for a disc we find

$$\mu(\phi) \sim \lambda_1\phi - \frac{\kappa}{3I_0}\phi^{1/2} + \mu_2(\phi) \quad \text{for } \phi \gg 1.$$

and $\Psi^{(1)}$ given asymptotically by (7.13) and (7.14).

7.3   Analysis of Onset in a General Bounded Domain

We turn now to the case where Ω is an arbitrary bounded simply connected domain in the plane. Again we denote by κ the curvature of $\partial\Omega$, though now $\kappa = \kappa(s)$ for $-L/2 < s < L/2$ where L is the arclength of the boundary. We will describe an asymptotic analysis of (7.1)–(7.2) in this setting for $\phi >> 1$. More details of this expansion can be found in [3] where Bernoff and the author carried out the program using a different scaling. (In [3], one finds an expansion for critical field at fixed temperature in the regime where the Ginzburg-Landau parameter is large.)

We again seek a boundary layer solution of the form

$$\Psi^{(1)} = \psi e^{i\phi G} e^{i\phi^{1/2}\beta^* s}.$$

where as before η is the distance to $\partial\Omega$ and $\nabla G = \mathbf{p}$. Again, near the boundary we decompose $\mathbf{A_N}$ as $\mathbf{A_N} = \mathbf{q} + \mathbf{p}$, where $\mathbf{q}$ and $\mathbf{p}$ are given by (7.7) and (7.8). We find that the eigenvalue problem (7.1) transforms near $\partial\Omega$ to:

$$\left(i\nabla - \phi\frac{(\eta(1 - \kappa(s)\eta/2) + \phi^{-1/2}\beta^*)}{1 - \kappa(s)\eta}\mathbf{t}\right)^2 \psi = \mu(\phi)\psi. \qquad (7.15)$$

As in the case of the disc, we expect the leading behavior in the normal direction to be $\psi_1(\xi)$, though we now expect this amplitude to be modulated by some transverse (tangential) amplitude. Let us again introduce $\xi = \phi^{1/2}\eta$. Then one might guess as before that

$$\psi \sim \psi_1(\xi)W_0(s), \qquad (7.16)$$

where the difference between this setting and a disc would be that W_0 would no longer be of modulus one. However, seeking a solution to (7.15) with an expansion such as (7.16) fails. In order to achieve the proper transverse behavior, i.e. to achieve a proper balance between the normal and tangential derivatives of ψ, we find we must introduce

$$\tau = \phi^{1/8}(s - s_0) \qquad \text{for some } s_0 \in (-L/2, L/2).$$

That is, we must *stretch the tangential variable s as well*, where s_0 remains to be determined. So we must now expand for the eigenvalue μ as

$$\mu = \mu_0\phi + \mu_1\phi^{7/8} + \mu_2\phi^{3/4} + \mu_3\phi^{5/8} + \dots \qquad (7.17)$$

and we must expand for $\psi = \psi(\xi, \tau)$ as

$$\psi = \Psi_0 + \frac{1}{\phi^{1/8}}\Psi_1 + \frac{1}{\phi^{1/4}}\Psi_2 + \dots \qquad (7.18)$$

We must also expand the curvature itself (via Taylor series) as

$$\kappa(\tau) = \kappa(s_0) + \frac{1}{\phi^{1/8}}\kappa_s(s_0)\tau + \frac{1}{2\phi^{1/4}}\kappa_{ss}(s_0)\tau^2 + \dots \qquad (7.19)$$

We then seek a boundary layer solution to (7.15) valid in a neighborhood of the point $s = s_0$ on $\partial\Omega$, i.e. valid for

$$0 < \xi << \phi^{1/2} \quad \text{and} \quad |\tau| << \phi^{1/8}.$$

Such a solution would then be matched to the normal state $\Psi^{(1)} \equiv 0$ outside of this neighborhood, so that necessarily the amplitude of the order parameter will be concentrated in a neighborhood of the point $s = s_0$. This represents a major departure from the behavior of eigenfunctions for the cases of the half-plane and disc in which the amplitude is constant along the boundary.

Substitution of (7.17), (7.18) and (7.19) into (7.15) yields at leading order ($\mathcal{O}(h)$) that $\mathcal{L}(\Psi_0, \mu_0) = 0$ so that again $\mu_0 = \lambda_1$ and

$$\Psi_0(\xi, \tau) = \psi_1(\xi) W_0(\tau)$$

with W_0 to be determined.

At order $\mathcal{O}(h^{7/8})$ and $\mathcal{O}(h^{3/4})$ we find:

$$\mathcal{L}(\Psi_i, \lambda_1) = -\mu_i \psi_1 W_0 \quad \text{for } i = 1, 2.$$

so that integration against ψ_1 gives $\mu_1 = \mu_2 = 0$. Indeed, one finds the next non-zero contribution to the eigenvalue expansion comes at order $\mathcal{O}(h^{1/2})$ where the solvability condition yields (as always, with extensive use of the moment identities):

$$\left(\kappa(s_0) + 3I_0\mu_4\right) W_0 = 0 \quad \text{so that}$$

$$\mu_4 = -\frac{\kappa(s_0)}{3I_0}.$$

Recalling that μ, and so μ_4 in particular, is to be the smallest eigenvalue, we have determined s_0 by the criterion

$$\kappa(s_0) = \max_s \kappa(s),$$

i.e. s_0 *corresponds to the location of maximum curvature along* $\partial\Omega$.

Since κ has its maximum at $s = s_0$ we have

$$\kappa_s(s_0) = 0 \quad \text{and} \quad \kappa_{ss}(s_0) \leq 0.$$

To keep things simple, we will assume:

(i) $\partial\Omega$ has a unique point of maximum curvature

(ii) $\kappa_{ss}(s_0) < 0$ (strict inequality).

(7.20)

So far we have

$$\mu \sim \lambda_1 \phi - \frac{\kappa(s_0)}{3I_0}\phi^{1/2} + \mathcal{O}(\phi^{3/8})$$

but we are compelled to continue the expansion since we still have not fully determined the leading order behavior of the eigenfunction. That is, we still have not determined $W_0(\tau)$.

Summarizing the remainder of the calculation, one find at order $\mathcal{O}(\phi^{3/8})$ that $\mu_5 = 0$ while at order $\mathcal{O}(\phi^{1/4})$, the solvability condition takes the form of an O.D.E. for W_0:

$$W_{0\tau\tau} - \left(\frac{\mu_6}{\beta^*} + \frac{\kappa_{ss}(s_0)}{6\beta^*}\tau^2\right)W_0 = 0$$

for $-\infty < \tau < \infty$ and we match to the normal state away from $s = s_0$ via the boundary conditions

$$W_0(-\infty) = 0 = W_0(\infty).$$

Through a simple change of variables, this problem can be converted into the standard harmonic oscillator problem:

$$f_{zz} + (\lambda - \frac{1}{4}z^2)f = 0, \quad f(-\infty) = 0 = f(\infty)$$

This can be solved explicitly with principal eigenfunction $e^{-z^2/4}$ and principal eigenvalue $\lambda = 1/2$. We then change back to the original variables to find

$$W_0(\tau) = e^{-c\tau^2} = e^{-c\phi^{1/4}(s-s_0)^2}$$

and

$$\mu_6 = \sqrt{\frac{\lambda_1 |\kappa_{ss}(s_0)|}{6\beta^*}}$$

where $c = c(\mu_6, \beta^*) > 0$.

We conclude that near the point of maximum curvature of $\partial\Omega$ we have

$$\left|\Psi^{(1)}\right| \sim \psi_1(\phi^{1/2}\eta)e^{-c\phi^{1/4}(s-s_0)^2}$$

while $\left|\Psi^{(1)}\right|$ is exponentially small elsewhere in Ω.

For the expansion in $\mu = \mu(\phi)$ we have found

$$\mu = \lambda_1\phi - \frac{\kappa(s_0)}{3I_0}\phi^{1/2} + \sqrt{\frac{\lambda_1 |\kappa_{ss}(s_0)|}{6\beta^*}}\phi^{1/4} + \dots \tag{7.21}$$

This completes the formal expansion of (7.1) for a general domain. The expansion succeeds in predicting an exponential localization of the superconducting state near the point of maximum curvature. On the other hand, in light of the exponential decay away from this point, it bears no information about the phase of the eigenfunction on the rest of the boundary, thus precluding any predictions about possible oscillations in the temperature/field transition curve which one might expect to arise from a phase winding type of argument as in the disc.

7.4   Rigorous Justification of Boundary Localization

Much of the formal argument just presented for a smooth bounded domain can be rigorously justified. With regard to the expansion (7.21), one can obtain a rigorous upper bound of this form after a straight-forward but tedious calculation by substituting the approximate eigenfunction constructed above into the Rayleigh quotient (7.3). (See [9], Proposition 4.1.).

To prove the corresponding lower bound is harder, and at this point we can only prove rigorously that

$$\lim_{\phi \to \infty} \frac{\mu(\phi)}{\phi} = \lambda_1. \tag{7.22}$$

(In this regard, see also [16].)

Concerning the exponential localization of the first eigenfunction, we have rigorously justified the decay away from the boundary and have so far obtained a preliminary result in the direction of showing decay *along* the boundary. We summarize this work here:

Theorem 1. *Let $\Omega \subset I\!\!R^2$ be a bounded, open, simply connected domain with $\partial\Omega \in C^{3,\alpha_0}$ for some $\alpha_0 \in (0,1)$. If $\{\Psi^\phi\}$ denotes a sequence of eigenfunctions corresponding to the first eigenvalue $\mu(\phi)$ given by (7.3), normalized so that $\left\|\Psi^\phi\right\|_{L^\infty(\Omega)} = 1$, then for ϕ sufficiently large we have*

$$\left|\Psi^\phi(z)\right| \le c_1 e^{-c_2 \phi^{1/2} \, \mathrm{dist}(z, \partial\Omega)} \text{ for all } z = (x, y) \in \Omega, \tag{7.23}$$

for constants c_1 and c_2 independent of ϕ. Moreover, if Ω is not a disc, then we have

$$\lim_{\phi \to \infty} \left(\min_{z \in \partial\Omega} \left|\Psi^\phi(z)\right| \right) = 0. \tag{7.24}$$

In order to see the implications of (7.24), it is interesting to consider the case of a smooth domain Ω whose boundary agrees with a circle except along a very small arc. The theorem above guarantees that the first eigenfunction will feel this 'defect' in the boundary and decay to zero somewhere when the flux ϕ is large. (Presumably the decay will be everywhere except where the curvature is maximized at the junction between the arc and the circular boundary.)

7.5   Infinite Curvature – Domains with Corners

Throughout this discussion, we have always assumed our boundary was smooth. It is natural to ask, however, about what happens to the eigenvalue problem (7.3) when the domain is not smooth. This question has been considered in numerous experiments but the question has also been considered from the standpoint of Ginzburg-Landau theory (see e.g. [15,13,19]).

The experiments and the theory predict a localization of the nucleating superconducting state in a vicinity of a corner, in keeping in spirit with the above discussion of localization near points of maximum curvature.

If one focuses on the case of a square, and again seeks an expansion for μ in the large flux regime, then one discovers a marked departure from the earlier cases. These cases all rely on the basic principle that if one 'blows up' the eigenvalue problem in a neighborhood of a boundary point–which is in effect the result of taking $\phi \gg 1$–then to leading order what one sees is a half-plane. From this perspective then, it is not surprising that one always obtains (7.22) as the leading order behavior of the eigenvalue.

If one 'blows up' the problem near the corner of a square, however, one obviously still feels the corner. A numerical computation on the problem (7.3) for Ω a square confirms this distinction and leads to a prediction of (cf. [15]):

$$\mu(\phi) \sim 0.55\phi \text{ for } \phi \gg 1,$$

whereas we recall that $\mu(\phi) \approx 0.59\phi$ for a smooth sample. Further work of a rigorous nature is now being done on this problem.

References

1. Bauman, P., Phillips, D. and Tang, Q., "Stable nucleation for the Ginzburg-Landau system with an applied magnetic field," Arch. Rat. Mech. Anal. **142** (1998), 1-43.
2. Berger, J. and Rubinstein, J., "Bifurcation analysis for phase transitions in superconducting rings with nonuniform thickness," SIAM J. Appl. Math. **58** (1998), 103-121.
3. Bernoff, A. and Sternberg, P., "Onset of superconductivity in decreasing fields for general domains," J. Math. Phys. **39** (1998), 1272-1284.
4. Bolley, C. and Helffer, B., " An application of semi-classical analysis to the asymptotic study of the supercooling field of a superconducting material," Ann. Inst. H. Poincare: Phys. Theor., **58** (1993), 189-233.
5. Buisson, O., Gandit, P., Rammal, R., Wang, Y.Y. and Pannetier, B., "Magnetization oscillations of a superconducting disk," Phys. Let. A **150** (1990), 36-42.
6. Benoist, R. and Zwerger, W., Z. Phys. **B103**, (1997).
7. Chapman, S.J., "Nucleation of superconductivity in decreasing fields, I," Euro. J. Appl. Math. **5** (1994), 449-468.
8. de Gennes, P.G. *Superconductivity in Metals and Alloys*, Addison Wesley (1989).
9. Del Pino, M., Felmer, P. and Sternberg, P., "Boundary concentration for eigenvalue problems related to the onset of superconductivity," Comm. Math. Phys. **210** (2000), 413-446.
10. Dauge, M. and Helffer, B., "Eigenvalues variation I, Neumann problem for Sturm-Liouville operators," J.D.E. **104** (1993), 243-262.
11. Du, Q., Gunzburger, M.D. and Peterson, J.S., "Analysis and approximation of the Ginzburg-Landau model of superconductivity," SIAM Review **34** (1992), 45-81.

12. Deo, P.S., Schweigert, V.A. Peeters, F.M. and Geim, A.K., "Magnetization of mesoscopic superconducting disks," Phys. Rev. Let. **79**, no. 23, (1997), 4653-4656.

13. Fomin, V.M., Devreese, J.T. and Moshchalkov, V.V., Europhys. Lett. **42**, 553 (1998).

14. Ginzburg, V.L. and Landau, L.D., "On the theory of superconductivity," J.E.T.P. **20** (1950), 1064.

15. Jadallah, H. Rubinstein, J. and Sternberg, P., "Phrase transition curves for mesoscopic superconducting samples," Phys. Rev. Let. **82**, no. 14 (1999), 2935-2938.

16. Lu, K. and Pan, X.B., "Eigenvalue problems of Ginzburg-Landau operator in bounded domains," J. Math. Phys. **40** (1999), 2647-2670.

17. Moschchalkov, V.V., Gielen, L. Strunk, C., Jonckheere, R., Qiu, X., Van Haesendonck, C. and Bruynseraede, Y., "Effect of sample topology on the critical fields of mesoscopic superconductors," Nature **373** (January 1995), 319-322.

18. Saint-James, D. and de Gennes, P.G., "Onset of superconductivity in decreasing fields," Phys. Let. **7** (1963), 306-308.

19. Schweigert, V.A. and Peeters, F.M., "Influence of the confinement geometry on surface superconductivity," (1999) preprint.

8 On the Numerical Solution of the Time-Dependent Ginzburg–Landau Equations in Multiply Connected Domains

Gustavo C. Buscaglia[1], Carlos Bolech[2], and Arturo López[1]

[1] Centro Atómico Bariloche and Instituto Balseiro, 8400 Bariloche, Argentina
[2] Center for Materials Theory, Department of Physics and Astronomy, Rutgers University, Piscataway, NJ 08854, USA

Abstract. A numerical method for the solution of the time-dependent Ginzburg-Landau equations is detailed. The method is based on the popular technique of gauge invariant variables. Extension of the method to multiply connected domains is addressed. An implementation of the method is made available through the Web.

8.1 Introduction

The numerical simulation of superconductivity has attracted increasing attention during the last years, specially due to the appearance of high-temperature superconductors. At the mesoscopic level, the governing equations are provided by the Ginzburg-Landau theory, and are frequently referred to as TDGL equations (for Time Dependent Ginzburg-Landau equations). These are coupled nonlinear partial differential equations for the (complex) order parameter ψ and for the electromagnetic vector potential $\mathbf{A}$ (the scalar potential is usually eliminated through an appropriate choice of gauge).

Numerical approximations to the TDGL equations have been derived using both finite element [1] and finite difference [2–7] methods. Most physical applications use a specific finite difference method that we will refer to as ψU-method. The unknowns in the ψU-method in two spatial dimensions are the order parameter ψ and two auxiliary fields, $\mathcal{U}^x$ and $\mathcal{U}^y$, that are related to $\mathbf{A}$ by

$$\mathcal{U}^x(x,y,t) = \exp\left(-\imath \int_{x_o}^{x} A_x(\xi,y,t)\,d\xi\right) \tag{8.1}$$

$$\mathcal{U}^y(x,y,t) = \exp\left(-\imath \int_{y_o}^{y} A_y(x,\eta,t)\,d\eta\right) \tag{8.2}$$

The point (x_o, y_o) is arbitrary, $\imath = \sqrt{-1}$. Such variables were first introduced in lattice gauge theories [8]. To our knowledge they were first applied to the TDGL equations by Liu *et al* [3]. The ψU-method has since proved useful in the numerical simulation of many superconductivity phenomena [3,5–7].

The TDGL equations coupled with Maxwell equations with the zero scalar potential gauge choice lead to the following mathematical problem:

$$\partial_t \psi = -\frac{1}{\eta} \left[(-i\nabla - \mathbf{A})^2 \psi + (1 - T) \left(|\psi|^2 - 1 \right) \psi \right] + \tilde{f} \qquad (8.3)$$

$$\partial_t \mathbf{A} = (1 - T) \, \mathrm{Re} \left[\overline{\psi} \left(-i\nabla - \mathbf{A} \right) \psi \right] - \kappa^2 \nabla \times \nabla \times \mathbf{A} \qquad (8.4)$$

where lengths have been scaled in units of $\xi(0)$, time in units of $t_0 = \pi\hbar/(96k_B T_c)$, $\mathbf{A}$ in units of $H_{c2}(0)\xi(0)$ and temperatures in units of T_c. It has been assumed that the coherence length ξ obeys $\xi(T) = \xi(0)(1-T)^{-1/2}$, where T is the temperature in units of the critical temperature T_c, and that the Ginzburg-Landau parameter κ is independent of temperature. η is a positive constant (a ratio of characteristic times for ψ and $\mathbf{A}$), $\tilde{f}$ a random force simulating thermal fluctuations, k_B the Boltzmann constant and H_{c2} the upper critical magnetic field for type-II superconductors (see [9]). Re stands for "real part of".

Eqs. (8.3)-(8.4) are to be solved in a bounded domain Ω, complemented with initial conditions for ψ and $\mathbf{A}$, together with the following boundary conditions:

- **Boundary condition for A:** A given applied magnetic field H_e in the z-direction, possibly time-dependent but **spatially uniform**, is assumed. Continuity of the field thus implies

$$B_z := \widehat{\mathbf{e}}_z \cdot \nabla \times \mathbf{A} = H_e \qquad (8.5)$$

- **Boundary condition for ψ:** Zero supercurrent density perpendicular to the boundary is imposed, namely,

$$\widehat{\nu} \cdot (-i\nabla - \mathbf{A}) \psi = 0 \qquad (8.6)$$

where $\widehat{\nu}$ denotes the unit normal to the superconductor-vacuum interface. This automatically implies that the normal current perpendicular to the boundary also vanishes, since the total current across the superconductor-vacuum interface is zero. To see this, let $\mathbf{J}_s = (1 - T)\mathrm{Re}[\overline{\psi}(-i\nabla - \mathbf{A})\psi]$ denote the supercurrent density and $\mathbf{J}_n = -\partial_t \mathbf{A}$ the normal current density. Rewrite Eq. 8.4 as

$$\mathbf{J}_n + \mathbf{J}_s = \kappa^2 \nabla \times \nabla \times \mathbf{A}$$

Projection of this equation along the normal $\widehat{\nu} = (\nu_x, \nu_y, 0)$ leads to

$$\widehat{\nu} \cdot \mathbf{J}_s + \widehat{\nu} \cdot \mathbf{J}_n = \kappa^2 (\nu_x \partial_y - \nu_y \partial_x) B_z \qquad (8.7)$$

Since $(\nu_x \partial_y - \nu_y \partial_x)$ is nothing but the tangential derivative, and considering that the applied field H_e is uniform, the right-hand side of (8.7) identically vanishes showing that the total current across the boundary is zero.

For later use, we define the magnetization M_z as (see, e.g., [9])

$$M_z(t) = \frac{\int (B_z(x,y,t) - H_e)\, dx\, dy}{4\pi \int dx\, dy}$$

8.2   Numerical Method

Consider a rectangular mesh such as that of Fig. 8.1, consisting of $N_x \times N_y$ cells, with mesh spacings a_x and a_y. Any numerical method is defined by the (finite) unknowns of the method plus the equations relating these unknowns. In the ψU method the fundamental unknowns are three complex arrays:

- $\psi_{i,j}$, with $1 \le i \le N_x + 1$, $1 \le j \le N_y + 1$, associated to the nodes (or vertices) of the mesh. The value of $\psi_{i,j}$ approximates that of the order parameter at position (x_i, y_j). In the program, the corresponding array is F(i,j).
- $U^x_{i,j}$ (link variable in the x-direction, with $1 \le i \le N_x$, $1 \le j \le N_y + 1$, associated to the horizontal links (cell edges) of the mesh. The value $U^x_{i,j}$ approximates that of $\exp(-\imath \int_{x_i}^{x_{i+1}} A_x(\xi, y_j)\, d\xi)$.
- $U^y_{i,j}$ (link variable in the y-direction, with $1 \le i \le N_x + 1$, $1 \le j \le N_y$, associated to the vertical links of the mesh. The value $U^y_{i,j}$ approximates that of $\exp(-\imath \int_{y_j}^{y_{j+1}} A_y(x_i, \eta)\, d\eta)$.

To derive the discrete equations it is useful to notice that, from the definition of the link variables, discrete analogs of $\mathcal{U}^x$ and $\mathcal{U}^y$ from (8.1)-(8.2) can be defined at the nodes as

$$\mathcal{U}^x_{i,j} = \prod_{k=1}^{i-1} U^x_{k,j}, \qquad \mathcal{U}^y_{i,j} = \prod_{k=1}^{j-1} U^y_{i,k} \tag{8.8}$$

which leads to

$$U^x_{i,j} = \overline{\mathcal{U}}^x_{i,j} \mathcal{U}^x_{i+1,j}, \qquad U^y_{i,j} = \overline{\mathcal{U}}^y_{i,j} \mathcal{U}^y_{i,j+1} \tag{8.9}$$

8.2.1   Discretization of the TDGL Equations

In the following, discrete approximations for each term of (8.3)-(8.4) are derived, maintaining second order accuracy in space.

Term $(-\imath \nabla - \mathbf{A})^2 \psi$: From the identity

$$(-\imath \nabla - \mathbf{A})^2 \psi = -\overline{\mathcal{U}}^x \partial^2_{xx}(\mathcal{U}^x \psi) - \overline{\mathcal{U}}^y \partial^2_{yy}(\mathcal{U}^y \psi)$$

a second order approximation at (x_i, y_j) reads

$$(-\imath \nabla - \mathbf{A})^2 \psi \big|_{(x_i, y_j)} = -\frac{U^x_{i,j}\psi_{i+1,j} - 2\psi_{i,j} + \overline{U}^x_{i-1,j}\psi_{i-1,j}}{a_x^2}$$

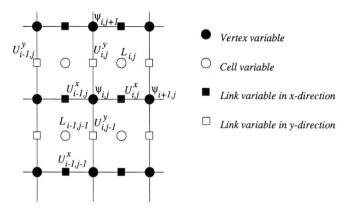

Fig. 8.1. Scheme of computational cells defining the numbering of discrete variables.

$$-\frac{U^y_{i,j}\psi_{i,j+1} - 2\psi_{i,j} + \overline{U}^y_{i,j-1}\psi_{i,j-1}}{a_y^2} + \mathcal{O}(a_x^2 + a_y^2) \qquad (8.10)$$

Term $(|\psi|^2 - 1)\psi$: It is readily approximated by

$$(\overline{\psi}_{i,j}\psi_{i,j} - 1)\psi_{i,j} \qquad (8.11)$$

Term $\mathrm{Re}\left[\overline{\psi}\left(-\imath\nabla - \mathbf{A}\right)\psi\right]$: From the identity

$$(-\imath\,\partial_x - A_x)\,\psi = -\imath\overline{U}^x\partial_x(U^x\psi)$$

it follows that

$$\mathrm{Re}\left[\overline{\psi}\left(-\imath\,\partial_x - A_x\right)\psi\right]\Big|_{x_i+\frac{a_x}{2},y_j} =$$

$$= \mathrm{Im}\left(\frac{\overline{U}^x_{i,j}\overline{\psi}_{i,j} + \overline{U}^x_{i+1,j}\overline{\psi}_{i+1,j}}{2}\,\frac{U^x_{i+1,j}\psi_{i+1,j} - U^x_{i,j}\psi_{i,j}}{a_x}\right) + \mathcal{O}(a_x^2)$$

$$= \frac{1}{a_x}\mathrm{Im}\left(\overline{\psi}_{i,j}\overline{U}^x_{i,j}U^x_{i+1,j}\psi_{i+1,j}\right) + \mathcal{O}(a_x^2) = \frac{1}{a_x}\mathrm{Im}\left(\overline{\psi}_{i,j}U^x_{i,j}\psi_{i+1,j}\right) + \mathcal{O}(a_x^2)$$

$$(8.12)$$

and analogously for the y component.

Term $\nabla \times \nabla \times \mathbf{A}$ $(= \nabla \times \mathbf{B})$: We introduce as auxiliary variable

$$L_{i,j} = U^x_{i,j}U^y_{i+1,j}\overline{U}^x_{i,j+1}\overline{U}^y_{i,j} \qquad (8.13)$$

In the program, the corresponding array is `bloop(i,j)`. From this and Stokes' identity it follows that

$$L_{i,j} = \exp\left(-\imath a_x a_y B_z(x_i + \frac{a_x}{2}, y_j + \frac{a_y}{2})\right)\left(1 + \mathcal{O}(a_x^4 + a_y^4)\right) \qquad (8.14)$$

so that, since $\mathbf{B} = (0, 0, B_z)$ and thus $\nabla \times \mathbf{B} = (\partial_y B_z, -\partial_x B_z, 0)$, we can use the approximations

$$\partial_y B_z(x_i + \frac{a_x}{2}, y_j) = \frac{\imath}{a_x a_y^2} \left(\overline{L}_{i,j-1} L_{i,j} - 1 \right) + \mathcal{O}(a_x^2 + a_y^2) \quad (8.15)$$

$$-\partial_x B_z(x_i, y_j + \frac{a_y}{2}) = \frac{\imath}{a_x^2 a_y} \left(\overline{L}_{i,j} L_{i-1,j} - 1 \right) + \mathcal{O}(a_x^2 + a_y^2) \quad (8.16)$$

Term $\partial_t \mathbf{A}$: From

$$\partial_t \left[\overline{\mathcal{U}}^x(x, y, t) \mathcal{U}^x(x + \delta, y, t) \right]$$

$$= -\imath \, \overline{\mathcal{U}}^x(x, y, t) \mathcal{U}^x(x + \delta, y, t) \int_x^{x+\delta} \partial_t A_x(\xi, y, t) \, d\xi$$

$$= -\imath \delta \, \overline{\mathcal{U}}^x(x, y, t) \mathcal{U}^x(x + \delta, y, t) \partial_t A_x(x + \frac{\delta}{2}, y, t) + \mathcal{O}(\delta^2) \quad (8.17)$$

it follows that

$$\partial_t A_x(x_i + \frac{a_x}{2}, y_j, t) = \frac{\imath}{a_x} \overline{U}_{i,j}^x \partial_t U_{i,j}^x + \mathcal{O}(a_x^2) \quad (8.18)$$

and similarly for $\partial_t A_y$.

Collecting the previous results, the numerical method for interior nodes reads:

$$\partial_t \psi_{i,j} = \frac{U_{i,j}^x \psi_{i+1,j} - 2\psi_{i,j} + \overline{U}_{i-1,j}^x \psi_{i-1,j}}{\eta a_x^2}$$

$$+ \frac{U_{i,j}^y \psi_{i,j+1} - 2\psi_{i,j} + \overline{U}_{i,j-1}^y \psi_{i,j-1}}{\eta a_y^2} - \frac{1-T}{\eta} (\overline{\psi}_{i,j} \psi_{i,j} - 1)\psi_{i,j} + \tilde{f}_{i,j} \quad (8.19)$$

$$\partial_t U_{i,j}^x = -\imath(1-T)U_{i,j}^x \mathrm{Im} \left(\overline{\psi}_{i,j} U_{i,j}^x \psi_{i+1,j} \right) - \frac{\kappa^2}{a_y^2} U_{i,j}^x \left(\overline{L}_{i,j-1} L_{i,j} - 1 \right) \quad (8.20)$$

$$\partial_t U_{i,j}^y = -\imath(1-T)U_{i,j}^y \mathrm{Im} \left(\overline{\psi}_{i,j} U_{i,j}^y \psi_{i,j+1} \right) - \frac{\kappa^2}{a_x^2} U_{i,j}^y \left(\overline{L}_{i,j} L_{i-1,j} - 1 \right) \quad (8.21)$$

Finally, a simple forward-Euler scheme is adopted to discretize the time variable with step Δt, namely

$$\psi_{i,j}(t + \Delta t) = \psi_{i,j}(t) + \Delta t \, \partial_t \, \psi_{i,j}(t) \quad (8.22)$$

and analogously for $U_{i,j}^x$ and $U_{i,j}^y$. Notice that the random force $\tilde{f}$ is also treated as a vertex variable. At each vertex it is selected from a Gaussian distribution with zero mean and standard deviation σ given by

$$\sigma = \sqrt{(\pi E_0/6\Delta t)(T/T_c)}$$

as done in Ref. [5].

8.2.2 External Boundary Conditions

Equations (8.19)-(8.21) are not defined for boundary nodes or links. We adopt the usual methodology of constraining boundary values of the unknowns to satisfy *first order* approximations of the boundary conditions.

If the boundary is aligned with the y-axis, the zero-current condition implies $(-\imath\partial_x - A_x)\psi = 0$ or, equivalently, $-\imath\overline{U}^x\partial_x(U^x\psi) = 0$. For the order parameter at $i = 1$ (West boundary) and $i = N_x + 1$ (East boundary) this is implemented as

$$\psi_{1,j} = U^x_{1,j}\psi_{2,j} \qquad \psi_{N_x+1,j} = \overline{U}^x_{N_x,j}\psi_{N_x,j} \qquad (8.23)$$

Similarly, for the South ($j = 1$) and North ($j = N_y + 1$) boundaries the expression is

$$\psi_{i,1} = U^y_{i,1}\psi_{i,2} \qquad \psi_{i,N_y+1} = \overline{U}^y_{i,N_y}\psi_{i,N_y} \qquad (8.24)$$

For the link variables, it remains to define how to update the values of those on the boundary. Notice that, for cells with two edges on the boundary only the product of the two link variables has numerical consequences, since it is the total circulation of the vector potential around the cell that propagates inside the computational domain. We have thus one unknown for each cell on the boundary, with the other three link variables already calculated from Eq. (8.20) or Eq. (8.21). Let H_e be the applied field and let the cell (i, j) be at the boundary. From

$$L_{i,j} = U^x_{i,j}U^y_{i+1,j}\overline{U}^x_{i,j+1}\overline{U}^y_{i,j} = \exp\left(-\imath a_x a_y H_e\right) \qquad (8.25)$$

the unknown link variable is readily obtained.

Remark: Notice that it is not difficult to obtain second order approximations to the boundary conditions that would preserve the accuracy of the scheme of Eqs. (8.19)-(8.21). Taking as example the East boundary ($i = N_x + 1$), a second order approximation of the zero-current condition leads to

$$\partial_t\psi_{i,j} = \frac{-2\psi_{i,j} + 2\overline{U}^x_{i-1,j}\psi_{i-1,j}}{\eta a_x^2} + \frac{U^y_{i,j}\psi_{i,j+1} - 2\psi_{i,j} + \overline{U}^y_{i,j-1}\psi_{i,j-1}}{\eta a_y^2}$$
$$-\frac{1-T}{\eta}(\overline{\psi}_{i,j}\psi_{i,j} - 1)\psi_{i,j} + \tilde{f}_{i,j} \qquad (8.26)$$

This coincides with (8.19) under the assumption $U^x_{i,j}\psi_{i+1,j} = \overline{U}^x_{i-1,j}\psi_{i-1,j}$, or, equivalently, $\mathcal{U}^x_{i+1,j}\psi_{i+1,j} = \mathcal{U}_{i-1,j}\psi_{i-1,j}$ (a second order approximation to $\partial_x(\mathcal{U}^x\psi) = 0$). One can proceed analogously with the link variables. This variant has not yet been adopted in practice, though it deserves at least a try.

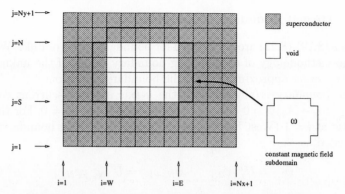

Fig. 8.2. Scheme of a computational mesh with a hole.

8.2.3 Boundary Conditions at Holes

If the domain is multiply connected, at the hole boundary the same boundary conditions as before apply, namely zero perpendicular supercurrent and magnetic field continuity. In what concerns the order parameter, one proceeds exactly as for the exterior boundary (Eqs. (8.23) and (8.24)). However, the magnetic field inside the hole is not known *a priori*. Our methodology has been implemented for the case of one rectangular hole, but it is easily generalized. Let H_i be the magnetic field inside the hole. Since there are no currents in the hole, H_i is uniform and only depends on time.

Consider the hole shown in Fig. 8.2. It is delimited by the vertex lines $i = W$, $i = E$, $j = S$ and $j = N$. Since the adjacent cells must also have $B_z = H_i$, the magnetic field is uniform in the subdomain ω depicted in Fig. 8.2. The key remark is that the link variables corresponding to the boundary of ω can be calculated using Eqs. (8.20) and (8.21).

The algorithm to update H_i and the link variables at the boundary of the hole is (with a notation similar to that in the program):

- Update all interior link variables according to (8.20)-(8.21).
- Compute

$$\mathcal{P}_{x1} = \overline{U}_{W,S-1}^y \left(\prod_{i=W}^{E-1} U_{i,S-1}^x \right) U_{E,S-1}^y \tag{8.27}$$

$$\mathcal{P}_{x2} = U_{E,N}^y \left(\prod_{i=W}^{E-1} \overline{U}_{i,N+1}^x \right) \overline{U}_{W,N}^y \tag{8.28}$$

$$\mathcal{P}_{y1} = U_{W-1,S}^x \left(\prod_{j=S}^{N-1} \overline{U}_{W-1,j}^y \right) \overline{U}_{W-1,N}^x \tag{8.29}$$

$$\mathcal{P}_{y2} = U_{E,S}^x \left(\prod_{j=S}^{N-1} U_{E+1,j}^y \right) \overline{U}_{E,N}^x \tag{8.30}$$

Notice that $\mathcal{P} = \mathcal{P}_{x1}\mathcal{P}_{x2}\mathcal{P}_{y1}\mathcal{P}_{y2}$ approximates $\exp(-\imath \int \mathbf{A} \cdot \mathbf{ds})$ where the circulation is calculated around ω. Consistently with the previous approximations, we take $\mathcal{P} = \exp(-\imath M a_x a_y H_i)$ where M is the number of cells in ω,

$$M = (N - S)(E - W) + 2(N - S) + 2(E - W)$$

- Update H_i using, as immediate from above,

$$\partial_t H_i = \frac{\imath}{M a_x a_y} \overline{\mathcal{P}} \partial_t \mathcal{P} \tag{8.31}$$

- Update all link variables at the hole boundary using (8.25), with H_i in the place of H_e.

The implementation uses the logical array Bulk (i,j) to determine if cell (i,j) is in the superconductor or in the hole.

8.2.4   Discretization of the Free Energy

Let, as before, Ω denote the domain occupied by the superconductor, and let Ω_H denote the hole (if any). The exact expression for the Gibbs' free energy is

$$\mathcal{G} = \int_{\Omega \cup \Omega_H} \left\{ |\psi|^2 \left(\frac{1}{2}|\psi|^2 - 1 \right) + \frac{1}{1 - T} |(-\imath \nabla - \mathbf{A})\psi|^2 \right.$$
$$\left. + \frac{\kappa^2}{(1 - T)^2} \left[|\nabla \times \mathbf{A}|^2 - 2\mathbf{H}_e \cdot (\nabla \times \mathbf{A}) \right] \right\} d\Omega \tag{8.32}$$

where the terms involving ψ must be taken as identically zero in Ω_H (the hole).

From the previous definitions $\mathcal{G}$ in cell (i,j) is approximated by

$$\mathcal{G}_{i,j} = \frac{a_x a_y}{4} \left(\frac{|\psi_{i,j}|^4}{2} - |\psi_{i,j}|^2 + \frac{|\psi_{i+1,j}|^4}{2} - |\psi_{i+1,j}|^2 \right.$$
$$+ \frac{|\psi_{i+1,j+1}|^4}{2} - |\psi_{i+1,j+1}|^2 + \frac{|\psi_{i,j+1}|^4}{2} - |\psi_{i,j+1}|^2 \right)$$
$$+ \frac{a_x a_y}{2(1 - T)} \left(\frac{|U_{i,j}^x \psi_{i+1,j} - \psi_{i,j}|^2}{a_x^2} + \frac{|U_{i,j+1}^x \psi_{i+1,j+1} - \psi_{i,j+1}|^2}{a_x^2} \right.$$
$$+ \frac{|U_{i,j}^y \psi_{i,j+1} - \psi_{i,j}|^2}{a_y^2} + \frac{|U_{i+1,j}^y \psi_{i+1,j+1} - \psi_{i+1,j}|^2}{a_y^2} \right)$$
$$+ \frac{\kappa^2 a_x a_y}{(1 - T)^2} \frac{\imath \ln L_{i,j}}{a_x a_y} \left(\frac{\imath \ln L_{i,j}}{a_x a_y} - 2H_e \right) \tag{8.33}$$

Notice that, since $L_{i,j}$ approximates $\exp(-\imath a_x a_y B_z)$ at cell (i,j), and since a_x, a_y and B_z are small (remember that distances are in units of $\xi(0)$ and

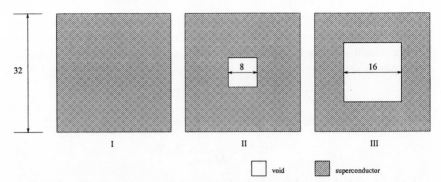

Fig. 8.3. Scheme of the cases considered.

fields in units of $H_{c2}(0)$), the imaginary part of $\ln L_{i,j}$ must lie in the branch that is closest to zero, i.e., between $-\pi$ and π. Again, the terms involving ψ must be taken as identically zero if `Bulk (i,j)` is false.

8.3 Examples

We detail in the following a few numerical examples. These depict some of the typical phenomena that occur in superconducting systems and how they are modeled by the described method. The reader is referred to [7] for an application of the model to the study of vortex arrays in superconducting thin films. It is also possible to extend the formulation to consider d-wave superconductors, as has been done in [10,11].

Consider first the case of a square sample with no hole (case I, see Fig. 8.3), with dimensions $32\,\xi(0) \times 32\,\xi(0)$. We use, with the units defined in Section 1, $a_x = a_y = 0.5$, $\kappa = 2$, $\eta = 1$ and $T = 0.5$, with a noise constant of $E_0 = 10^{-5}$. With these definitions $H_{c1} = 0.04$ and $H_{c2} = 0.5$ for the bulk material, while $H_{c3} = 0.85$ for a semi-infinite domain. Numerical limitations arise in the choice of the time step due to the forward-Euler treatment of the equations. A practical rule for time step selection is

$$\Delta t \leq \min\left\{\frac{h^2\eta}{4}, \frac{h^2}{4\kappa^2}\right\} \tag{8.34}$$

where

$$h^2 = \frac{2}{\frac{1}{a_x^2} + \frac{1}{a_y^2}}$$

In this case we choose $\Delta t = 0.015$, which satisfies stability. External field begins at $H_e = 0$ and is linearly incremented to $H_e = 1$ along the 10^6 simulated time steps. Variables are homogeneously initialized to a perfect Meissner state, $\psi(t = 0) = 1$, $\mathbf{A}(t = 0) = 0$.

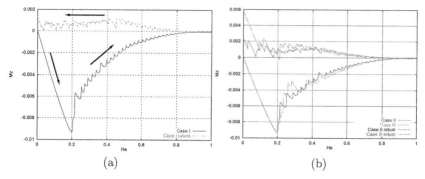

Fig. 8.4. Magnetization curves for the different cases considered.

This case is particularly simple and runs at about 2500 steps per minute on a personal computer. The magnetization curve that is obtained can be seen in Fig. 8.4 (a). For applied fields smaller than $H_e = 0.203$ the sample remains in a Meissner state, but at this field an instability develops that leads to the entrance of four vortices with the consequent jump in the magnetization. Similar vortex-entrance events occur at $H_e = 0.216$ (8 vortices), 0.252 (4 vortices), 0.270 (4 vortices), 0.288 (4 vortices), 0.312 (4 vortices), etc. In Figs. 8.5 we show the distribution of the modulus of the order parameter on the sample for several values of the applied field. It can be observed that the arrangement of the vortices is strongly affected by the finite size of the sample and its square symmetry. Larger samples allow for the obtention of hexagonal vortex lattices (an example can be seen in Fig. 8.6). Also notice that for $H_e > 0.4$ the superconductivity is strongly depressed in the sample's interior, but recovers near the boundary. This gradually leads, for $H_e \sim 0.5$ or greater, to surface superconductivity in a layer a few coherence lengths thick. The corners always remain the points where the order parameter is maximal.

The series of minima in the magnetization curves deserve special attention. Similar extrema were measured by Guimpel *et al* [12] and by Brongersma *et al* [13]. A related phenomenon was reported by Hünnekes *et al* [14]. In [7] it was argued that the minima reflect the magnetization behavior of the superconducting sheet at the sample surface and not rearrangements in the vortex lattice (which do occur). The simple case reported here confirms this argument, since minima extend far beyond H_{c2} ($H_e = 0.5$) and must thus come from a surface effect.

In Fig. 8.4 (a) the hysteretic behavior of the system is clearly observed. Beginning at the calculated solution for $H_e = 1$, negative increments in the applied field were imposed to show the hysteresis. Some snapshots of $|\psi|$ can be seen in Fig 8.7. For any given field, there are many more vortices in the sample when the applied field is decreasing than increasing. The lower (absolute) values in the magnetization suggest a smaller barrier for vortices

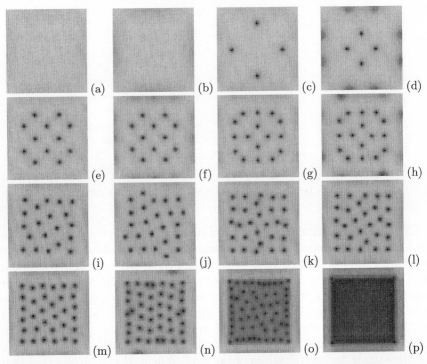

Fig. 8.5. Colour graph of the modulus of the order parameter for different applied fields. Case I. $H_e = 0.160$ (a), 0.200 (b), 0.205 (c), 0.215 (d), 0.220 (e), 0.250 (f), 0.260 (g), 0.270 (h), 0.280 (i), 0.290 (j), 0.320 (k), 0.330 (l), 0.350 (m), 0.400 (n), 0.500 (o), 0.700 (p). Colours run from yellow ($|\psi| \sim 1$) to blue ($|\psi| \sim 0$).

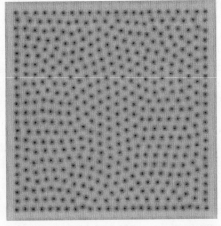

Fig. 8.6. Vortex arrangement at $H_e = 0.4$ in a square sample with dimensions $96\xi(0) \times 96\xi(0)$.

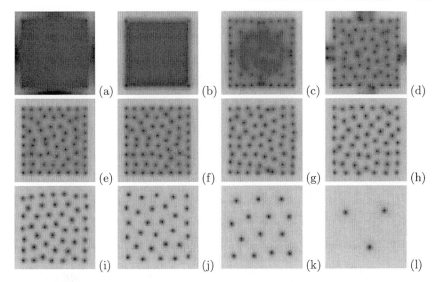

Fig. 8.7. Colour graph of the modulus of the order parameter for different applied fields. Case I, return (applied field descending from 1 to 0). $H_e = 0.900$ (a), 0.600 (b), 0.460 (c), 0.450 (d), 0.445 (e), 0.430 (f), 0.400 (g), 0.350 (h), 0.300 (i), 0.200 (j), 0.100 (k), 0.000 (l).Colours run from yellow ($|\psi| \sim 1$) to blue ($|\psi| \sim 0$).

to leave the system than the barrier for vortices to enter the system. The results are however not conclusive since in the simulation the rate of change of the applied field is rather high and time-dependent terms play a role. The three vortices in Fig. 8.7 (l), for example, leave the sample if the field is kept constant at $H_e = 0$ during another 35000 time steps.

Consider now a hollow sample. Let the hole be a centered square, with dimensions $8\xi(0) \times 8\xi(0)$ (case II), or $16\xi(0) \times 16\xi(0)$ (case III). The same process of increasing and decreasing the applied field as before is conducted. The magnetization curves can be observed in Fig. 8.4 (b). Essentially the same structure as before arises, but at zero field in case II and III there remain fluxoids "trapped" by the sample. These fluxoids (5 for case II and 13 for case III) are localized in the hole and do *not* leave the sample if the simulation is continued keeping $H_e = 0$ for as many as 10^6 additional time steps.

Vortex arrangements are also different, as shown in Figs. 8.8-8.9. It is interesting to remark how the first four vortices that enter the system are "captured" by the hole in case II, and similarly for the first sixteen ones in case III. In Fig. 8.8 (b) the instant before the capture has been depicted, and similarly in Fig. 8.9 (f). The dynamics can be better grasped looking at animated sequences. Some are available at the same webpage mentioned above, or can be reproduced with the program in a few hours of CPU time.

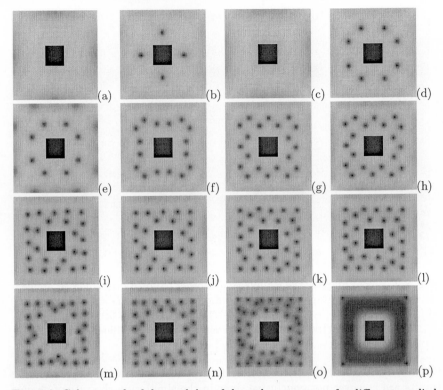

Fig. 8.8. Colour graph of the modulus of the order parameter for different applied fields. Case II. $H_e = 0.202$ [0] (a), 0.206 [4] (b), 0.210 [4](c), 0.230 [12](d), 0.250 [12] (e), 0.270 [20] (f), 0.300 [24] (g), 0.310 [24] (h), 0.320 [28] (i), 0.330 [28] (j), 0.340 [32] (k), 0.350 [32] (l), 0.37 [36] (m), 0.400 [40] (n), 0.450 [52] (o), 0.550 (p).Colours run from yellow ($|\psi| \sim 1$) to blue ($|\psi| \sim 0$). Between brackets the winding number along the external boundary is reported (above H_{c2} the winding number is not evaluated because round-off pollutes the results).

We end up here the example section, since it is not the purpose of this article to discuss particular applications of the proposed method but rather to illustrate some physically interesting cases that can be simulated with the program.

8.4   Concluding Remarks

A numerical method for solving the 2D TDGL equations in multiply connected domains has been described, together with the underlying term-by-term approximation formulae. An implementation of the method is available at http://cabmec1.cnea.gov.ar/~gustavo, and is readily extendable to more than one hole to consider more complex mesoscopic systems. A few examples showing the main phenomena that can be simulated with the

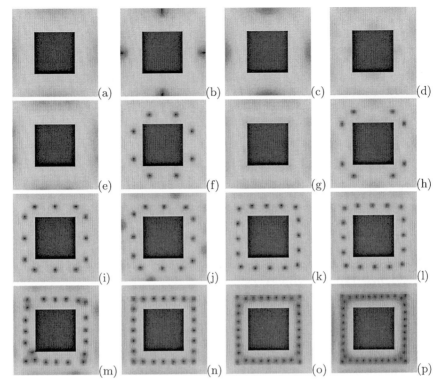

Fig. 8.9. Colour graph of the modulus of the order parameter for different applied fields. Case III. $H_e = 0.200$ [0] (a), 0.202 [0] (b), 0.218 [4](c), 0.220 [8](d), 0.230 [8] (e), 0.234 [16] (f), 0.250 [16] (g), 0.270 [24] (h), 0.300 [28] (i), 0.330 [28] (j), 0.340 [32] (k), 0.350 [32] (l), 0.400 [40] (m), 0.420 [44] (n), 0.450 [52] (o), 0.550 (p).Colours run from yellow ($|\psi| \sim 1$) to blue ($|\psi| \sim 0$). Between brackets the winding number along the external boundary is reported.

method have been reported. The reader is encouraged to download the program and test other configurations, such as eccentric holes (as considered in Section 5.4.7) or squid-like systems. It is also possible to introduce defects, simply by changing $|\psi|^2 - 1$ in Eq. 8.3 by $|\psi|^2 - r$, with $r < 1$ at regions with defects. Numerically, this amounts to a straightforward modification of Eq. 8.19. Extension to d-wave superconductors is more involved, but the term-by-term approximations of Section 2 lead to an appropriate scheme (some results can be found in [10,11]).

References

1. Q.Du., M.D.Gunzburger and J.S.Peterson, Phys. Rev. B **46**, 9027-9034 (1992), SIAM Rev. **34**, 54-81 (1992). Z.Chen and K.-H.Hoffmann, Report No. 433 (1993), Inst. Angew. Math. Stat., Techn. Univ. München.

2. R. Kato, Y. Enomoto and S. Maekawa, Phys. Rev. B. **44**, 6916-6920 (1991). Y. Enomoto and R. Kato, J. Phys.: Condens. Matter **3**, 375-380 (1991).
3. F. Liu, M. Mondello and N. Goldenfeld, Phys. Rev. Lett. **66**, 3071-3074 (1991).
4. H. Frahm, S. Ullah and A. Dorsey, Phys. Rev. Lett. **66**, 3067-3070 (1991).
5. R. Kato, Y. Enomoto and S. Maekawa, Phys. Rev. B **47**, 8016-8024 (1993). See also R. Kato *et al*, Physica C **227**, 387-394 (1994).
6. M. Machida and H. Kaburaki, Phys. Rev. Lett. **71**, 3206-3209 (1993).
7. C. Bolech, G. C. Buscaglia and A. López, Phys. Rev. B **52** (RC), 15719-15722 (1995).
8. K. G. Wilson, Phys. Rev. D **10**, 2445-2459 (1974).
9. M. Tinkham, *Introduction to Superconductivity*, Mc Graw Hill, New York, 1975.
10. J. J. Vicente Álvarez, G. C. Buscaglia and C. Balseiro, Phys. Rev. B **54**, 16168-16170 (1996).
11. J. J. Vicente Álvarez, C. Balseiro and G. C. Buscaglia, Phys. Rev. B **58**, 11181-11184 (1998).
12. J. Guimpel, L. Civale, F. de la Cruz, J. M. Murduck and I. K. Schuller, Phys. Rev. B **38**, 2342-2344 (1988).
13. S. H. Brongersma, E. Verweij, N. J. Koeman, D. G. de Groot and R. Griessen, Phys. Rev. Lett. **71**, 2319-2322 (1993).
14. C. Hünnekes, H. G. Bohn, W. Schilling and H. Schulz, Phys. Rev. Lett. **72**, 2271-2273 (1994). These authors do not measure the magnetic moment of the film, but instead its internal friction at 250 Hz, and attribute the peaks they obtain to rearrangements of the flux line lattice. A comment on this paper by S. H. Brongersma, B. I. Ivlev and R. Griessen and the authors' reply can be found in Phys. Rev. Lett. **73**, 3329-3330 (1994).

9 Formation of Vortex–Antivortex Pairs

Sanatan Digal[1], Rajarshi Ray[2], Supratim Sengupta[2], and Ajit M. Srivastava[2]

[1] Fakultat fur physik, Universitat Bielefeld, Universitatsstrasse, D-33615 Bielefeld, Germany
[2] Institute of Physics, Sachivalaya Marg, Bhubaneswar-751005, India

Abstract. We discuss formation of vortices and antivortices in field theory systems. We first describe conventional models, where such topological defects are produced either via thermal fluctuations, or via a non-equilibrium mechanism, known as *the Kibble mechanism*, during a phase transition. We then describe a new mechanism, recently proposed by us, where defect-antidefect pairs are formed due to strong oscillations, and subsequent *flipping* of the order parameter field. We also describe a novel phenomenon in which defect-antidefect pairs are produced via this *flipping mechanism* due to resonant oscillations of the order parameter field which is driven by a periodically varying temperature T, with T remaining much below the critical temperature T_c. Also, in a rapid heating of a localized region to a temperature *below* T_c, far separated vortex and antivortex can form. We discuss possible experimental tests of our predictions of defect-antidefect pair production *without* ever going through a phase transition.

9.1 Introduction

The subject of topological defects is a highly interdisciplinary area in physics. Topological defects occur in many different branches of physics, and are of interest to the researchers in these fields for varied reasons. In condensed matter physics topological defects arise in a variety of systems, such as vortex filaments in superfluid helium, flux tubes in superconductors, defects in liquid crystals etc., and have been studied, both theoretically and experimentally, for a long time. In the last two decades or so, certain developments have taken place in the fields of particle physics as well as in condensed matter physics, which have made it possible for the study of topological defects in one field to directly relate to the developments in the other field [1,2]. An important example of this is provided by the study of the processes of formation of topological defects. First serious attempts in this direction were made in particle physics models of the early universe [3] where the possibility of occurrence of topological defects provided the long sought source of density fluctuations which could lead to the formation of structure in the universe, such as galaxies, clusters of galaxies etc.

The entire picture of the formation of these defects in phase transitions in the early universe is very similar to what has been experimentally observed in condensed matter physics for a long time. However, the issue of details of the processes of formation of these defects and their evolution had not attracted much attention. It turns out that not only is the general picture of topological defects in both subjects similar; even the theoretical framework for discussing these defects, and their formation is almost identical. Thus, the processes of defect formation which were proposed for the early universe, are completely valid for describing the formation of topological defects in a condensed matter system. Recognition of this has made it possible to carry out direct experimental investigation of such aspects of theories of topological defects using suitable condensed matter systems [4–8]. Needless to say that such experimental verification would be out of question for particle physics since these defects are supposed to have formed during the very early stages of the evolution of the universe, at extreme densities and temperatures, which are not present anywhere in the present universe (and not conceivable in any laboratory experiments in foreseeable future).

In this article we will discuss various theories of topological defect formation. We will first describe the conventional models which are proposed for topological defect formation in a phase transition. Then we discuss a new model of defect formation, recently proposed by some of us, where defect-antidefect pairs form due to strong oscillations, and subsequent *flipping* of the order parameter (OP) field [9,10]. Topological defects can form via this mechanism, even without the system ever going through a phase transition. We demonstrate this by showing that vortices are produced via this new *flipping mechanism* due to resonant oscillations of the field which is driven by a periodically varying temperature T, with T remaining much below the critical temperature T_c [11]. Also, in a rapid heating of a localized region to a temperature *below* T_c, far separated vortex and antivortex can form (or, a large string loop can form for a three dimensional sample). We discuss how one can experimentally test the predictions of this new mechanism in systems such as superfluid helium and superconductors.

9.2   Theoretical Framework for Discussing Defect Formation

We will confine our discussion to the formation of vortices and antivortices arising due to spontaneous breaking of a global U(1) symmetry. Nevertheless, much of our discussion is general and applies to other defects, and to particle physics as well as systems in condensed matter physics. The differences between the relativistic case (as appropriate for particle physics) and condensed matter systems primarily arise from the equations of motion. We will discuss these differences at appropriate places.

Our model Lagrangian describes a system with a spontaneously broken global U(1) symmetry in 2+1 dimensions. The Lagrangian is expressed in terms of scaled, dimensionless variables (by choosing suitable units),

$$\mathcal{L} = \frac{1}{2}(\partial_\mu \Phi^\dagger)(\partial^\mu \Phi) - V(\phi) \qquad (9.1)$$

Here $\Phi = \Phi_1 + i\Phi_2$ is a complex scalar order parameter field with magnitude ϕ. $V(\phi)$ is the effective potential, which is the analog of the Ginzburg-Landau (GL) free energy (for uniform Φ) in the context of condensed matter systems and describes the symmetry breaking pattern of the theory. Just like for GL free energy, we choose the form of $V(\phi)$ to describe spontaneous breaking of U(1) symmetry.

$$V(\phi) = \frac{1}{4}(\phi^2 - 1)^2 + \frac{\alpha}{8}T^2\phi^2. \qquad (9.2)$$

T is the temperature of the system and α is a dimensionless parameter. We take $\alpha = 4$. (Here the critical temperature $T_c = 1$ with our choice of units and parameters.) The temperature dependence of the effective potential, given above, is motivated from the one-loop finite temperature corrections. It will be more appropriate to use linear temperature dependence for condensed matter systems. For much of the discussion this difference will not be relevant, except for the discussion at the end about the flipping mechanism. There, defect formation is found to be more efficient for linear temperature dependence.

$V(\phi)$ in Eq.(2) describes the symmetric phase of the system at high temperatures, with $\phi = 0$ being the vacuum state (the state of lowest free energy). For low temperatures $(T < T_c)$, the ground state occurs at a non-zero value of ϕ, and is not symmetric under U(1) transformation, thereby describing the spontaneously broken phase. The order parameter space (the vacuum manifold) is a circle S^1 here, characterized by the phase θ of Φ for fixed magnitude ϕ. Topological defects (U(1) vortices and antivortices), with non-trivial windings of θ, will form during the phase transition, which for this case is a second order phase transition.

9.3   Conventional Models of Formation of Vortices and Antivortices

Conventionally, defect production has been studied in two different situations. In certain condensed matter systems it is possible to produce defects by external influence, such as in an external magnetic field (which leads to formation of flux tubes in type II superconductors), or due to moving boundaries, such as production of vortices in a rotating vessel containing superfluid helium, (or nucleation of vortex rings in superflow through a small orifice [12]). The only other method of producing topological defects is either due to thermal fluctuations at temperatures close to the transition temperature (with defect density suppressed by the Boltzmann factor), or in a phase transition via the *Kibble mechanism* where defect production happens due to formation of a kind of domain structure during phase transition with defects forming at the junctions of these domains. As we mentioned above, the Kibble mechanism was originally proposed for studying defect formation in the early universe. However, the mechanism as such has complete general applicability and in fact has been recently verified by studying defect formation in certain condensed matter systems [5,6].

In the Kibble mechanism, the domain like structure arises from the fact that during phase transition, the orientation of the OP field can only be correlated within a finite region. The order parameter can be taken to be roughly uniform within a correlation region (domain), while varying randomly from one domain to the other. This situation is very natural to expect in a first order phase transition, where the transition to the spontaneous symmetry broken phase happens via nucleation of bubbles (for which case $V(\phi)$ will have a different form than given in Eq.(2)). Inside a bubble the orientation of the order parameter (say, phase θ of Φ in Eq.(1)) will be uniform, while θ will vary randomly from one bubble to another. Eventually bubbles grow and coalesce, leaving a region of space where θ varies randomly at a distance scale of the inter-bubble separation, thereby leading to a domain like structure. The same situation happens for a second order transition where the orientation of the OP field is correlated only within a region of the size of the correlation length. This again results in a domain like structure, with domains being the correlation volumes. There are non-trivial issues in this case in determining the appropriate correlation length for determining the initial defect density. This is due to the effects of critical slowing down of the dynamics of the OP field near the transition temperature. For a discussion of these issues, see ref.[8] and references therein.

In order to determine whether a topological defect (vortex) has formed in a region, one needs the information about θ at every point on a closed path in that region. For this one needs to know how θ will vary in between any two domains, once θ is known inside the two domains. An important ingredient in the Kibble mechanism is the assumption that in between any two adjacent domains, the OP field is supposed to vary with least gradient. This is usually called the *geodesic rule* and arises naturally from the consideration of minimizing the gradient energy.

One can take the example of a superfluid phase transition in ^{4}He corresponding to spontaneous breaking of U(1) symmetry and characterized by a complex order parameter. The GL free energy for this will be given by an expression as in Eq.(2) (with linear temperature dependence for the coefficient of the ϕ^2 term). In the superfluid phase, the magnitude ϕ of the order parameter gives the degree of superfluidity, while its phase θ can vary spatially over distances larger than the correlation length. In this case, string defects (superfluid vortices) arise at the junctions of domains if θ winds non trivially around a closed path going through adjacent domains. Simple arguments show [13,5] that the probability of string formation at a junction of three domains (in two space dimensions) is equal to 1/4.

It is important to mention that in the above argument, no use is made of the field equations. Thus, whether the system is a relativistic one, or a non-relativistic one appropriate for condensed matter physics with the dynamics of the order parameter being given by time dependent Ginzburg Landau (TDGL) equations, there is no difference in the defect production (per domain). It is this universality of the prediction of defect density (number of defects per domain) in the Kibble mechanism which has been utilized to experimentally verify this prediction (which was originally given for cosmic defect production) in liquid crystal systems [5]. Note that defect density as such is not universal since it depends on the domain size. For a second order transition, domains are not easily identified (as in the case of superfluid ^{4}He [7]), making it difficult to utilize the universality of defect density (defects per domain) for a clean experimental verification of the theoretical prediction. For a first order transition proceeding via bubble nucleation, bubbles are the domains, which may be easily identified facilitating the measurement of defect density (defects per domain), as in the case of liquid crystals, see [5]. See, also, [6] for the experimental verification of another universal prediction of the Kibble mechanism relating to the defect-antidefect correlations.

As the Kibble mechanism is formulated only in terms of the order parameter field (phase θ of Φ for Eq.(1)), one may expect that presence of other fields, such as gauge fields, will also not alter the predictions of this mechanism. Thus, one should expect the same predictions to apply for superconductors as for superfluid ^{4}He, both described by the same free energy (similar to that in Eq.(2)). There is, however, a subtlety here [14] when gauge fields are present, due to gauge invariance. As far as the spatial variation of θ alone is concerned, any possible continuous variation of θ is allowed on an open path in the region intermediate to any two domains, all such possibilities being related to each other by gauge transformations and hence physically equivalent. As the geodesic rule concerns the variation of θ from one point to another different point, we conclude that for gauge theories the geodesic rule for θ becomes unjustified. Without a criterion like the geodesic rule, one can not specify the configuration of θ at every point on a closed path (say, going through three adjacent domains) and hence can not determine whether the path encloses any string or not. In view of this, it becomes important to carry out experimental tests of the Kibble mechanism (such as those in refs. [4–7]) for superconductors which are the only experimentally accessible systems with spontaneously broken gauge symmetries which support topologically nontrivial defects, namely string defects, see ref.[15] in this context. The issue of geodesic rule for gauge defects has been investigated in the literature and it has been shown that in certain situations (such as in the presence of strong damping) the geodesic rule arises dynamically [16,17]. However, in the presence of strong field oscillations, the geodesic rule may not hold, see [18].

9.4   A New Mechanism
of Defect-Antidefect Production

An important aspect of the Kibble mechanism is that it does not crucially depend on the dynamical details of the phase transition. For example, the number of defects (per domain) produced via the Kibble mechanism depends only on the topology of the order parameter space and on spatial dimensions. Dynamics plays a role here only in determining the relevant correlation length [8], which in turn determines the domain size, affecting net number of defects produced in a given region. Still, the number of defects per domain is entirely independent of the dynamics. (Apart from some special situations, e.g. in a very slow first order transition, see [17].)

Recently, a new mechanism for defect production has been proposed by some of us [9,10] where the dynamics of the OP field plays a very important role. Here, defect-antidefect pairs are produced due to strong oscillations of the field. Whenever the field passes through zero magnitude, while oscillating (in a region where the field is non-uniform), a defect-antidefect pair gets created.

The essential physical picture underlying this mechanism is as follows [10]. We will discuss the formation of $U(1)$ global vortices in $2+1$ dimensions, with the Lagrangian given in Eq.(1). Consider a region of space in which the phase θ of the OP field Φ varies uniformly from α to β as shown in Fig.1a. At this stage there is no vortex or anti-vortex present in this region. Now suppose that the magnitude of the OP field undergoes oscillations, resulting in the passage of Φ through zero, in a small region in the center enclosed by the dotted loop, see Fig.1b. (We will discuss later how such oscillations can arise.) From the effective potential $V(\phi)$ (Eq.(2), for $T < T_c$) it is easy to see that oscillation of Φ through zero magnitude amounts to a change in Φ to the diametrically opposite point on the order parameter space S^1. This process, which causes a discontinuous change in θ by π, was termed as the flipping of Φ in [10].

For simplicity, we take θ to be uniform in the flipped region. Consider now the variation of θ along the closed path AOBCD (shown by the solid curve in Fig.1b) and assume that θ varies along the shortest path on the order parameter space S^1 (as indicated by the dotted arrows), as we cross the dotted curve, i.e. the variation of θ from the unflipped to the flipped region follows the geodesic rule. [Even if θ varies along the longer path on S^1, we still get a pair, with the locations of the vortex and the antivortex getting interchanged.] It is then easy to see that θ winds by 2π as we go around the closed path, showing that a vortex has formed inside the region. As the net winding surrounding the flipped region is zero, it follows that an anti-vortex has formed in the other half of the dotted region. One can also see it by explicitly checking for the (anti)winding of θ.

Another way to see how flipping of Φ results in the formation of a vortex-antivortex pair is as follows. Consider the variation of θ around the closed path AOBCD in Fig.1b before flipping of Φ in the dotted region. Such a variation of θ corresponds to a shrinkable loop on the order parameter space S^1. After flipping of Φ in the dotted region, a portion in the center of the arc connecting $\theta=\alpha$ to $\theta=\beta$ on S^1 moves to the opposite side of S^1. If the midpoint of the arc originally corresponded to $\theta=\gamma$, flipping of Φ changes γ to $\gamma+\pi$. We assume that different points on the arc move to the opposite side of S^1 maintaining symmetry about the mid point of the arc (and say, also maintaining the orientation of the arc). Then one can see that the loop on the order parameter space S^1 becomes non-shrinkable, and has winding number one, see [10] for details. Thus a vortex has formed inside the region enclosed by the solid curve. Obviously, an anti-vortex will form in the left half of Fig.1b.

Every successive passage of Φ through zero will create a new vortex-antivortex pair. Density waves generated during field oscillations lead to further separation of a vortex-antivortex pair created earlier. The attractive force between the vortex and anti-vortex lead to their eventual annihilation. We now discuss how such oscillations of the OP field can arise. As shown in [9,10], such strong oscillations can naturally arise in a first order transition during coalescence of expanding bubbles (when dissipation is not too strong, or when explicit symmetry breaking is also present). In a strong first order transition, bubble walls acquire large kinetic energies before bubbles start coalescing. This kinetic energy leads to strong oscillations, and subsequent flipping of the OP field, resulting in production of defect-antidefect pairs [9,10]. This mechanism is valid even for second order phase transitions, where oscillations of the OP field can result during a quench from very high

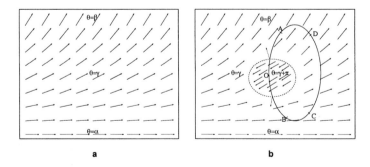

Fig. 9.1. (a) A region of space with θ varying uniformly from α at the bottom to some value β at the top. **(b)** Flipping of Φ in the center (enclosed by the dotted loop) has changed $\theta = \gamma$ to $\theta = \gamma+\pi$ resulting in a vortex-antivortex pair production

temperatures [10]. Further, this mechanism is also applicable for the formation of other topological defects. For example, it was shown in [10] that this mechanism also applies to the production of monopoles as well as textures. For string production in 3+1 dimensions, above arguments can easily be seen to lead to the production of a string loop enclosing the flipped region [10]. Production of defect-antidefect pairs due to field oscillations has also been seen for gauge defects, see [18].

9.5   Resonant Production of Vortex-Antivortex Pairs

We now discuss a new phenomenon where defect-antidefect pairs can form without any external influence (like external fields or moving boundaries etc.) and at temperatures much below the critical temperature, without ever going through a phase transition [11]. In our model, defects form due to resonant oscillations of the OP field at a temperature which is periodically varying, but remains much below the critical temperature T_c. (Periodic variation of the temperature will certainly be externally controlled. In the same way, a system undergoing phase transition has externally controlled temperature.) We also show that in rapid heating of a localized region to a temperature *below* T_c, far separated vortex and antivortex (or a large string loop in 3+1 dimensions) can form. We study the system described by Eq.(1). As mentioned above, the temperature dependence of the effective potential in Eq.(2) is motivated from the one-loop finite temperature corrections. We get similar results with linear temperature dependence in the effective potential, which is more appropriate for condensed matter systems. (In fact, for linear temperature dependence we get defect production for even lower temperatures.)

Another important point we mention is that the basic physics of our model resides in the time dependence of the effective potential. In the present case we achieve this by using a time dependent temperature. In many condensed matter systems one can equally well do this by periodically varying some other thermodynamic variable such as pressure (which may even be experimentally more feasible for the periodic variation case), or even a time-dependent external electric or magnetic field (say, for liquid crystals, though there are subtleties in that case [10].) One can even have coupling to an oscillating uniform background field leading to resonance in the OP field, as in the case of reheating after inflation in the early universe. There has been much discussion of non-thermal symmetry restoration, and defect production during subsequent symmetry breaking, in the context of post-inflationary reheating [19,20]. The oscillatory temperature dependence in our model leads to field equations which are similar to those with an oscillating inflaton field coupled to another scalar field (with the oscillating temperature playing the role of the inflaton field). However, there are also crucial differences between our results and those in the literature, see ref. [11] for detailed discussion.

We now discuss the dynamics of the field, using the following equations for field evolution.

$$\partial^2 \Phi_i/\partial t^2 + \eta \partial \Phi_i/\partial t - \nabla^2 \Phi_i + V'(\phi) = 0. \tag{9.3}$$

Here $V(\phi)$ is the effective potential in Eqn.(2). We solve Eq.(3) using a second order staggered leapfrog algorithm. η is the dimensionless dissipation coefficient. As the production of vortex-antivortex pairs happens here due to the flipping mechanism, the dynamics of the field plays a crucial role. Here it makes a difference whether the dynamics is given by Eq.(3), or by the TDGL equations. We find that for a system whose dynamics is completely dominated by the dissipative term (i.e. η is very large), we do not get any defect production (which is natural to expect since for large dissipation, oscillations and subsequent flippings of the OP field will be suppressed). However, for small dissipation, defects are produced via the flipping mechanism. We have not included any noise term in the above equations. The basic physics we discuss above does not depend on that. Further, as discussed above, time dependence of the effective potential could be thought to arise from some source other than the temperature (with temperature kept very low to suppress any thermal fluctuations). Thus here we concentrate on discussing this phenomenon without any noise term. (It will be interesting to explore the effects of noise on defect production in our model and explore connections with the well studied phenomenon of *stochastic resonance* in condensed matter systems [21].)

We first discuss a situation in which a central patch in the system is instantaneously heated from zero temperature to a temperature $T_0 < T_c$ and maintained at that temperature, while the region outside the patch is kept at zero temperature. One could easily produce such a situation by rapidly heating a small portion of the system (such as superfluid ^{4}He or superconductor), which is kept in contact with a low temperature reservoir. Eventually the temperature of the system may approach a uniform value, but the production of defects will happen on the time scale of the relaxation of the OP field

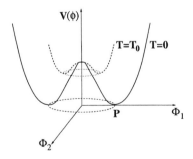

Fig. 9.2. Solid curve shows the effective potential at $T = 0$. The system is instantaneously heated to $T = T_0$ with effective potential shown by the dashed curve

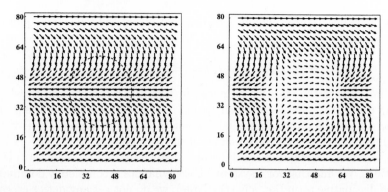

Fig. 9.3. The central region is instantaneously heated to a temperature $T_0 = 0.72$ (left figure). A far separated vortex-antivortex pair is produced due to flipping of Φ in the heated region (right figure)

(depending on heating rate). We first give qualitative arguments explaining the physics underlying the process of defect formation in this case. Assume that the system initially is at zero temperature with the OP field pointing in some direction on the order parameter space, denoted by point P in Fig.2. Thick solid curve in Fig.2 represents the effective potential at this stage. Now suppose that the temperature is instantaneously increased to a value T_0 and kept there, where T_0 is less than the critical temperature T_c. The dashed curve in Fig.2 denotes the effective potential at $T = T_0$. As the temperature has been increased suddenly, the OP field will still have the value shown by point P in Fig.2. However, now the effective potential being given by the dashed curve, the OP field at point P becomes unstable and starts rolling down towards the minimum of the new effective potential (dashed curve). It is clear that if the value of the potential energy at point P, with respect to the new effective potential, is larger than the height of the central bump then the field will overshoot the bump and roll down to the opposite side (ignoring dissipation). That is, the OP field will be *flipped*, leading to vortex-antivortex pair production via the flipping mechanism (for non-uniform θ, as discussed above).

We study the evolution of the field by Eqn.(3), with T in the effective potential taken as the local value of temperature. The temperature profile of the wall separating the central region from the surrounding region at zero temperature is taken to be of the form $T(r) = (T_0/2)(1 - \tanh((r - R)/\triangle))$, where R is the radius of the (circular) central patch and $\triangle$ is the wall thickness. (Our results do not depend on the profile used for this boundary region.) We use a 500×500 lattice with the physical size of the lattice equal to 80×80. The time step Δt was taken to be 0.008. We take $R = 24$ and $\triangle = 0.32$. The initial magnitude of Φ was taken to be uniform, equal to the zero temperature vacuum expectation value (vev), while its phase varied linearly from 0 to 2π along the Y-axis, being uniform along the X-axis, as shown in the left figure in Fig.3. (This choice facilitated the use of periodic boundary conditions to evolve the field. Defect production happens with much smaller phase variation also.)

The instantaneous heating of the central region (interior of the circle shown in the left figure in Fig.3) with $T_0 = 0.72$ (with $\eta = 0.4$), lead to destabilization of the OP field in that region. The field, while trying to relax to its new equilibrium value, overshot the central barrier and ended up on the opposite side of the order parameter space. For the above field configuration, this lead to formation of a vortex at one boundary of the *flipped* region, with an anti-vortex forming at the opposite boundary (as discussed above). The figure on right in Fig.3 shows the field configuration at $t = 21.6$. A vortex-antivortex pair is seen here at the boundary of the heated region. Note that such a pair could not have been formed via any other mechanism. Defects could not form here via the Kibble mechanism as that requires the system to go through a symmetry breaking phase transition. In our case the temperature never crosses the transition temperature. Similarly, such defects could never be produced by thermal fluctuations, as that process can not directly produce a pair in which a defect and an antidefect are so far separated.

These arguments show that the above scenario can be used to provide an unambiguous experimental verification of the flipping mechanism described in [10] for defect production. For example, one can use superfluid ^{4}He (or superconductor) to experimentally test this scenario by locally heating a central region to a temperature $T_0 < T_c$ (T_0 should be large enough to allow for overshooting). In $3 + 1$ dimensions, this would lead to the formation of a string loop, with a size of order of the size of the heated central region. The condition of small phase variation across the system can be easily achieved for superfluid helium by allowing for small, uniform, superflow across the system. (For superconductors, one can allow small supercurrent.) The fact that the formation of such a large string loop does not require a phase transition, makes this a clean signal, unpolluted by the presence of smaller string loops. One can even consider a film of superfluid helium (or a superconductor). Again, by allowing small uniform superflow, and rapidly heating a localized portion of the system, one should observe a far separated vortex-antivortex pair.

We now describe the resonant production of vortex-antivortex pairs, induced by periodically varying the temperature T. T (in Eqn.(2)) is taken to be spatially uniform, with its periodic variation given by $T(t) = T_0 + T_a \sin(\omega t)$. The choice of frequency ω was guided by the range of frequency required to induce resonance for the case of spatially uniform field (evolved by Eqn.(3)). We find that resonance happens when ω lies in a certain range. (We are assuming that for the relevant range of ω here, the system can be considered in quasi-equilibrium so that the use of temperature dependent effective potential makes sense.) This frequency range for which resonant defect production occurs depends on the average temperature (T_0) as well as on the amplitude of oscillation T_a, with the range becoming larger as T_0 approaches T_c.

Initially the magnitude of Φ is taken to be equal to the vev at $T = T_0$, over the entire lattice. The phase θ of Φ was chosen to have small random variations at each lattice point, with θ lying between 20^0 and 40^0 (left figure in Fig.4). We have carried out simulations for other widely different initial configurations as well. In one case we took the domain structure developing at late stages in the above simulations, and used it as the initial data for evolution with a different set of parameters. In another case, we took the lattice to consist of only four domains with small variations of θ between domains. Defects are always produced as long as there is some *non-uniformity* in the initial phase distribution. Because of the coupling between the phase and magnitude of the OP field, even a slight non-uniformity in phase is amplified due to magnitude fluctuations brought about by the periodically varying temperature. Although we find that at early times defect formation depends on the initial configuration, the *asymptotic average defect density* seems to remain unchanged.

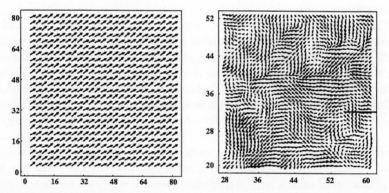

Fig. 9.4. Left figure shows the initial configuration of Φ for the entire lattice. The figure on the right shows the plot of Φ at t = 908.0 for the resonant oscillation case for a portion of lattice, showing randomly oriented domains, with 4 vortex-antivortex pairs. $T_0 = 0.31$ and $T_a = 0.125$ for this case

The figure on the right in Fig.4 shows a portion of the lattice at time $t = 908.0$, for $\omega = 1.19$, $T_0 = 0.31$, $T_a = 0.125$ and $\eta = 0.013$, showing four vortex-antivortex pairs. We count only those pairs which are separated by a distance larger than $2m_H^{-1}$. Here m_H^{-1} $(= 1$ at $T = 0)$ is the relevant length scale (correlation length) in Eq.(1). We emphasize again, formation of these vortices can not be accounted for by Kibble mechanism, or by the thermal production of defects. Although, the basic mechanism of defect production here is completely different from the production of defects via thermal fluctuations, it is interesting to check what sort of defect densities are expected from thermal fluctuation mechanism at the temperatures under consideration. For this, we have estimated the energy of a vortex-antivortex pair, using the numerical techniques in ref.[22], to be about 2.5 (for $T_0 = 0.31$) when the separation between the vortex and the antivortex is about $2m_H^{-1}$ (with m_H being temperature dependent). If we take thermally produced defect density to be of order $\sim T^2 \exp(-E_{\text{pair}}/T)$, then even with $T = T_0 + T_a = 0.435$, the defect density is only about 6×10^{-4}, which is less than the defect densities we find. Thus detection of defects via resonant production should not have any difficulty from the thermal contribution to the defect density (at least for values of T which are not too close to T_c).

We have carried out simulations for various values of T_0 (with $\omega = 1.19$, $T_a = 0.125$, and $\eta = 0.013$ kept fixed), starting with similar initial configurations. As a function of time, defect density rises from zero, and eventually fluctuates about an average value. We carry out the evolution till $t \sim 1000$, when this average defect density becomes reasonably constant. For this set of parameters, the smallest value of T_0 for which we observe resonant defect production was equal to 0.31. (For linear temperature dependence in $V(\phi)$ in Eqn.(2) we get defect production at much lower temperatures, with smallest value of $T_0 = 0.10$ and $T_a = 0.08$, $\omega = 1.19$.) Fig.5 shows temporal variation of defect density at large times. The fluctuations in defect density are due to pair annihilations and pair creations of vortices which keep happening periodically because of localized flipping of Φ. The number of vortices is equal to the number of antivortices, as the flipping mechanism creates only vortex-antivortex pairs.

The asymptotic average defect density (for $900 \le t \le 1000$) shows nontrivial variation as a function of T_0. We find that the defect density peaks at a

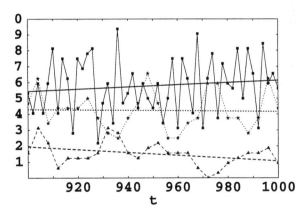

Fig. 9.5. Evolution of density n of vortices and antivortices at large times. Straight lines show linear fits to the respective curves. n for $\omega = 1.19$ and 1.31, are shown by the solid and the dotted curves respectively (with $T_0 = 0.5$ and $T_a = 0.125$). Dashed curve shows n for $T_0 = 0.31$, $T_a = 0.125$ and $\omega = 1.19$

value of $T_0 \simeq 0.5$. This drop in average defect density for larger temperatures can be due to a decrease in m_H. Due to flatter V_{eff} near the true vacuum, Φ does not gain too much potential energy when V_{eff} changes due to change in temperature (with $T_a = 1.0$), which may make flipping of Φ difficult (even though the barrier height is smaller now). We also find that there is an optimum value of frequency ω for which the defect density is maximum. It is reasonable to expect that if ω is too large, it would be difficult for the field to overshoot the central barrier, since larger frequency will lead to averaging out of the force on Φ due to the rapidly changing effective potential. On the other hand, if ω is too small, the field may have enough time to relax to its equilibrium value without being sufficiently destabilized, making the overshooting of Φ more difficult. By keeping all other parameters fixed ($T_0 = 0.5$, $T_a = 0.13$ and $\eta = 0.013$), we have explored the behavior of defect density for different values of frequency. (For the above parameters, resonant defect production happens for $0.94 < \omega < 1.31$). For $\omega = 0.94$, 1.0, 1.13, 1.19, and 1.31, we find the average defect density to be 0.0048, 0.0061, 0.0059, 0.0058, and 0.0042 respectively.

We have made some check on the dependence of the asymptotic average defect density on T_a. For $T_0 = 0.5$, $\omega = 1.19$ and $\eta = 0.013$, the average defect density is larger for $T_a = 0.13$ compared to the case when $T_a = 0.25$. (Again, note that we are focusing on asymptotic defect density.) When we increase η, it results in suppression of the defect density since damping makes it difficult for Φ to overshoot the barrier. Moreover, as η is gradually increased, the minimum value of T_0 required to induce resonant oscillations also increases and for $\eta > 0.63$, resonant production of vortices is not possible for $T_0 < T_c$. We also find that the range of frequency for which resonant defect production occurs, becomes narrower on increasing η.

9.6   Conclusion

Study of the formation of topological defects is an active field. There have been many interesting developments recently in this area. New theories have been proposed which lead to different predictions about the defect distributions. It is possible to experimentally test many of these predictions. It will be very interesting to verify the prediction that topological defects can form via the flipping mechanism, without ever going through phase transition. Superfluid helium, or superconductors are some systems where these experimental tests can be carried out. Resonant production of defects also presents novel possibilities for defect production. Experimental verification of this will be very interesting, especially from the point of view of its implications for the defect production in the inflationary theories of the universe.

References

1. A. Vilenkin and E.P.S. Shellard, "Cosmic strings and other topological defects", (Cambridge University Press, Cambridge, 1994).
2. *Formation and interactions of topological defects*, Edited by, A.C. Davis and R. Brandenberger, Proceedings of NATO Advanced Study Institute, 1994, (Plenum, New York, 1995).
3. T.W.B. Kibble, J. Phys. A **9**, 1387 (1976).
4. I. Chuang, R. Durrer, N. Turok and B. Yurke, Science **251**, 1336 (1991).
5. M.J. Bowick, L. Chandar, E.A. Schiff and A.M. Srivastava, Science **263**, 943 (1994).
6. S. Digal, R. Ray, and A.M. Srivastava, Phys. Rev. Lett. **83**, 5030 (1999).
7. P.C. Hendry et al., J. Low. Temp. Phys. **93**, 1059 (1993); G.E. Volovik, Czech. J. Phys. **46**, 3048 (1996) Suppl. S6; M.E. Dodd et al., Phys. Rev. Lett. **81**, 3703 (1998); G. Karra and R.J. Rivers, Phys. Rev. Lett. **81**, 3707 (1998).
8. W.H. Zurek, Phys. Rep. **276**, 177 (1996), see also A. Yates and W.H. Zurek, Phys. Rev. Lett. **80**, 5477 (1998).
9. S. Digal and A.M. Srivastava, Phys. Rev. Lett. **76**, 583 (1996).
10. S. Digal, S. Sengupta, and A.M. Srivastava, Phys. Rev. D **55**, 3824 (1997); Phys. Rev. D **56**, 2035 (1997); Phys. Rev. D **58**, 103510 (1998).
11. S. Digal, R. Ray, S. Sengupta, and A.M. Srivastava, Phys. Rev. Lett. **84**, 826 (2000).
12. M. Stone and A.M. Srivastava, J. Low Temp. Phys. **102**, 445 (1995).
13. T. Vachaspati and A. Vilenkin, Phys. Rev. D **30**, 2036 (1984).
14. S. Rudaz and A.M. Srivastava, Mod. Phys. Lett. A **8**, 1443 (1993).
15. S. Rudaz, S. Varma, and A.M. Srivastava, Int. J. Mod. Phys. A **14**, 1591 (1999).
16. M. Hindmarsh, A.C. Davis and R.H. Brandenberger, Phys. Rev. D **49**, 1944 (1994); R. H. Brandenberger and A.C. Davis, Phys. Lett. B **332**, 305 (1994); T.W.B. Kibble and A. Vilenkin, Phys. Rev. D **52**, 679 (1995).
17. J. Borrill, T.W.B. Kibble, T. Vachaspati, and A. Vilenkin, Phys. Rev. D **52**, 1934 (1995).
18. E.J. Copeland and P.M. Saffin, Phys. Rev. D **54**, 6088 (1996); E.J. Copeland and P.M. Saffin, Phys. Rev. D **56**, 1215 (1997).
19. J.H. Traschen and R.H. Brandenberger, Phys. Rev. D **42**, 2491 (1990); Y. Shtanov, J. Traschen, and R. Brandenberger, Phys. Rev. D **51**, 5438 (1995); L. Kofman, A. Linde, and A.A. Starobinsky, Phys. Rev. D **56**, 3258 (1997).
20. M.F. Parry and A.T. Sornborger, Phys. Rev. D **60**, 103504 (1999); I.Tkachev, S. Khlebnikov, L. Kofman, and A. Linde, Phys. Lett. B **440**, 262 (1998); S. Kasuya and M. Kawasaki, Phys. Rev. D **58**, 083516 (1998).
21. L. Gammaitoni, et.al. "Stochastic Resonance", Rev. Mod. Phys. **70**, 223 (1998)
22. A.M. Srivastava, Phys. Rev. D **47**, 1324 (1993).

10 The Order Parameter as a Macroscopic Quantum Wavefunction

Antony J. Leggett

Department of Physics, University of Illinois at Urbana-Champaign, 1110 West
Green Street, Urbana, IL 61801-3080, USA

Abstract. Although it is only in the very simplest cases that the order parameter
of a superfluid or superconductor can be regarded literally as a Schrödinger wave
function, it is quite generally "very like" one, and thinking of it in this way can
often be quite helpful to one's intuition, particularly in cases involving internal
degrees of freedom. In this chapter I shall briefly sketch the basis for this point of
view, starting with the very simplest case, that of a noninteracting Bose gas, and
proceeding through progressively more complicated examples, to end with a Fermi
superfluid with "exotic" pairing.

10.1 The Free Bose Gas

Consider a set of N identical noninteracting atoms without internal degrees
of freedom and subject to Bose statistics, moving freely in a volume V with
periodic boundary conditions. It is a standard result, originally due to Einstein, that in the thermodynamic limit $N, V \to \infty$, $N/V \to$ const. $\equiv n$, there
exists a temperature $T_c(n)$ such that below T_c the thermal equilibrium single-
particle distribution over plane-wave states $\mathbf{k}$, $\langle a_{\mathbf{k}}^{+} a_{\mathbf{k}} \rangle(T) \equiv \langle n_{\mathbf{k}}(T) \rangle$ has the
property

$$\langle n_0(T) \rangle \equiv N_0(T) \sim N, \qquad \langle n_{\mathbf{k} \neq 0} \rangle(T) \sim 1 \qquad (10.1)$$

i.e. below T_c a finite fraction $f(T) \equiv N_0(T)/N$ of all atoms occupy the zero-
momentum (lowest-energy) single-particle state. This phenomenon is known
as *Bose-Einstein condensation* (BEC), and the atoms which occupy the zero-
momentum state are known as the *condensate*. The specific form of the condensate fraction $f(T)$ for the free gas is $1 - (T/T_c(n))^{3/2}$, so that f tends to
0 for $T \to T_c(n)$ and to 1 for $T \to 0$.

Suppose now we perturb the system with some space- and time-dependent
c-number external potential $V(\mathbf{r}, t)$. Under the action of V the original single-
particle wave functions, which are time-independent ($e^{i\mathbf{k} \cdot \mathbf{r}}$) except for the
overall phase factor $\exp[-i(\hbar^2 k^2/2m)t]$, will evolve into a set of functions
$\psi_{(\mathbf{k})}(\mathbf{r})$ which have a nontrivial time dependence; the label $(\mathbf{k})$ is attached to
indicate the state from which they evolved. By unitarity, the states $\psi_{(\mathbf{k})}(\mathbf{r}t)$
for different $\mathbf{k}$ are at all times mutually orthogonal; moreover, because of the
product nature of the wave functions which were represented in the original
(thermal-equilibrium) density matrix, the average occupation number $\langle n_{(\mathbf{k})} \rangle$
of the state $\psi_{(\mathbf{k})}(\mathbf{r}t)$ is just equal to the original $\langle n_{\mathbf{k}} \rangle$. In particular, the state

$\psi_{(0)}(\mathbf{r}t)$, and *only* that state, has a macroscopic occupation N_0. For notational convenience, let us denote this state by $\chi_0(\mathbf{r}t)$ and call it the "wave function of the condensate"; note that it is no more and no less than a particular (time- and space-dependent) single-particle wave function. Then we can define the order parameter $\Psi(\mathbf{r}t)$ by the simple prescription

$$\Psi(\mathbf{r}t) \equiv \sqrt{N_0}\chi_0(\mathbf{r}t) \tag{10.2}$$

where for this simple case $N_0 \equiv N_0(T) = N(1 - (T/T_c)^{3/2})$. Thus, up to its overall normalization the order parameter is indeed nothing but a simple one-particle Schrödinger wave function.

The state of the system considered above is rather special, in that it is specified to have evolved from the thermal equilibrium state under the action of an external perturbation, and one might suspect that a definition similar to (10.2) might actually be useful more generally. This is true, but at this point we might as well go directly to the case of an interacting Bose gas.

10.2   The Interacting Bose Gas

We consider now the general case of an interacting Bose gas (but still without internal degrees of freedom) subject to some external potential $V(\mathbf{r}t)$ and not necessarily in thermal equilibrium. Quite generally, we can define the single-particle density matrix

$$\rho(\mathbf{r}, \mathbf{r}' : t) \equiv \langle \psi^+(\mathbf{r})\psi(\mathbf{r}') \rangle$$
$$\equiv \sum_i p_i \int dr_2...dr_N \Psi_N^{(i)*}(\mathbf{r}, \mathbf{r}_2...\mathbf{r}_N : t)\Psi_N^{(i)}(\mathbf{r}', \mathbf{r}_2...\mathbf{r}_N : t) \tag{10.3}$$

where the $\Psi_N^{(i)}$ are functions which diagonalize the N-body density matrix with eigenvalues p_i (i.e., crudely speaking, "the N-particle system is in state $\Psi_N^{(i)}$ with probability p_i"). The quantity $\rho(\mathbf{r}, \mathbf{r}' : t)$ is clearly Hermitian with respect to its arguments $(\mathbf{r}, \mathbf{r}')$ and thus can be diagonalized with real eigenvalues which we denote $\langle n_i \rangle(t)$

$$\rho(\mathbf{r}, \mathbf{r}' : t) = \sum_i \langle n_i \rangle(t)\chi_i^*(\mathbf{r}t)\chi_i(\mathbf{r}'t) , \tag{10.4}$$

where for any given time the eigenfunctions $\chi_i(\mathbf{r}t)$ constitute a complete orthonormal set. Intuitively, the quantity $\langle n_i \rangle(t)$ is the average occupation, at time t, of the single-particle state $\chi_i(\mathbf{r} : t)$. Note that the choice of the χ_i is simply dictated by the requirement of absence of "off-diagonal" terms in (10.4), and they need not be eigenfunctions of any particular quantity (e.g. the kinetic energy) other than $\hat{\rho}$ itself.

We now define the system of interest as possessing (at time t) "BEC" if *at least one* of the eigenvalues $\langle n_i \rangle(t)$ is of order of the total particle number N, and as possessing "simple BEC" if *one and only one* of the eigenvalues has this property. It is impossible to overemphasize, first, that the conventional definition of an order parameter (see below) makes sense *only* if the BEC is "simple", i.e., if the "condensate" is uniquely defined, and second, that in the general case (as distinct from the simple special case discussed in Section 10.1) we have no a priori reason to believe that in all states of physical interest any BEC occurring need be simple. In fact, the reasons why "simple BEC" tends to be the norm, at any rate in spinless systems, are quite subtle and have to do with the effects of the interatomic interaction, which for most cases of practical interest is repulsive;[1] they are discussed in detail in [7], section 4 (cf. also [8]).

Given that the system of interest indeed shows simple BEC, we can define an order parameter as in (10.2), the only difference being that the quantity $N_0 \equiv \langle n_0 \rangle(t)$ is in general a function of time:

$$\Psi(\mathbf{r}t) = \sqrt{N_0(t)}\chi_0(r,t) \ . \tag{10.5}$$

Note that, even in thermodynamic equilibrium at $T = 0$ the quantity N_0 need not be equal to the total number of particles N (in fact, in a case of practical interest – liquid ^{4}He – the condensate fraction $f(0) \equiv N_0(T = 0)/N$ is believed to be less than 10%). We see that, just as in the special case of Section 10.1, the order parameter is essentially a particular single-particle wave function; the only "macroscopic" thing about it is that the corresponding single-particle state is occupied by a macroscopic number of atoms.

10.3   Bose Gas with Internal Degrees of Freedom

Although the Bose system in which BEC was originally realized, liquid ^{4}He, has no (relevant) internal degree of freedom, this is no longer the case with a system of more recent interest, namely the ultracold atomic alkali gases (for review, see e.g. [4]); these systems are invariably characterized by hyperfine and Zeeman quantum numbers, and the energy scales involved are such that it is often quite easy to produce situations in which two or more different hyperfine-Zeeman states coexist. The general description of BEC in such systems is discussed in detail in [7]; in the present context the simplest approach is to define a fixed (space- and time-independent) set of basic vectors α in the "internal" space (which need not necessarily correspond to the eigenfunctions of the single-particle Hamiltonian, though this choice is usually convenient where it is possible) and to treat the label α on a par with the coordinate

---

[1] Bose systems with attractive effective interactions are generally unstable against collapse in real space, though in a "trap" geometry there are some complications: cf. [2], section III.c.

vector $\mathbf{r}$. Thus, one defines a creation operator $\psi(\mathbf{r}, \alpha)$ ($\equiv \psi_\alpha(\mathbf{r})$ in a more familiar notation) and generalizes equations (10.3) and (10.4) to read

$$\rho(\mathbf{r}, \alpha, \mathbf{r}', \alpha' : t) \equiv \langle \psi^+(\mathbf{r}, \alpha)\psi(\mathbf{r}, \alpha')\rangle(t)$$
$$= \sum_{i\eta} \langle n_{i\eta}\rangle(t)\chi^*_{i\eta}(\mathbf{r}, \alpha : t)\chi_{i\eta}(\mathbf{r}', \alpha' : t) \quad , \qquad (10.6)$$

where the eigenfunctions $\chi_{i\eta}(\mathbf{r}, \alpha : t)$ are "spinors" (multicomponent wave functions) in the space of the internal degrees of freedom; note that the spinor structure of those eigenfunctions may depend on the index i. As above, we define the system to show simple BEC if one and only one eigenvalue $\langle n_{0,\eta_0}\rangle(t) \equiv N_0(t)$ is of order N, and can then generalize the definition (10.5) of the order parameter to read

$$\Psi(\mathbf{r}, \alpha' : t) \equiv \sqrt{N_0(t)}\chi_{0,\eta_0}(\mathbf{r}, \alpha : t) \qquad (0 \equiv i, \eta) \qquad (10.7)$$

The question of the justification for the assumption that any BEC occurring is simple is even more delicate in this case, and in fact one not infrequently has to deal with cases where it is not: see [7], section 7, and cf. also below, Section 10.5.

10.4   Cooper Pairing: s-Wave Case

As is well known, the basis of the BCS theory of superconductivity is that electrons near the Fermi surface of the metal in question pair off to form a bound state with total spin and orbital angular momentum zero (Cooper pair), in such a way that all pairs occupy the same two-particle state. It is thus tempting to think of the Cooper pairs as bosons, and the formation of the superconducting state as a sort of BEC of these bosons. Were the radius of the pairs much less than the mean distance between them ("Bose limit"), this picture would be at least qualitatively valid, and one could then simply apply the analysis developed in Section 10.2 above (and moreover, generalize it to the case of "exotic" pairing along the lines of Section 10.3). However, the pre-1986 superconductors are certainly in the opposite limit: the "pair radius" is much *greater* than the average inter-pair separation ("BCS limit") and as a result one certainly cannot neglect the fermion nature of the electrons composing the pair. (The situation is a bit more ambiguous for the cuprates, but they too are probably closer to the BCS than to the Bose limit.)

A simple schematic representation of the N-electron wave function corresponding to the ground state assumed by Bardeen et al. is [1], apart from normalization,

$$\Psi_N(\mathbf{r}_1\sigma_1...\mathbf{r}_N\sigma_N) = \qquad (10.8)$$
$$\mathcal{A}\varphi(\mathbf{r}_1 - \mathbf{r}_2)\chi_s(1, 2)\varphi(\mathbf{r}_3 - \mathbf{r}_4)\chi_s(3, 4)...\varphi(\mathbf{r}_{N-1} - \mathbf{r}_N)\chi_s(N - 1, N)$$

where $\chi_s(1,2)$ denotes the spin singlet combination of the spins of particles 1 and 2 and $\mathcal{A}$ is the operator of antisymmetrization between any two electrons. Because the spin singlet function is already antisymmetric in the interchange, the orbital function $\varphi(\mathbf{r}_1 - \mathbf{r}_2)$ must be symmetric. At finite temperature (less than the superconducting transition T_c), a typical wave function represented in the density matrix has some fraction $f(T)$ of all the electrons paired off as in (10.8), with the rest occupying independent single-particle states. Now, just as in Section 10.1, we can envisage generalizing (10.8) by applying a space- and time-dependent external potential, so that $\varphi(\mathbf{r}_1\mathbf{r}_2 : t)$ now becomes a function of time and of the center-of-mass coordinate $\mathbf{R} \equiv (\mathbf{r}_1\mathbf{r}_2{:}t)$ as well as of $\mathbf{r}_1$-$\mathbf{r}_2$. One might therefore think that it would be natural to attempt to define an "order parameter" $\Psi(\mathbf{r}_1, \mathbf{r}_2 : t)$ by the prescription

$$\Psi_{\mathrm{trial}}(\mathbf{r}_1, \mathbf{r}_2 : t) = \sqrt{N_0}\varphi(\mathbf{r}_1, \mathbf{r}_2 : t) \qquad (?) \qquad (10.9)$$

where $N_0 \equiv Nf(T)$.

However, this definition turns out to be rather awkward to implement for the more general case, and in any case the quantity $\varphi(\mathbf{r}_1\mathbf{r}_2 : t)$ turns out not to have much direct physical significance. Rather, following the classic discussion of [11], let us define for an arbitrary state the two-particle density matrix

$$\rho(\mathbf{r}_1\sigma_1, \mathbf{r}_2\sigma_2, \mathbf{r}_3\sigma_3, \mathbf{r}_4\sigma_4 : t) \equiv \langle \psi_{\sigma_1}^+(\mathbf{r}_1)\psi_{\sigma_2}^+(\mathbf{r}_2)\psi_{\sigma_3}(\mathbf{r}_3)\psi_{\sigma_4}(\mathbf{r}_4)\rangle(t) \quad (10.10)$$

and study the special case $\sigma_1 = -\sigma_2 = \uparrow$, $\sigma_3 = -\sigma_4 = \downarrow$. Then, using the Hermitian property of ρ, we can write it in a form analogous to (10.4):

$$\rho(\mathbf{r}_1 \uparrow, \mathbf{r}_2 \downarrow, \mathbf{r}_3 \downarrow, \mathbf{r}_4 \uparrow: t) = \sum_i \langle n_i \rangle(t)\chi_i^*(\mathbf{r}_1\mathbf{r}_2)\chi_i(\mathbf{r}_4\mathbf{r}_3) . \qquad (10.11)$$

As in the Bose case, let us *assume* that one and only one of the eigenvalues $\langle n_i \rangle(t)$ is extensive, i.e. of order of the total number N ("simple pseudo-BEC"); this is certainly true in the BCS ground state, and there are general arguments similar to those applying in the Bose case which indicate that the (attractive) interaction between fermions tends to disfavor macroscopic occupation of more than one (orbital) two-particle state, see e.g. [6], pp. 160–163. Then we can define a (two-particle) order parameter $\Psi(\mathbf{r}_1, \mathbf{r}_2 : t)$ by the prescription (where $N_0(t)$ is the unique macroscopic eigenvalue and χ_0 the corresponding eigenfunction)

$$\Psi(\mathbf{r}_1\mathbf{r}_2 : t) \equiv \sqrt{N_0(t)}\chi_0(\mathbf{r}_1\mathbf{r}_2 : t) . \qquad (10.12)$$

In the BCS formalism the constraint of fixed particle number is relaxed, so that we can assign a finite value to so-called "anomalous" averages such as $\langle \psi_\alpha(\mathbf{r})\psi_\beta(\mathbf{r}')\rangle$; then we have for $\Psi(\mathbf{r}_1\mathbf{r}_2 : t)$ the simple expression

$$\Psi(\mathbf{r}_1\mathbf{r}_2 : t) = \langle \psi_\uparrow(\mathbf{r}_1)\psi_\downarrow(\mathbf{r}_2)\rangle(t) \equiv F(\mathbf{r}_1, \mathbf{r}_2 : t) \qquad (10.13)$$

where we note an alternative notation frequently used in the literature. One advantage of defining the order parameter by (10.12) [or (10.13)] rather than (10.9) is that the formula for the change in the expectation values of "two-particle" quantities such as the potential energy which is explicitly[2] due to pair formation, when expressed in terms of the quantity defined by (10.12), is formally identical to that for the total potential energy in a two-particle system: for example we have

$$(\delta\langle V \rangle)_{\text{pairing}} = \int \int d\mathbf{r}_1 d\mathbf{r}_2 V(\mathbf{r}_1 - \mathbf{r}_2)|\Psi(\mathbf{r}_1, \mathbf{r}_2 : t)|^2 \qquad (10.14)$$

By contrast, if we were to use the definition (10.9) the expression for $(\delta\langle V \rangle)_{\text{pairing}}$ would be very messy.

The question of the normalization of Ψ is a little trickier than in the Bose case. If we use the definition (10.13) (or equivalently (10.12) with χ_0 normalized to unity) and calculate the right-hand side for the BCS thermal equilibrium state, we find, using the standard BCS notation, the result

$$N_0(T) \equiv \int \int |\Psi(\mathbf{r}_1 \mathbf{r}_2)|^2 d\mathbf{r}_1 d\mathbf{r}_2 = V\left(\frac{dn}{d\epsilon}\right) \int d\epsilon \left(\frac{\Delta(T)}{2E}\right)^2 \tanh^2 E/2T \qquad (10.15)$$

The expression on the right can thus be regarded, in an intuitive sense, as a measure of the "number of Cooper pairs" formed in the system; it is of order $N(\Delta(0)/\epsilon_F)$ at $T = 0$ and of order $N\Delta^2(T)/T_c\epsilon_F$ in the limit $T \to T_c$. (By contrast, in the "Bose" limit the number of pairs at $T = 0$ would of course be simply $N/2$).

By this time the reader may be distinctly puzzled, since he or she is probably used to thinking of the order parameter of a superconductor as a quantity $\Psi(\mathbf{r}t)$ which has only a single space argument, whereas the quantity introduced in (10.12) and discussed above is a function of two space arguments $\mathbf{r}_1, \mathbf{r}_2$. Actually, the traditional order parameter $\Psi(\mathbf{r}t)$, which was originally introduced by Ginzburg and Landau in the context of a phenomenological approach before the microscopic BCS theory was developed, is straightforwardly related to the $\Psi(\mathbf{r}_1\mathbf{r}_2 : t)$ of (10.12). Let's write the latter in terms of the center-of-mass coordinate $\mathbf{R} \equiv \frac{1}{2}(\mathbf{r}_1 + \mathbf{r}_2)$ and the relative coordinate $\rho \equiv \mathbf{r}_1 - \mathbf{r}_2$:

$$\Psi(\mathbf{r}_1, \mathbf{r}_2 : t) \equiv \Psi(\mathbf{R}, \rho : t) . \qquad (10.16)$$

In the simple s-wave case considered in this section, the dependence of Ψ on the relative coordinate ρ is fixed, in thermodynamic equilibrium, by the energetics and is a function only of $| \rho |$, and more generally cannot deviate far from its equilibrium form without a major cost in energy (i.e. the internal "structure" of the Cooper pairs is fixed to a good approximation). Thus it

---

[2] As distinct from possible "indirect" changes due to pair formation, e.g. in the Fock term in the energy.

is convenient to eliminate ρ from the description. The most common way of doing this is simply to write

$$\Psi(\mathbf{r}, t) \equiv \text{const.} \times \Psi(\mathbf{R}, \mathbf{0} : t)_{\mathbf{r}=\mathbf{R}} , \qquad (10.17)$$

but it would also be possible, for example, to define $\Psi(\mathbf{r}t)$ as the integral of $\Psi(\mathbf{R}, \rho: t)$ over ρ for $\mathbf{r} = \mathbf{R}$.[3] Thus, the traditional order parameter is, up to a (possibly temperature-dependent) constant, simply the "effective" wave function of the center of mass of the Cooper pairs. Note that, just as in the Bose case, the only "macroscopic" thing about this wave function is that it is occupied by a macroscopic $[O(N)]$ number of pairs.

A final detail concerns the choice of the constant in (10.17). It turns out that the original choice of Ginzburg and Landau, when interpreted in terms of BCS theory with a contact interaction $-g\delta(\mathbf{r})$, actually identifies the order parameter with the local (complex) energy gap $\Delta(\mathbf{r}t)$ [3]. Since in such a theory the gap is just $g\Psi(\mathbf{R}, \mathbf{0} : t)$, this choice is equivalent to setting the constant in (10.17) equal to g. However, it should be emphasized that this choice is simply a matter of convention and nothing depends on it; this remains equally true in the case of a more general form of interaction.

10.5   Cooper Pairs with Internal Degrees of Freedom

An important simplifying factor in the case of the s-wave pairing, believed to describe the "classic" superconductors, is that the spin structure of the pairs is a singlet and thus is fixed once and for all, while the dependence of the two-particle order parameter $\Psi(\mathbf{R}, \rho)$ on ρ is only on the magnitude $|\rho|$ and is fixed, in equilibrium, by the energetics. Thus the pairs have no "orientation" or adjustable internal structure. By contrast, there exists a class of Fermi superfluids – the superfluid phases of liquid ^3He, some more recently discovered superconductors and, probably, neutron stars – where the pairs possess a nontrivial spin and/or orbital structure which is not completely determined by the (main part of the) Hamiltonian.[4]

In such cases, the simplest procedure consists in first generalizing (10.13) to arbitrary spin components α, β:

$$\Psi_{\alpha\beta}(\mathbf{r}_1, \mathbf{r}_2 : t) \equiv \langle \psi_\alpha(\mathbf{r}_1)\psi_\beta(\mathbf{r}_2)\rangle (\equiv F_{\alpha\beta}(\mathbf{r}_1, \mathbf{r}_2 : t)) \qquad (10.18)$$

and writing it in analogy to (10.17) in the form

$$\Psi_{\alpha\beta}(\mathbf{r}_1, \mathbf{r}_2 : t) \equiv \Psi_{\alpha\beta}(\mathbf{R}, \rho : t) \qquad (10.19)$$

---

[3] To lowest nontrivial order in the standard Ginzburg-Landau gradient expansion these definitions are effectively equivalent; at higher order they may produce minor differences.

[4] For an exhaustive discussion of the topic of this section, see [10], chs. 3 and 5.

In the case of spin singlet (but possibly orbitally anisotropic) pairing we have, as in Section 10.4, $\Psi_{\alpha\beta}(\mathbf{R},\boldsymbol{\rho}: t) = -\Psi_{\beta\alpha}(\mathbf{R},\boldsymbol{\rho}: t) = \frac{1}{2}(1 - \delta_{\alpha\beta})\Psi(\mathbf{R},\boldsymbol{\rho}: t)$. In the case of triplet pairing, we could of course characterize the spin state by the amplitudes for each of the three Zeeman substates, but it is often more convenient to represent it by an equivalent spin-space vector $\mathbf{d}(\mathbf{R},\boldsymbol{\rho}:t)$ defined in terms of the 2×2 matrix $\Psi_{\alpha\beta}$ by

$$\mathbf{d}(\mathbf{R}, \boldsymbol{\rho} : t) \equiv Tr(-i\sigma_2\boldsymbol{\sigma}\Psi(\mathbf{R}, \boldsymbol{\rho} : t)) \ . \tag{10.20}$$

We still need to characterize the "orientation" of the orbital part of the wave function. For simplicity, I specialize to the case of p-wave pairing, which occurs in ^{3}He and probably some superconductors:[5] in this case one can for example define a quantity $d_{i\mu}(\mathbf{R} : t)$ (where i denotes the component of $\mathbf{d}$ in spin space) by

$$d_{i\mu}(\mathbf{r}, t) \equiv \int d^3\rho\, \rho_\mu d_i(\mathbf{R}, \boldsymbol{\rho} : t)_{\mathbf{r}=\mathbf{R}} \tag{10.21}$$

or by any of a number of definitions which are essentially equivalent in the usual Ginzburg-Landau limit. The quantity $d_{i\mu}$ defined by (10.21) is a bivector, that is, it is simultaneously a vector in spin space with components $i=1,2,3$ and a vector in orbital space with components $\mu=1,2,3$. This description is qualitatively similar, though not identical, to that of a Bose gas with an internal ("spin") degree of freedom, see Section 10.3; it can be generalized in a fairly obvious way to the case of pairing with $\ell \neq 1$.

In a simple system such as liquid ^{3}He, where the interatomic interactions are to a first approximation invariant under separate rotations of the spin and orbital coordinates, the free energy F is to the same approximation invariant under an overall rotation of the order parameter $d_{i\mu}$ in spin and orbital space separately. However, this does *not* mean, as one might perhaps think, that F is a function only of the quantity $|d_{i\mu}|^2$; in general, forms of the order parameter which are equivalent in this respect but cannot be related by any rotation [e.g. $d_{i\mu} = \delta_{\mu z}\delta_{iz}$ and $d_{i\mu} = 2^{-1/2}\delta_{\mu z}(\delta_{ix} + i\delta_{iy})$] correspond to different values of free energy even in the above approximation. What it does mean is that it is possible for the order parameter at one point in space and time to be rotated, in spin and/or orbital space, relative to its value elsewhere at a cost only of the relevant gradient terms in the Ginzburg-Landau free energy. For a detailed discussion in the specific context of ^{3}He, see [10] or [9].

One caveat should be made about the above analysis (which is standard in the literature): The definition (10.18) (like (10.13) in the s-wave case) implicitly relies on the assumption that one and only one eigenvalue of the two-particle density matrix (10.10) is "of order N", i.e. we have simple, rather than general, "pseudo-BEC". If there is an exact or approximate conservation

---

[5]   In real-life superconductors it is necessary to classify the internal structure of the pairs by its behavior under the operations of the crystal symmetry group rather than under the rotation group O(3) so that the s-, p-classification is only approximate.

law in the system, it is not at all obvious that this must be true. For example, in the absence of the small spin-nonconserving nuclear dipole interaction, the ground state wave function of superfluid ^{3}He in the A phase would actually be a product of two independent wave functions for the up- and down-spin atoms, and would then have two different two-particle eigenfunctions with occupation $\sim N$. In this and similar cases, it may be shown that in the thermodynamic limit an arbitrarily weak spin-nonconserving interaction will enforce the "simplicity" condition, but it is not guaranteed a priori in general that this limit corresponds to the actual conditions of a real-life experiment. For a detailed discussion of this point, see [5].

References

1. Ambegaokar, V., in R. D. Parks, ed., *Superconductivity*, Vol.I, Marcel Dekker, New York, 1969.
2. Dalfovo, F., S. Giorgini,L. P. Pitaevskii and S. Stringari, Rev. Mod. Phys. **71**, 463 (1999).
3. Gor'kov, L. P., Zh. Eksp. Teor. Fiz. **36**, 1918:translation, Soviet Physics JETP **36**, 1364 (1959).
4. Ketterle, W., D. S. Durfee andD. M. Stamper-Kurn, in Proc. Intl. School "Enrico Fermi", course CXL, ed. M. Inguscio, S. Stringari and C. E. Wieman, IOS Press, Amsterdam, 1999.
5. Leggett, A. J., in *Bose Einstein Condensation*, ed. A. Griffin, D. W. Snoke and S. Stringari, Cambridge University Press, Cambridge, U.K., 1995, p.452.
6. Leggett, A. J.,in *Electron*, ed. M. Springford, Cambridge University Press, Cambridge, U.K., 1997, p. 148.
7. Leggett, A. J.,in preparation for submission to Rev. Mod. Phys.
8. Nozières, P., in *Bose Einstein Condensation* (see above), p. 15.
9. Volovik, G. E. , *Exotic Properties of Superfluid ^{3}He*, World Scientific, Singapore, 1992.
10. Vollhardt, D., and P. Wölfle, *The Superfluid Phases of Helium-3*, Taylor and Francis, London, 1990.
11. Yang, C. N., Rev. Mod. Phys. **34**, 694 (1962).

11 The Ehrenberg–Siday–Aharonov–Bohm Effect

Charles G. Kuper

Department of Physics, Technion — Israel Institute of Technology, 32000 Haifa, Israel

11.1 Introduction

In classical electrodynamics, the scalar and vector potentials are mathematical artifacts, designed to simplify calculation, without any real physical significance. However, in quantum theory this is no longer true; potentials enter Schrödinger's equation in an intimate way, and can produce directly observable effects. As far as I am aware, this was first pointed out by Ehrenberg and Siday [1] in a widely neglected paper. Aharonov and Bohm [2] gave a much more explicit account, illustrated by two striking *gedanken* experiments, and brought the issue to the notice of the community.

Although the Ehrenberg–Siday–Aharonov–Bohm (ESAB) effect is a necessary consequence of quantum mechanics, it seemed at first sight to be counter-intuitive. It was therefore deemed necessary to find *direct* experimental confirmation of its existence. Experiments by Chambers [3] and by Boersch and Lischke [4] appeared to provide this confirmation, but a series of articles by Loinger and coworkers [5] showed that these experiments were open to an alternative interpretation. Roy [6] drew attention to the importance of topological considerations, and it was these topological constraints which led the present author [7] to suggest that an experiment using a toroidal magnet could yield an unambiguous test for the reality of the effect. The subsequent work of Tonomura and his coworkers [8,9] used such a toroidal geometry, and thus gave an elegant and unambiguous experimental demonstration of the reality of the ESAB effect.

11.2 The Role of the Electromagnetic Potentials in Quantum Mechanics

The potentials $\mathbf{A}, \Phi$ are defined (in Gaussian units) by

$$\mathbf{B} = \nabla \times \mathbf{A},$$

$$\mathbf{E} = -\nabla\Phi - \frac{1}{c}\frac{\partial \mathbf{A}}{\partial t}, \tag{11.1}$$

together with the constraints

$$\nabla^2\Phi + \frac{1}{c}\frac{\partial}{\partial t}(\nabla \cdot \mathbf{A}) \qquad = -4\pi\rho,$$

$$\nabla^2 \mathbf{A} - \frac{1}{c^2}\frac{\partial^2 \mathbf{A}}{\partial t^2} - \nabla\left(\nabla\cdot\mathbf{A} + \frac{1}{c}\frac{\partial \Phi}{\partial t}\right) = -\frac{4\pi}{c}\mathbf{J}. \tag{11.2}$$

and the Lorentz condition

$$\nabla\cdot\mathbf{A} + \frac{1}{c}\frac{\partial \Phi}{\partial t} = 0 \tag{11.3}$$

Here $\mathbf{E}$, $\mathbf{B}$ are the electric and magnetic fields respectively, and ρ, $\mathbf{J}$ are the charge and current densities. These potentials automatically satisfy Maxwell's equations

$$\nabla\cdot\mathbf{E} = 4\pi\rho, \qquad \nabla\times\mathbf{E} + \frac{1}{c}\frac{\partial \mathbf{B}}{\partial t} = 0,$$

$$\nabla\cdot\mathbf{B} = 0, \qquad \nabla\times\mathbf{B} - \frac{1}{c}\frac{\partial \mathbf{E}}{\partial t} = \frac{4\pi}{c}\mathbf{J}. \tag{11.4}$$

However, the constraints (11.2) and (11.3) do not fix $\mathbf{A}$, Φ uniquely since, if χ is an arbitrary scalar field satisfying the wave equation,

$$\nabla^2\chi = \frac{1}{c}\frac{\partial^2 \chi}{\partial t^2} \tag{11.5}$$

then the gauge transformation

$$\Phi' = \Phi - \frac{1}{c}\frac{\partial \chi}{\partial t}$$

$$\mathbf{A}' = \mathbf{A} + \nabla\chi \tag{11.6}$$

yields an equally good set of potentials. Since the "real" physically observable quantities are $\mathbf{E}$, $\mathbf{B}$, and not $\mathbf{A}$, Φ, all observable effects must be gauge invariant.

It is easily shown that the Lagrangian

$$\mathcal{L} = mc^2\sqrt{1 - v^2/c^2} + (e/c)\mathbf{v}\cdot\mathbf{A} - e\Phi \tag{11.7}$$

yields the usual classical equations for a charged particle in an electromagnetic field. Here m is the rest mass of the particle, $\mathbf{v}$ its velocity, and $\mathbf{p}_N$ its Newtonian momentum. This Lagrangian gives the canonical momentum

$$\mathbf{p} \equiv \frac{\partial \mathcal{L}}{\partial \mathbf{v}} = \mathbf{p}_N + (e/c)\mathbf{A} \tag{11.8}$$

and Hamiltonian

$$\mathcal{H} \equiv \mathbf{p}\cdot\mathbf{v} - \mathcal{L} = \sqrt{(c\mathbf{p} - e\mathbf{A})^2 + m^2 c^4} + e\Phi. \tag{11.9}$$

Note that neither the Lagrangian, the Hamiltonian, nor the canonical momentum are gauge invariant.

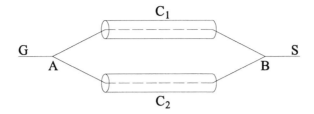

Fig. 11.1. The first Aharonov-Bohm *gedanken* experiment. The beam, emerging from an electron gun G, is split at A. One half passes through the Faraday cage C_1, and the other half, through C_2. The two beams are reunited at B and produce interference fringes on a screen at S

Making the usual transition to quantum mechanics by interpreting the canonical momentum $\mathbf{p} = -i\hbar\nabla$ yields the non-relativistic Schrödinger equation

$$i\hbar\frac{\partial\Psi}{\partial t} = \left\{\frac{1}{2m}\left(\mathbf{p} - e\mathbf{A}/c\right)^2 + e\Phi\right\}\Psi \qquad (11.10)$$

in which the state vector Ψ depends explicitly on the potentials, and is therefore also not gauge invariant. However, gauge transformations only affect the *phase* of Ψ. $|\Psi^2|$, which is observable, *is* gauge invariant.

11.3   The Aharonov–Bohm *Gedanken* Experiments

In the first *gedanken* experiment, let us consider a coherent electron beam split into two parts, each of which passes through a long cylindrical Faraday cage (Fig. 11.1). After leaving the cages, the beams recombine, to produce interference fringes in the usual manner. The beam is pulsed; the duration of a pulse is short compared to the transit time. While the electrons are inside the cages, an electrostatic potential Φ is applied to the cage C_1 and then removed, while the cage C_2 is grounded throughout. The electrons never experience any electric (or magnetic) field, but nevertheless the phase of the beam in C_1 is changed relative to C_2, leading to a displacement of the interference fringes.

Fig. 11.2 illustrates the second *gedanken* experiment. M is an infinitely long magnet, perpendicular to the plane of the diagram. The electron beam is split at A into two coherent beams, going around the magnet along paths C_1 and C_2, before being reunited at B. The phase difference between the two paths is

$$\Delta \equiv \delta_{C_1} - \delta_{C_2} = \frac{e}{\hbar c}\left\{\int_{C_1}\mathbf{A}\cdot d\mathbf{r} - \int_{C_2}\mathbf{A}\cdot d\mathbf{r}\right\}$$

$$= \frac{e}{\hbar c}\oint_C\mathbf{A}\cdot d\mathbf{r} = \frac{e}{\hbar c}\int_\Sigma\mathbf{B}\cdot d\mathbf{S}, \qquad (11.11)$$

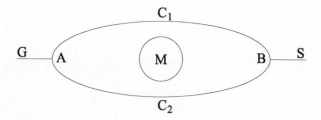

Fig. 11.2. The second Aharonov-Bohm *gedanken* experiment. The electron beam emerging from G is split at A. The two halves travel along the paths C_1 and C_2, on opposite sides of the infinitely long magnet M. The two beams are reunited at B and produce interference fringes on a screen at S

by Stokes's theorem. Here C is a closed contour around M and Σ is a surface of integration bounded by C. Eq. (11.11) demonstrated explicitly the gauge invariance of Δ, but again illustrates that the *inaccessible* field **B** will move the interference fringes.

11.4   Early Experimental "Verification"

Many experiments have been reported, which are basically just variants of the second Aharonov–Bohm *gedanken* experiment. However, infinitely long magnets or solenoids are not obtainable, and the experiments used short finite magnets. I will describe briefly two examples, taken from a large set.

In Chambers's experiment [3], an electron beam was split by an electrostatic "biprism", and the magnetic field **B** was contained within a magnetic whisker ($\sim 1\mu$m in diameter, and ~ 0.5mm long). Boersch and Lischke [4] used a solenoid, consisting of a hollow superconducting tube (of outer diameter 1.44 μm, inner diameter 0.76μm, and length $\sim 100\mu$m), which could contain either an odd or even number of flux quanta $\Phi_0 = \pi\hbar c/e$.

In all experiments of this class, the expected fringe shifts were observed. However, Bocchieri *et al.* [5] have remarked that another interpretation is possible. Since the magnet is always quite short compared with the electron trajectories, the electrons will always experience a "leakage" field, so it was not true that the electron trajectories went through a field-free region of space.

11.5   Roy's Theorems

Roy [6] gave mathematically sufficient conditions for the "non-existence" of the ESAB effect, *i.e.* sufficient conditions for all observable effects to be derivable from the action of *accessible* fields only. His theorems establish that in any singly-connected region of space, the field strengths **E, B** will determine all physical effects on the particle. Hence for an unequivocal demonstration that the effect is real, a multiply-connected geometry is essential.

Roy defines paths in space-time $z_\mu(x,\xi)$, $\mu = 1, 2, 3, 4$, $z_4 = iz_0$, where ξ is a real parameter, satisfying $-\infty < \xi < 0$, and where z_μ are single-valued differentiable functions of ξ and of the coordinates x_ν. The paths are required to satisfy the conditions

$$z_\mu(x, 0) = x_\mu \, ,$$
$$\lim_{\xi \to -\infty} z_\mu(x, \xi) = \infty \, . \qquad (11.12)$$

The electromagnetic potentials $A_\mu(x)$ are assumed to be single-valued functions, satisfying the conditions

$$\int_{-\infty}^{0} d\xi \, A_\mu(z) \frac{\partial z_\mu}{\partial \xi} < \infty \, ,$$
$$\lim_{\xi \to -\infty} A_\nu(z) \frac{\partial z_\nu}{\partial x_\mu} = 0 \qquad (11.13)$$

and the fields $F_{\mu,\nu} \equiv \partial_\mu A_\nu - \partial_\nu A_\mu$ satisfy

$$\int_{-\infty}^{0} d\xi \, F_{\mu\nu} \frac{\partial z_\mu}{\partial \xi} \frac{\partial z_\nu}{\partial x_\rho} < \infty. \qquad (11.14)$$

Since the integrand is continuous and the integral is uniformly convergent, we may differentiate under the integral sign, and then integrate by parts, to get

$$\frac{\partial}{\partial x_\mu} \left(\int_{-\infty}^{0} d\xi \, A_\nu(z) \frac{\partial z_\nu}{\partial \xi} \right) = A_\mu(x) + \int_{-\infty}^{0} d\xi \, F_{\rho\nu} \frac{\partial z_\nu}{\partial \xi} \frac{\partial z_\rho}{\partial x_\mu}. \qquad (11.15)$$

Roy's first theorem states:

If a charged particle is confined within a region $\mathcal{R}$, and if for every x_μ in $\mathcal{R}$ there exists a path $z_\mu(x, \xi)$ satisfying the conditions (11.12), then all physical effects on the particle are determined by the $F_{\mu\nu}$ in $\mathcal{R}$ alone.

The proof follows by gauge invariance, noting that the left-hand side of (11.15) is single valued; the potential $\mathbf{A}_\mu$ is equivalent to

$$\mathbf{A}'_\mu \equiv \int_{-\infty}^{0} d\xi \, F_{\nu\rho} \frac{\partial z_\nu}{\partial \xi} \frac{\partial z_\rho}{\partial x_\mu}, \qquad (11.16)$$

which contains only the field strengths in $\mathcal{R}$.

The second theorem states:

If two potentials $A_\mu^{(1)}$ and $A_\mu^{(2)}$ in $\mathcal{R}$, satisfy

$$\partial_\mu(\Delta A_\nu) - \partial_\nu(\Delta A_\mu) = 0 \, ;$$
$$\int_{-\infty}^{0} d\xi \, \Delta A_\mu(z) \frac{\partial z_\mu}{\partial \xi} < \infty \, ;$$
$$\lim_{\xi \to -\infty} \frac{\partial z_\nu}{\partial x_\mu} \Delta A_\nu(z) = 0 \qquad (11.17)$$

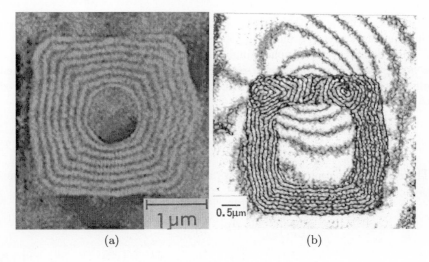

(a)   (b)

Fig. 11.3. Interferograms at normal incidence, from a thin-film toroidal permalloy magnet (Tonomura *et al.* [8].)**(a)** "Ideal" situation, with the flux well confined within the magnet; **(b)** there is flux leakage

(where $\Delta A_\mu \equiv A_\mu^{(1)} - A_\mu^{(2)}$) for at least one path $z_\mu(x, \xi)$ satisfying the conditions (11.12) and lying entirely within $\mathcal{R}$, then the two potentials are equivalent.

The proof follows from (11.15), replacing A_μ by ΔA_μ; it is the derivative with respect to x_μ of a single-valued function, *i.e.* it is equivalent to zero potential.

From Roy's theorems, it follows that in any singly-connected region of space, all observable effects result from *accessible* fields; a non-trivial ESAB effect can only be tested using a topology which is not singly connected.

11.6   Crucial Experiments

In the light of Roy's theorems, it is clear that a real direct experimental verification of the ESAB effect requires a topologically different setup from that of the early experiments. The present author [7] suggested that a real experiment could be performed in which the magnetic field is confined within a hollow superconducting torus. The walls can be thick enough, compared with the London penetration depth, so that there need be no significant stray or fringe fields.

Tonomura's group in Japan have performed a series of very elegant experiments [8,9] which establish the reality of the ESAB effect in a totally convincing manner. In these experiments, the phase of the electron beam is followed by using holographic interferometry. The method is designed to produce contour lines of constant phase. The electron beam is split coherently

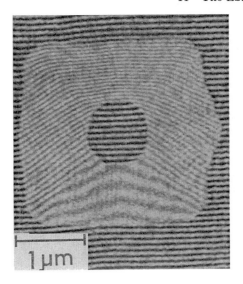

Fig. 11.4. Interferograms taken at oblique incidence (Tonomura *et al.* [8]), showing contour lines of magnetic flux, and demonstrating the refraction of the electron phase by the magnetic medium

into a reference beam and a beam which passes through the specimen — a toroidal magnet. The interference between the reference beam and the beam which irradiates the specimen produces a hologram. The photographic plate containing the hologram is then placed in an optical system to reconstruct a map of the interference fringes.

In the first series of experiments, Tonomura *et al.* [8] used a thin-film permalloy toroidal magnet. When the electron beam is at normal incidence, the fringes of constant phase are concentric loops around the torus. The interferograms of many specimens showed that the magnetic flux was indeed well confined within the torus. (Fig. 3a). Specimens where the flux was not confined were easily identified by the fact that some of the contour lines have to exit from the torus (Fig. 11.3b).

At oblique incidence, interference fringes can be seen in all three regions: outside the specimen, inside the central hole, and inside the magnetic material itself (Fig. 11.4). Note that the fringes *inside* the material are bent by refraction, but are bent back again inside the hole (*cf.* the refraction of light in a parallel-sided glass block). The experiment verifies the effect quantitatively, to an accuracy of better than 20%.

The criticism of Bocchieri *et al.* [5] is still not completely answered by this experiment, since the electron beam is not totally excluded from the region where the flux sits. To answer this possible criticism, Tonomura *et al.* varied the energy of the electrons (80, 100, and 125 keV), in order to vary the penetrability of the magnet to the electrons. Varying the beam energy

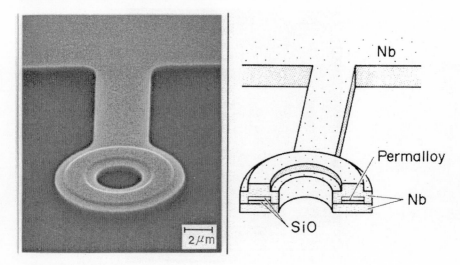

Fig. 11.5. The design of Tonomura's [9] niobium-sheathed toroidal magnets. Left: scanning electron micrograph; right: schematic

led to no change in the displacement of the phase between the hole and the exterior.

To counter Bocchieri's further assertion that even the *slightest* penetration can "explain" the effect, Tonomura [9] made a series of toroidal magnets ($\lesssim 10\mu$m in size), completely sheathed by superconducting Nb of thickness $> \lambda_{\mathrm{L}} \sim 10^2$ nm (Fig. 11.5). The sheath serves both to confine the flux within the magnet, and to exclude the electron beam from the magnet.

When the Nb is superconducting, only two different interferograms are found, as expected (Fig. 11.6a, b). They correspond to the two cases where the number n of flux quanta within the magnet is even or odd. (The total flux within the torus is $\Phi = n\Phi_0$.) However, when the temperature T of a specimen is $> T_c$, Φ is no longer quantized, and intermediate phase shifts are seen (Fig. 11.6c). This experiment of Tonomura thus gives a very elegant and direct confirmation of the reality of the ESAB effect.

Acknowledgment

I wish to thank A. Tonomura for Figs. 11.3–11.6.

References

1. W. Ehrenberg and R.E. Siday, *Proc. Phys. Soc. (Lond.)* **B 62**, 8 (1949).
2. Y. Aharonov and D. Bohm, *Phys. Rev.* **115**, 485 (1959); **123**, 1511 (1961).
3. R.G. Chambers, *Phys. Rev. Lett.* **5**, 3 (1960).

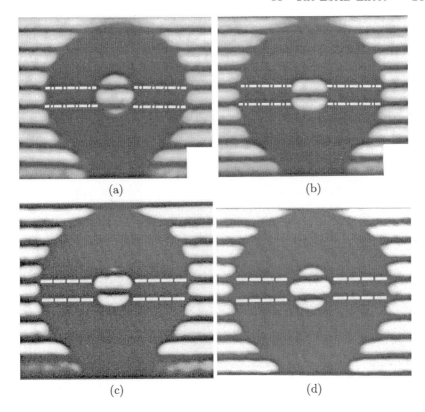

(a)   (b)

(c)   (d)

Fig. 11.6. Interferograms of two of Tonomura's [9] samples. At 4.5 K (when the Nb is superconducting) only two different interferograms are seen, corresponding to a phase shift of **(a)** zero, and **(b)** π respectively. But when $T > T_c$ (=9.2 K), the phase changes abruptly; in **(c)**, at 15 K, it has changed from π to 0.8π, and in **(d)**, at 300 K, it is 0.3π (due to the change in the magnetization of the permalloy)

4. H. Boersch and B. Lischke, *Z. Phys.* **237**, 449 (1979).
5. P. Bocchieri and A. Loinger, *Nuovo Cimento* **A 47,** 475 (1978); P. Bocchieri, A. Loinger and G. Siragusa *ibid.* **A 51,** 1 (1979); —, *ibid.* **A 56,** 55 (1980); —, *Lett. Nuovo Cimento* **35,** 79 (1982); A. Loinger, *Revista del Nuovo Cimento* **A 10,** 1 (1987).
6. S.M. Roy, *Phys. Rev. Lett.* **44,** 111 (1980).
7. C.G. Kuper, *Phys. Lett.* **A 79,** 413 (1980).
8. A. Tonomura, T. Matsuda, R. Suzuki, A. Fukuhara, N. Osakabe, H. Umezaki, J. Endo, K Shinagawa, Y. Sugita, and H. Fujiwara, *Phys. Rev. Lett.* **48,** 1443 (1982).
9. A. Tonomura, *Physica* **B 151,** 206 (1988).

12 Connectivity and Superconductivity in Inhomogeneous Structures

Guy Deutscher

School of Physics and Astronomy, Tel Aviv University, 69879 Tel Aviv, Israel

Abstract. Inhomogeneous superconductors, made up of superconducting and non superconducting regions, or of regions having different critical temperatures, have properties that may differ considerably from those of ideally homogeneous superconductors. We review here some examples of interest, such as superconductor/normal metal contacts, superconducting networks embedded in insulating matrices, and granular superconductors.

12.1 Introduction

A superfluid is in general described by a macroscopic wave function:

$$\psi = |\psi|e^{i\varphi} \tag{12.1}$$

where the amplitude $|\psi|$ and the phase φ may be space, and also time dependent. The superfluid density $|\psi|^2$ expresses the macroscopic occupation of a quantum state. The particles that have condensed in that state must evidently be bosons. In the case of superfluid helium, these particles are the individual atoms. In the case of a superconductor, they are pairs of electrons of opposite momentum and spins, the famous "Cooper pairs". In a BCS superconductor, pairs form and condense at the same temperature, the critical temperature T_c. But other kinds of condensation leading to superconductivity are in principle possible, for instance with pairs forming at a higher temperature than the condensation temperature. In any case, the superfluid density is defined only below the condensation temperature T_c.

In an homogeneous superconductor, the superfluid density determines the value of the London penetration depth λ_L that characterizes the Meissner effect:

$$\lambda_L^{-2} = 16\pi e^2 |\psi|^2 (mc^2)^{-1} \tag{12.2}$$

where m is the single electron effective mass. In homogeneous superconductors described by the theory of Bardeen, Cooper and Schrieffer (BCS), the concepts of the superfluid density and of the pair potential Δ, or order parameter (identical to the energy gap in the homogeneous case) are intimately related:

$$|\psi|^2 = \frac{2mC}{\hbar}\Delta^2 \tag{12.3}$$

The value of the coefficient C can be determined experimentally by comparing values of λ and of Δ (the latter obtained by a tunneling experiment that detects the energy of the first excited states). The value of C has also been calculated within the framework of the BCS theory. In a pure metal, $C \approx N(0)\xi_0^2$, where $N(0)$ is the normal state density of states at the Fermi level, and $\xi_0 = \frac{\hbar v_F}{\pi \Delta}$ is the coherence length. At $T = 0$ (i.e. in the absence of any excitations), the superfluid density is then equal to the free electron density: the penetration depth is in fact independent from the pair potential. It will retain the same value for infinitesimally small values of the pair potential, or of the critical temperature T_c, related to Δ by $\Delta = 1.75kT_c$. The limit $\Delta \to 0$ corresponds to a vanishingly small interaction parameter V, $\Delta \approx \Theta_D \exp(-1/NV)$, where Θ_D is the Debye temperature of the metal. We shall see below that the indirect relationship between Meissner currents and the order parameter allows an interesting situation where these currents can flow in an inhomogeneous superconductor, in regions where $V = 0$ (and therefore $\Delta = 0$).

Another important quantity determined by the order parameter is the maximum velocity v_c that the condensate particles can acquire without being excited out of the condensate. In a BCS superconductor

$$v_c = \frac{\Delta}{p_F} \tag{12.4}$$

where p_F is the Fermi momentum. In general, what is being measured experimentally is the critical current density:

$$j_c = 2e|\psi|^2 \frac{\Delta}{p_F} \tag{12.5}$$

In the Ginzburg Landau formulation of the free energy of the condensed state

$$F_s = F_n + a|\psi|^2 + \frac{b}{2}|\psi|^4 + \frac{1}{2m}|(-i\hbar\nabla - \frac{2eA}{\hbar c})|^2 + \frac{h^2}{8\pi} \tag{12.6}$$

— where the coefficient a is a linear function of the temperature T that goes to zero at T_c, the coefficient b is taken as constant, and h is a local field — the critical current density is given by:

$$j_e = \frac{4e\psi_0^2}{3\sqrt{3}} \frac{\hbar}{m\xi(T)} \tag{12.7}$$

where the coherence length $\xi(T)$ is related to the coefficient a by

$$\xi(T)^2 = \frac{\hbar^2}{2m|a|} \tag{12.8}$$

and ψ_0^2 is the equilibrium value of the superfluid density, $\psi_0^2 = -\frac{a}{b}$.

In inhomogeneous superconductors, the definition of an order parameter is not obvious. They may display some or all of the manifestations of super-conductivity described above — Meissner effect, critical current density, gap in the excitation spectrum. But the relationship between these quantities will in general be different from what they are in a BCS superconductor.

12.2   The Case
of a Normal-Metal/Superconductor Interface

We consider an interface between a semi-infinite BCS superconductor having an interaction parameter V and a gap Δ, with a normal metal in which the interaction parameter is zero, or extremely small compared to that of the superconductor. There exists a proximity effect between the two sides. Two theoretical approaches have been proposed to describe it. In the dirty limit — where the electron mean free path is substantially smaller than the coherence length — the proximity effect can be understood as due to the diffusive leakage of Cooper pairs from the S side into N. In the clean limit, one can, following Saint James, consider the case where S is semi-infinite, and N is a slab of finite thickness d . An incoming electron from the N side, having an energy counted from the Fermi level that is smaller than the energy gap in the S side, cannot penetrate inside S. It creates an electron-hole pair in S, combines with the electron to create a Cooper pair in S, while the hole is reflected back into N. This process is usually known as an Andreev reflection. After being again reflected at the outer surface of N, this time as a normal reflection, the hole hits again the S/N interface, undergoing a second Andreev reflection that re-establishes the charge and spin of the original quasi-particle. After a second normal reflection at the outer surface of N, the cycle starts again. This description is equivalent to the behavior of a particle in a potential well of thickness $2d$. Its energy is quantized. When $2d > \xi_s$, the separation between the energy levels is given by:

$$\Delta\varepsilon = \frac{\hbar^2}{2m} \frac{\pi k_F}{d} \tag{12.9}$$

where k_F is the Fermi wave vector in N. This separation is half that of a normal particle in a potential well of width d. Since there are no states available below the first level, it plays the role of an effective energy gap, if one probes only excitations propagating perpendicular to the interface. Note that this gap comes about because normal quasi-particles cannot penetrate inside S: it is a proximity effect. Another way to describe it, is to remark that the reflected hole has a momentum that is almost equal and opposite to that of the incoming electron (to within $k_F \frac{\Delta}{E_F}$): together, they are almost a Cooper pair, that has been induced in N by the proximity of S.

Let us now apply a magnetic field parallel to the interface, on the N side of the bi-layer. Will there be a Meissner effect in N, namely will the applied field be screened inside N and have a very small value at the interface – or will it be essentially unscreened in N? In other terms, is there a finite superfluid density in N? And if yes, what is its value? The answer to this question is not obvious. On the one hand, the leakage of Cooper pairs from S to N implies the existence of a finite superfluid density. On the other hand, if the interaction parameter is strictly zero in N, so is the local pair potential (the gap in an homogeneous superconductor). This comes about because in the linear perturbation approximation, valid for small values of the pair potential, it is given by an expression of the form:

$$\Delta(r) = V(r) \int K(r, r')\Delta(r')dr' \tag{12.10}$$

where $K(r, r')$ is a kernel having a range that depends on the normal state properties. If the local value of V is zero, so is the local pair potential. However, if the pair potential is non zero somewhere in the sample, the local value of Δ/V is non zero everywhere. Δ/V is the pair amplitude. What is required to have finite Meissner currents is a finite pair amplitude and not necessarily a finite pair potential. These currents are given in the dirty limit by

$$j(x) = -\chi(x)A(x) \tag{12.11}$$

where $A(x)$ is the local value of the vector potential, and $\chi(x)$ is given by

$$\chi(x) = \frac{4\pi T\sigma}{cD^2} \sum_n \int \Delta(x')K_n(x - x')dx' \tag{12.12}$$

Here σ is the normal state conductivity and D is the coefficient of diffusion.

Although the Meissner current is determined by a non-local relation, it may be described by a London like equation with an effective local superfluid density (or order parameter), or in other terms a local penetration depth. But this is not quite in the spirit of the London equation, which is strictly local.

Because the kernels K_n decrease exponentially away from the S/N interface, so does the function $\chi(x)$ inside N ($x¿0$). This means that the effective local penetration depth diverges exponentially away from the interface. Although the Meissner currents are finite everywhere, they become exponentially weak away from the interface. Let us define ρ as the distance in N measured from the interface. We can divide N into two regions. Far from the interface, $\lambda(\rho) > \rho$: the field penetrates essentially freely. For all practical purposes, the superfluid density is zero. Close to the interface, it may or may not be that $\lambda(\rho) < \rho$: the field may or may not be screened. It can be shown that the condition

$$\lambda(\rho) = \rho \tag{12.13}$$

is equivalent to:

$$\lambda(\rho) = K^{-1} \tag{12.14}$$

where K^{-1} is the maximum range of the kernels K_n. The parameter λK plays the role of an effective Ginzburg Landau parameter. If it is everywhere in N larger than 1, there is in practice no Meissner effect in N. If, close to the interface, it becomes smaller than 1, then the field will not penetrate in that region.

The effective GL parameter is a strong function of temperature, contrary to the case of homogeneous superconductors. Near the critical temperature of S, it is larger than 1 everywhere in N, because of the decrease of the pair potential in S. At low temperatures, because the Kernel range varies as $T^{-1/2}$, the screening condition (12.13) is eventually fulfilled, first near the interface, then progressively further away. For instance, in the case of a contact between Pb, a. superconductor below 7.2K, and Cu, a non-superconductor, a Meissner effect is seen on the Cu side below about 2K. At very low temperatures, the screening distance from the interface can reach several microns.

It turns out that the study of the Meissner effect is well suited for a study of the superfluid density on the N side of S/N contacts. The screening length, because it is measured *from the interface*, and not from the outer surface of N, gives a good evaluation of the practical superfluid properties induced by the proximity effect.

An interesting question is that of the determination of the local value of the order parameter, in an inhomogeneous situation. Suppose for instance that one performs a tunneling experiment at the surface of the N side of an S/N contact. One may naively think that what will be measured is the value of the order parameter at that surface, as is usually the case when performing such an experiment on a superconductor. It turns out that this guess is usually wrong. The reason is, that for a thick N layer, the order parameter increases exponentially away from the surface, towards the S/N interface. Two effects then cancel each other: the exponentially decreasing sensitivity of the tunneling current away from the surface, and the exponentially increasing order parameter. For thick N layers (compared to the decay length of the order parameter in N), the energy scale of the tunneling characteristic is in fact determined by its value near the interface. The measurement is in fact highly *non-local*. There is one special case where the tunneling measurement is a local probe. This is when the N layer is at a temperature close to its own T_c. The decay length in N is then considerably increased beyond the range of the normal state kernel, and to first approximation what is being measured is the local value of the order parameter at the surface.

12.3   Superconducting Networks

We consider a system composed of superconducting and insulating regions, mixed at random. A lattice model for such a system would be composed of superconducting and insulating bonds. An experimental realization is a composite of a superconducting metal, such as Indium, and of an insulator or semiconductor, such as Germanium. The important points for the choice of the constituents are that they should not be soluble in each other, nor form definite compounds by some chemical reaction, and that they should both be in crystalline form under the conditions of preparation. Such systems can be prepared by co-evaporation of the two elements unto a substrate held at the appropriate temperature (such that both elements crystallize upon condensing on the substrate). The scale of the superconducting and insulating regions are asumed to be large enough so that their electronic properties are essentially identical to that of the bulk. For instance, the superconducting regions have the same critical temperature as the bulk, and the same local superfluid density. In practice this requires that their smaller dimension be of the order of 10 nm or more.

The macroscopic superfluid properties of such a system are entirely dependent on its geometry, which is itself strongly dependent on the respective volume fractions of the constituents. There exists a critical metal volume fraction x_c below which the metal forms only finite clusters, and above which it forms both finite clusters and an infinite one. At $x < x_c$ the compound is macroscopically an insulator. Its resistance is finite at any temperature and for all current values. An applied magnetic field will penetrate throughout any thickness of the composite, although it will be excluded from the volume of the finite clusters in a way that depends on their geometry. Since the resistance is finite and the macroscopic London penetration depth is infinite, the macroscopic superfluid density is zero.

A more interesting situation arises at $x > x_c$. The infinite cluster then displays the basic properties of a superfluid. Below T_c, it has zero resistance, a finite critical current density, and a finite London penetration depth. Describing the network in terms of a percolation model, it is characterized by a length scale, called the percolation correlation length ξ_p. This length scale diverges at x_c as

$$\xi_p = \xi_{p0}(x - x_c)^{-\nu} \tag{12.15}$$

The infinite cluster is composed of blobs connected by thin links that contain a small number of parallel paths. ξ_p is the typical distance that separates these links. The number of parallel paths that constitute a link remains finite as x_c is approached. In a d dimensional medium, the density of links determines the macroscopic critical current density, which varies as:

$$j_c = j_{c0}(x - x_c)^{(d-1)\nu} \tag{12.16}$$

Here j_{c0} is the critical current density of the bulk. This relation is well followed by experiment. A measurement of the critical current density of random superconductor/insulator composites is in fact the most direct way to determine the critical exponent of the percolation correlation length. Values obtained experimentally are in good agreement with those obtained from Monte Carlo simulations. It is worth to note that the behavior of the normal state conductivity of the composite is not, contrary to the superconducting critical current density, determined uniquely by the density of weak links in the network. It is also determined by the length of these weak links, which has its own critical exponent. For instance, in 3 dimensions, the critical exponent for the critical current density is equal to 1.7, while that for the conductivity is equal to 2.0.

12.4   Granular Superconductors

We now consider another kind of mixture of superconducting and insulating regions. Instead of being distributed at random, the constituents have different morphologies. The superconductor is in the form of grains, and the insulator coats these grains, forming dielectric regions that act as tunneling barriers between the grains. Ideally, the grains have all the same size, and the barriers all the same thickness and transmission coefficient. In an experimental realization, the grains are made of Aluminium, and the dielectric barriers of Germanium. A granular composite is formed by vacuum deposition of the constituents unto a glass substrate held at room temperature. They hit the substrate at random, but because under the conditions of deposition, only Al crystallizes, Ge atoms are rejected from the crystallization regions and form the described dielectric coating (amorphous Ge) around the grains. As the grains grow, Ge atoms accumulate around them, until they form a continuous coating. At that stage, grain growth stops. Micrographs show that the Al grain size distribution is fairly narrow, with an average size that decreases as the Ge content increases (the Ge coating reaching continuity faster). A typical Al grain size is 10 nm for the concentrations of interest.

In the normal state, electrical conduction occurs by electron tunneling from grain to grain. Because of the small grain size, Coulomb effects induce a repulsive potential E_c for the transfer of one electron between two initially neutral grains. This potential is effective when the relaxation time τ is such that:

$$\frac{\hbar}{\tau} < E_c \qquad (12.17)$$

the composite is then an insulator. In the opposite case, it is in fact a metal, namely the transfer of electrons from grain to grain does not require an activation energy. As the Ge content is increased, the grains become smaller, hence $E_c \approx \frac{e^2}{\varepsilon r}$ becomes larger. Here ε is an effective dielectric constant of the medium, and r is the grain radius. At the same time, the Ge barrier becomes thicker and its cross section smaller, so τ becomes longer. At some point, condition (12.17) is met and the composite becomes and insulator.

Below a temperature T_{c0}, the grains become superconducting. What will then be the properties of the granular compound? If, above T_{c0}, the composite is metallic, we expect it to become itself superconducting. This is indeed the case if the grains are sufficiently large to have a well defined T_{c0}, as we have assumed here. A more delicate situation arises if the composite is insulating in its normal state. Can the composite still become superconducting? And, if not, how will its insulating properties be affected by the individual grains superconductivity?

At the time of writing, we do not have a full theoretical understanding of these questions. But experiments have revealed two interesting phenomena.

The first one is that there does exist a narrow window of concentration where the composite is insulating in the normal state, but nevertheless becomes superconducting below a temperature T_c, lower than T_{c0}. This proves that the Josephson coupling between superconducting grains, E_J, can be larger than the Coulomb potential E_c, although the normal state coupling energy, $\hbar/\tau$, is smaller than E_c. The net result is that an insulator — a material in which all electrons are localized in the normal state — can become a superconductor, due to inhomogeneity effects.

The second one is that, at slightly larger Ge concentrations, the Al-Ge composite becomes a *super-insulator* below T_c. In other words, as the Ge concentration is increased, the composite state at $T = 0$ goes from superconducting to super-insulating. The super-insulating state is characterized by a faster increase of the resistance below T_c than above T_c, as the temperature is lowered. By applying, below T_c, a magnetic field sufficiently strong to quench superconductivity in the grains, it is possible to compare directly the resistance in the super-insulating state to that in the normal state. At low temperatures, the former can be several orders of magnitude larger than the latter.

The origin of the super-insulating properties below T_c lies clearly in the opening up of the superconducting gap in the individual grains. This gap prevents quasi-particle tunneling (Giaever tunneling) — the only one allowed when the intergrain Josephson coupling is weak. What is remarkable is that the vast majority of intergrain Ge barriers act essentially as ideal Giaever junctions. This experiment proves that transport in the composite does indeed occur through inter-grain tunneling — and not, for instance, through electron hoping, as would be the case in an amorphous material. The granular structure does control the properties of the system.

Acknowledgements

This work was supported in part by the Heinrich Hertz Minerva Center for Superconductivity, and by the Oren Family Chair of Solid State Phyiscs.

Further Reading

On the proximity effect G. Deutscher and P. G. de Gennes, in *Treatise on Superconductivity*, ed. R. D. Parks, Marcel Dekker Inc., N.Y. (1969), p.1005, and references therein. For non-local tunneling effects, see in particular the reference to S. Mauro and P. G. de Gennes.

On non-local effects in proximity induced Meissner currents G. Deutscher, Solid State Comm. **9**, 891 (1971).

On percolation and superconductivity G. Deutscher, in *Chance and Matter*, eds. J. Souletie, J. Vannimenus and R. Stora, Les Houches session XLVI, NATO ASI, North Holland, Amsterdam (1987), pp. 4–66.

On granular superconductors B. Abeles, Adv. Phys. **24**, 407 (1975); A. Gerber et al., Phys. Rev. Lett. **78**, 4277 (1997).

Lecture Notes in Physics

For information about Vols. 1–522
please contact your bookseller or Springer-Verlag

Vol. 523: B. Wolf, O. Stahl, A. W. Fullerton (Eds.), Variable and Non-spherical Stellar Winds in Luminous Hot Stars. Proceedings, 1998. XX, 424 pages. 1999.

Vol. 524: J. Wess, E. A. Ivanov (Eds.), Supersymmetries and Quantum Symmetries. Proceedings, 1997. XX, 442 pages. 1999.

Vol. 525: A. Ceresole, C. Kounnas, D. Lüst, S. Theisen (Eds.), Quantum Aspects of Gauge Theories, Supersymmetry and Unification. Proceedings, 1998. X, 511 pages. 1999.

Vol. 526: H.-P. Breuer, F. Petruccione (Eds.), Open Systems and Measurement in Relativistic Quantum Theory. Proceedings, 1998. VIII, 240 pages. 1999.

Vol. 527: D. Reguera, J. M. G. Vilar, J. M. Rubí (Eds.), Statistical Mechanics of Biocomplexity. Proceedings, 1998. XI, 318 pages. 1999.

Vol. 528: I. Peschel, X. Wang, M. Kaulke, K. Hallberg (Eds.), Density-Matrix Renormalization. Proceedings, 1998. XVI, 355 pages. 1999.

Vol. 529: S. Biringen, H. Örs, A. Tezel, J.H. Ferziger (Eds.), Industrial and Environmental Applications of Direct and Large-Eddy Simulation. Proceedings, 1998. XVI, 301 pages. 1999.

Vol. 530: H.-J. Röser, K. Meisenheimer (Eds.), The Radio Galaxy Messier 87. Proceedings, 1997. XIII, 342 pages. 1999.

Vol. 531: H. Benisty, J.-M. Gérard, R. Houdré, J. Rarity, C. Weisbuch (Eds.), Confined Photon Systems. Proceedings, 1998. X, 496 pages. 1999.

Vol. 532: S. C. Müller, J. Parisi, W. Zimmermann (Eds.), Transport and Structure. Their Competitive Roles in Biophysics and Chemistry. XII, 400 pages. 1999.

Vol. 533: K. Hutter, Y. Wang, H. Beer (Eds.), Advances in Cold-Region Thermal Engineering and Sciences. Proceedings, 1999. XIV, 608 pages. 1999.

Vol. 534: F. Moreno, F. González (Eds.), Light Scattering from Microstructures. Proceedings, 1998. XII, 300 pages. 2000

Vol. 535: H. Dreyssé (Ed.), Electronic Structure and Physical Properties of Solids: The Uses of the LMTO Method. Proceedings, 1998. XIV, 458 pages. 2000.

Vol. 536: T. Passot, P.-L. Sulem (Eds.), Nonlinear MHD Waves and Turbulence. Proceedings, 1998. X, 385 pages. 1999.

Vol. 537: S. Cotsakis, G. W. Gibbons (Eds.), Mathematical and Quantum Aspects of Relativity and Cosmology. Proceedings, 1998. XII, 251 pages. 1999.

Vol. 538: Ph. Blanchard, D. Giulini, E. Joos, C. Kiefer, I.-O. Stamatescu (Eds.), Decoherence: Theoretical, Experimental, and Conceptual Problems. Proceedings, 1998. XII, 345 pages. 2000.

Vol. 539: A. Borowiec, W. Cegła, B. Jancewicz, W. Karwowski (Eds.), Theoretical Physics. Fin de Siècle. Proceedings, 1998. XX, 319 pages. 2000.

Vol. 540: B. G. Schmidt (Ed.), Einstein's Field Equations and Their Physical Implications. Selected Essays. 1999. XIII, 429 pages. 2000.

Vol. 541: J. Kowalski-Glikman (Ed.), Towards Quantum Gravity. Proceedings, 1999. XII, 376 pages. 2000.

Vol. 542: P. L. Christiansen, M. P. Sørensen, A. C. Scott (Eds.), Nonlinear Science at the Dawn of the 21st Century. Proceedings, 1998. XXVI, 458 pages. 2000.

Vol. 543: H. Gausterer, H. Grosse, L. Pittner (Eds.), Geometry and Quantum Physics. Proceedings, 1999. VIII, 408 pages. 2000.

Vol. 544: T. Brandes (Ed.), Low-Dimensional Systems. Interactions and Transport Properties. Proceedings, 1999. VIII, 219 pages. 2000.

Vol. 545: J. Klamut, B. W. Veal, B. M. Dabrowski, P. W. Klamut, M. Kazimierski (Eds.), New Developments in High-Temperature Superconductivity. Proceedings, 1998. VIII, 275 pages. 2000.

Vol. 546: G. Grindhammer, B. A. Kniehl, G. Kramer (Eds.), New Trends in HERA Physics 1999. Proceedings, 1999. XIV, 460 pages. 2000.

Vol. 547: D. Reguera, G. Platero, L.L. Bonilla, J.M. Rubí(Eds.), Statistical and Dynamical Aspects of Mesoscopic Systems. Proceedings, 1999. XII, 357 pages. 2000.

Vol. 548: D. Lemke, M. Stickel, K. Wilke (Eds.), ISO Surveys of a Dusty Universe. Proceedings, 1999. XIV, 432 pages. 2000.

Vol. 549: C. Egbers, G. Pfister (Eds.), Physics of Rotating Fluids. Selected Topics, 1999. XVIII, 437 pages. 2000.

Vol. 550: M. Planat (Ed.), Noise, Oscillators and Algebraic Randomness. Proceedings, 1999. VIII, 417 pages. 2000.

Vol. 551: B. Brogliato (Ed.), Impacts in Mechanical Systems. Analysis and Modelling. Lectures, 1999. IX, 273 pages. 2000.

Vol. 552: Z. Chen, R. E. Ewing, Z.-C. Shi (Eds.), Numerical Treatment of Multiphase Flows in Porous Media. Proceedings, 1999. XXI, 445 pages. 2000.

Vol. 553: J.-P. Rozelot, L. Klein, J.-C. Vial Eds.), Transport of Energy Conversion in the Heliosphere. Proceedings, 1998. IX, 214 pages. 2000.

Vol. 554: K. R. Mecke, D. Stoyan (Eds.), Statistical Physics and Spatial Statistics. The Art of Analyzing and Modeling Spatial Structures and Pattern Formation. Proceedings, 1999. XII, 415 pages. 2000.

Vol. 555: A. Maurel, P. Petitjeans (Eds.), Vortex Structure and Dynamics. Proceedings, 1999. XII, 319 pages. 2000.

Vol. 556: D. Page, J. G. Hirsch (Eds.), GTO Lectures on Astrophysics. Proceedings, 1999. X, 330 pages. 2000.

Vol. 557: J. A. Freund, T. Pöschel (Eds.), Stochastic Processes in Physics, Chemistry, and Biology. X, 330 pages. 2000.

Vol. 558: P. Breitenlohner, D. Maison (Eds.), Quantum Field Theory. Proceedings, 1998. VIII, 323 pages. 2000

Vol. 559: H.-P. Breuer, F. Petruccione (Eds.), Relativistic Quantum Measurement and Decoherence. Proceedings, 1999. X, 140 pages. 2000.

Monographs

For information about Vols. 1–21
please contact your bookseller or Springer-Verlag